A Graduate Course in NMR Spectroscopy

Ramakrishna V. Hosur
Veera Mohana Rao Kakita

A Graduate Course in NMR Spectroscopy

 Springer

Ramakrishna V. Hosur
UM-DAE Centre for Excellence in Basic
Sciences
University of Mumbai
Mumbai, Maharashtra, India

Veera Mohana Rao Kakita ⓘD
UM-DAE Centre for Excellence in Basic
Sciences
University of Mumbai
Mumbai, India

ISBN 978-3-030-88771-1   ISBN 978-3-030-88769-8   (eBook)
https://doi.org/10.1007/978-3-030-88769-8

Cover illustration: Front cover art created by the authors

This Springer imprint is published by the registered company Springer Nature Switzerland AG
The registered company address is: Gewerbestrasse 11, 6330 Cham, Switzerland

Preface

Nuclear magnetic resonance (abbreviated as NMR) has come a long way since its first observation in the condensed phase in 1946, independently by two groups, one at Stanford led by Felix Bloch and the other at Massachusetts Institute of Technology led by E. M. Purcell. The ensuing years saw many new discoveries such as chemical shifts, coupling constants, spin echo, nuclear double resonance, Fourier transform (FT) NMR, multidimensional NMR, magnetic resonance imaging (MRI), to name a few, which have led to widespread application of NMR in different areas of science, namely physics, chemistry, biology, and medicine. Present-day research in NMR is highly interdisciplinary drawing expertise from all the different branches of science and technology. The frontiers are continually expanding at a breathtaking pace and it is impossible to foresee the limits, if there are any, for many more years to come. It has thus become necessary to learn this technique, no matter in which branch of science one is pursuing research. To be successful in such a situation, the students will have to be taught mathematics and physics at the undergraduate level, no matter which branch the students will eventually graduate in, namely physics, chemistry, or biology. Therefore, many universities are designing curricula keeping these factors in mind.

NMR can be taught at various levels of complexity depending upon the background of the students in terms of their training in physics, chemistry, and mathematics. The mathematical rigor can sometimes be compromised for ease of explanation, and the depth can also be different depending upon the specific area of research. There are several books already available to cater to those needs. These are written to address particular categories of students, and accordingly, the styles, coverage, and technical details vary. As a result, in most modern books, the very early developments and the fundamental concepts are not covered sufficiently to provide space for more modern developments. However, for early learners entering the research field, those details are crucial, otherwise, those will never get clarified and there will be gaps in the understanding of the students. With this view, this book is intended for students who wish to pursue PhD in chemistry or biophysics or structural biology, and who have had reasonable exposure in physics and mathematics. It deals exclusively with solution state NMR which is also termed as high-resolution NMR. There is no intention to be exhaustive, but the essential basic principles are covered in sufficient detail, which will enable the students to read more specialized or advanced books at a later stage comfortably to follow more involved developments.

The book is organized in six chapters: (i) "Basic Principles," (ii) "Analysis of High-Resolution NMR Spectra," (iii) "Fourier Transform NMR," (iv) "Polarization Transfer," (v) "Density Matrix description of NMR," and (vi) "Multidimensional NMR." Mathematical rigor has been kept to the level necessary to explain the concepts, and the content is designed in such a way that the book can be used as a textbook for a one-semester graduate course in NMR (60 h). The aim has been to present the very basic concepts and also some of the exciting modern advances for applications in chemistry and biology at one place; presently, those who have been taking my course have to search through various books, many of which are not easily available or accessible. An appendix is provided at the end to cover some advanced topics to be more inclusive. For better absorption of the concepts, exercises are given at the end of each chapter.

The chapters and the contents in the book are organized in such a way as to indicate chronological evolution of the technique, on one hand, and increasing level of complexity and mathematical rigor, on the other. The book progresses systematically in terms of concepts and developments. The students are assumed to have certain background with regard to mathematics and basic quantum mechanics. The mathematics background would include calculus, matrices and determinants, differential equations, and vector algebra. Notwithstanding, some explanations are also given at relevant places in the book for the benefit of those who did not have exposure to quantum mechanics.

This book makes no attempt to cover applications of the technique in any great detail. Since the applications are too wide ranging, we believe no justice can be made in a book which is intended to be a graduate course to teach the principles of the technique. Nevertheless, the illustrations chosen to drive the concepts will indicate the possible applications of the technique. The course will enable the students to understand the experiments well and interpret the experimental data in a reliable and meaningful manner. The final chapter, namely "Multidimensional NMR" covers some experiments developed in my laboratory for application to proteins. These have not been covered in any book so far, although they have been described in some review articles.

This book has evolved from teaching over a period of more than 20 years. The students who have taken this course have established themselves as independent scientists at reputed institutions and have been pursuing excellent research in biophysics and structural biology.

Recommended Books for Further Reading

High Resolution NMR by J. A. Pople, W. G. Schnieder, H. S. Bernstein, McGraw Hill, 1959

Principles of Magnetic Resonance, C. P. Slichter, 3rd ed., Springer, 1990

High Resolution NMR, E. D. Becker, 3rd ed., Elsevier, 2000

Principles of NMR in one and two dimensions, R. R. Ernst, G. Bodenhausen,
  A. Wokaun, Oxford, 1987

Spin Dynamics, M. H. Levitt, 2nd ed., Wiley 2008

High Resolution NMR Techniques in Organic Chemistry, T. D. W. Claridge, 3rd
  ed., Elsevier, 2016

NMR Spectroscopy: Basic Principles, Concepts and Applications in Chemistry,
  H. Günther, 3rd ed., Wiley, 2013

Understanding NMR Spectroscopy, J. Keeler, Wiley, 2005

Protein NMR Spectroscopy, J. Cavanagh, N. Skelton, W. Fairbrother, M. Rance, A.
  Palmer III, 2nd ed., Elsevier, 2006

The Nuclear Overhauser Effect in Structural and Conformational Analysis,
  D. Neuhaus, M. P. Williamson, 2nd ed., Wiley 2000

Nuclear Overhauser Effect: Chemical Applications, J. H. Noggle, R. E. Schirmer,
  Academic Press, 1971

Introduction to Solid-State NMR Spectroscopy, M. Duer, Wiley 2010

Mumbai, Maharashtra, India   Ramakrishna V. Hosur
Mumbai, India   Veera Mohana Rao Kakita

Contents

1   **Basic Concepts**   1
     1.1   Nuclear Spin and Magnetic Moments   2
     1.2   Nuclear Spins in a Magnetic Field   4
     1.3   Spin-Lattice Relaxation   10
     1.4   Spin Temperature   13
     1.5   Resonance Absorption of Energy and the NMR Experiment   14
     1.6   Kinetics of Resonance Absorption   18
     1.7   Selection Rules   21
     1.8   Line Widths   22
     1.9   Bloch Equations   24
     1.10   More About Relaxation   28
     1.11   Sensitivity   36
     1.12   Summary   38
     1.13   Further Reading   38
     1.14   Exercises   39

2   **High-Resolution NMR Spectra of Molecules**   41
     2.1   Introduction   42
     2.2   Chemical Shift   43
     2.3   Spin-Spin Coupling   50
     2.4   Analysis of NMR Spectra of Molecules   54
     2.5   Dynamic Effects in the NMR Spectra   76
     2.6   Summary   85
     2.7   Further Reading   86
     2.8   Exercises   86

3   **Fourier Transform NMR**   89
     3.1   Introduction   90
     3.2   Principles of Fourier Transform NMR   91
     3.3   Theorems on Fourier Transforms   97
     3.4   The FTNMR Spectrometer   100
     3.5   Practical Aspects of Recording FTNMR Spectra   101
     3.6   Data Processing in FT NMR   112

3.7   Phase Correction   117
3.8   Dynamic Range in FTNMR   122
3.9   Solvent Suppression   122
3.10   Spin Echo   126
3.11   Measurement of Relaxation Times   129
3.12   Water Suppression Through the Spin Echo: Watergate   133
3.13   Spin Decoupling   134
3.14   Broadband Decoupling   136
3.15   Bilinear Rotation Decoupling (BIRD)   138
3.16   Summary   139
3.17   Further Reading   139
3.18   Exercises   139

4   **Polarization Transfer**   143
4.1   Introduction   144
4.2   The Nuclear Overhauser Effect (NOE)   144
4.3   Origin of NOE   147
4.4   Steady-State NOE   152
4.5   Transient NOE   155
4.6   Selective Population Inversion   157
4.7   INEPT   159
4.8   INEPT$^+$ 162
4.9   Distortionless Enhanced Polarization Transfer (DEPT)   163
4.10   Summary   165
4.11   Further Reading   166
4.12   Exercises   166

5   **Density Matrix Description of NMR**   169
5.1   Introduction   169
5.2   Density Matrix   170
5.3   Elements of Density Matrix   172
5.4   Time Evolution of Density Operator ρ   175
5.5   Matrix Representations of RF Pulses   179
5.6   Product Operator Formalism   184
5.7   Summary   198
5.8   Further Reading   198
5.9   Exercises   198

6   **Multidimensional NMR Spectroscopy**   203
6.1   Introduction   204
6.2   Two-Dimensional NMR   205
6.3   Two-Dimensional Fourier Transformation in NMR   207
6.4   Peak Shapes in Two-Dimensional Spectra   209
6.5   Quadrature Detection in Two-Dimensional NMR   211
6.6   Types of Two-Dimensional NMR Spectra   212
6.7   Three-Dimensional NMR   249

6.8   Summary 270
6.9   Further Reading 270
6.10   Exercises 271
Reference 276

7   Appendix 277
7.1   Appendix A1: Dipolar Hamiltonian 278
7.2   Appendix A2: Chemical Shift Anisotropy 280
7.3   Appendix A3: Solid-State NMR Basics 282
7.4   Appendix A4: Selection of Coherence Transfer Pathways
        by Linear Field Gradient Pulses 286
7.5   Appendix 5: Pure Shift NMR 290
7.6   Appendix A6: Hadamard NMR Spectroscopy 298

Correction to: Multidimensional NMR Spectroscopy C1

Solutions to Exercises 301
Chapter 1 301
Chapter 2 305
Chapter 3 309
Chapter 4 311
Chapter 5 311
Chapter 6 313

Basic Concepts

1

Contents

1.1	Nuclear Spin and Magnetic Moments	2
1.2	Nuclear Spins in a Magnetic Field	4
1.3	Spin-Lattice Relaxation	10
1.4	Spin Temperature	13
1.5	Resonance Absorption of Energy and the NMR Experiment	14
	1.5.1 The Basic NMR Spectrometer	17
1.6	Kinetics of Resonance Absorption	18
1.7	Selection Rules	21
1.8	Line Widths	22
1.9	Bloch Equations	24
1.10	More About Relaxation	28
1.11	Sensitivity	36
1.12	Summary	38
1.13	Further Reading	38
1.14	Exercises	39

Learning Objectives
- Introducing the concepts of nuclear spin, nuclear magnetic moment, and spin angular momentum
- Interactions of nuclear magnetic moments with magnetic field
- Resonance absorption of energy by nuclear spins
- Relaxation phenomena of the nuclear magnetization
- Essential elements of NMR spectrometer

© The Author(s), under exclusive license to Springer Nature Switzerland AG 2022
R. V. Hosur, V. M. R. Kakita, *A Graduate Course in NMR Spectroscopy*,
https://doi.org/10.1007/978-3-030-88769-8_1

1.1   Nuclear Spin and Magnetic Moments

Atomic nuclei are composed of "protons" and "neutrons" which, in turn, are made up of fundamental elementary particles called "quarks." There are two types of quarks termed as up-quarks and down-quarks. These point-like elementary particles have *intrinsic* properties of electric charge and spin angular momentum (or simply spin). Up-quarks have a charge of +2e/3, while down-quarks have a charge of −e/3; e refers to the electronic charge. Addition of charges of the quarks making up the proton (two up-quarks and one down-quark) leads to an electric charge of +1 for the proton. Similarly, for the neutron (one up-quark and two down-quarks), one gets a charge of 0. Spin is an intrinsic property of elementary particles, and its direction is an important degree of freedom. It is sometimes visualized as the rotation of an object around its own axis (hence the name "spin"), though this notion is somewhat misguided at subatomic scales because elementary particles are believed to be point like. However, spin obeys the same mathematical laws as quantized angular momenta do. Spin angular momentum is represented by a vector, and for either of the quarks, the component of the spin angular momentum along any axis is always either $+\hbar/2$ or $-\hbar/2$, where $\hbar = (h/2\pi)$ is the reduced Planck's constant. Hence, these particles are said to be spin ½ particles. The spin values for the proton and neutron are derived by addition of the spins of the constituent quarks as per the principles of quantum mechanics, and it turns out that both proton and neutron have spin values of ½. These are again vector quantities, and the addition of the contributions of these in every nucleus which contain different numbers of protons and neutrons, according to the rules of quantum mechanics, leads to different nuclear states with different spin quantum numbers, and these can have different energies. The lowest energy state or the so-called ground state represents the most stable state of the isotope of the element in question. Any further discussion on these aspects of nuclear structure and theory is beyond the scope of this book. The spin quantum number of the ground state mentioned above is often referred to as its *intrinsic* nuclear spin, which is represented by the letter *I*. The corresponding spin angular momentum is represented by the vector **I** and has a magnitude given by $\hbar\sqrt{I(I+1)}$.

For the most stable states of the nuclei (ground state or the lowest energy state), some empirical rules seem to be valid for the calculation of nuclear spin values:

(i) If the nuclear charge represented by the atomic number is even and if the isotope number is even, then the spin I takes the value 0.
(ii) If the nucleus has odd mass number (the same as the isotope number), then its spin will have half-integer values such as 1/2, 3/2, 5/2, etc.
(iii) For odd atomic number and even mass number, the nuclear spin takes the integral values 1, 2, 3, etc.

Associated with the spin angular momentum, **I**, the nuclei have another *intrinsic* property, namely, magnetic moment, which is a consequence of a complex interplay of the motions of the particles inside the nucleus. This is also a vector, which is represented by the symbol μ. Magnetic moment and spin angular momentum of a nucleus are linearly related.

$$\mu = \gamma I \qquad (1.1)$$

The constant of proportionality, γ, is called *gyromagnetic ratio* or *magnetogyric ratio*, which is a characteristic of the given nucleus. This constant can be positive or negative; some nuclei have positive γ and some others have negative γ; depending upon this sign, the magnetic moment and the angular momentum vectors are oriented in the same direction or in opposite directions. The magnitude of the magnetic moment is clearly proportional to the magnitude of the spin angular momentum. The *magnetogyric ratios* have to be determined experimentally by resonance absorption techniques to be described later.

Table 1.1 lists the nuclear spin values, natural abundance of the isotopes, and magnetogyric ratios of some of the most commonly used nuclei in chemistry and biology.

From the general principles of quantum mechanics, the magnitude of the spin angular momentum is given by the square root of the eigenvalue of the operator, I^2. The x, y, and z Cartesian components of the angular momentum are represented by the operators I_x, I_y, and I_z, respectively. The I^2 operator commutes with all the three Cartesian components, I_x, I_y, and I_z. Therefore, I^2 can have common eigenfunctions with any of the Cartesian components. Although the I_x, I_y, and I_z operators can have discrete eigenvalues and corresponding eigenfunctions, they do not have common eigenfunctions since the operators for the Cartesian components themselves do not commute with each other. This means that all the three components and the magnitude of the spin angular momentum cannot be well-defined at the same time. Therefore, the magnitude and orientation of the spin angular momentum vector should be described using any one pair $[I^2, I_z]$ or $[I^2, I_x]$ or $[I^2, I_y]$. Conventionally, it is the $[I^2, I_z]$ pair. Once the eigenfunctions for this pair are chosen, the other Cartesian components I_x and I_y remain undefined. The discrete eigenvalues of I_z and the corresponding eigenfunctions form the basis for all NMR discussions. While the eigenvalue of I^2 yields the square of the magnitude of the angular momentum, the eigenvalues of I_z are used to describe the orientation of the angular momentum vector.

Table 1.1   Nuclear spin and related data on some chosen nuclei

Isotope	Nuclear spin	Natural abundance	Magnetogyric ratio rad $s^{-1}T^{-1}$	NMR frequency at 11.7433 T (MHz)
^1H	½	~100%	267.522×10^6	-500
^2H	1	0.015%	41.066×10^6	-76.753
^{13}C	1/2	1.1%	67.283×10^6	-125.725
^{15}N	1/2	0.37%	-27.126×10^6	50.684
^{14}N	1	99.6%	19.338×10^6	-36.132
^{19}F	1/2	~100%	251.815×10^6	-470.470
^{31}P	1/2	~100%	108.394×10^6	-202.606

Specifically, for a nucleus with spin I, the I_z component can have only $(2I + 1)$ values, namely, $I\hbar$, $(I - 1)\hbar$, $(I - 2)\hbar$, $(I - 3)\hbar$,...$(-I + 1)\hbar$, $-I\hbar$. All these states have the same total energy and thus are degenerate. The component values are often represented as $m\hbar$, where the symbol m taking the values $-I$ to $+I$ is called the "magnetic quantum number" of the state. Thus, for example, for a nucleus with $I = 1/2$, the $\widehat{I}_z$ operator would have the eigenvalues $(-1/2)$ and $(+1/2)$. For a nucleus with $I = 1$, m can have values $-1, 0, +1$, and so on. These values fix the magnitude of the z-component of the angular momentum when the nucleus is in one of the eigenstates. Since the magnitudes of the angular momentum and its component along the z-axis are not equal in any of the eigenstates, it follows that the angular momentum vector in the individual eigenstates will be oriented at an angle to the z-axis, and this angle is different for different states: $\cos^{-1}\left(\frac{m}{\sqrt{I(I+1)}}\right)$. In the x-y plane, there are no restrictions on the orientations for any of the above states. Consequently, the x and the y components can lie anywhere between $-P$ and $+P$, where P is the projection of the angular momentum in the x-y plane for the particular state.

A concept which is exclusive to quantum mechanics is the so-called superposition of states. At any point in time, in the absence of any external forces, the nucleus must be represented by a state which is a superposition of the eigenstates with similar energies. Presently, all the eigenstates having different m values are degenerate in terms of energy and thus contribute to the spin state. In other words, the nucleus will behave as though it has the properties of all the eigenstates. For example, for a spin ½ nucleus which has two eigenstates represented by α and β, the spin state can be written as

$$\psi(t) = C_\alpha \mid \alpha > +C_\beta \mid \beta > \tag{1.2}$$

where C_α and C_β are time-dependent complex coefficients. Due to normalization, they obey the relation:

$$|C_\alpha|^2 + |C_\beta|^2 = 1 \tag{1.3}$$

The modulus squares in Eq. 1.3 represent the probabilities of the spin being in the respective states. These coefficients vary from spin to spin across the ensemble. The z-component of the angular momentum of the nuclear spin in such a superposition state will depend on the coefficients and thus can take any value between the limiting values $(-I\hbar$ to $I\hbar)$ across the ensemble. In other words, the angular momentum vector can have any orientation in space. This is a very general result and holds for any value of the nuclear spin.

1.2   Nuclear Spins in a Magnetic Field

Let us first look at the behavior of the spins in a classical physics picture. When an ensemble of nuclear magnetic moments is placed in a static magnetic field $\mathbf{H_o}$, the magnetic moments tend to orient themselves with respect to the magnetic field direction. The magnetic moments experience a torque given by $(\mathbf{\mu} \times \mathbf{H_o})$. According to the laws of motion, the torque is equal to the rate of change in angular momentum. Since the magnitude of angular momentum is fixed, the torque causes a change in the orientation of the angular momentum. This is described by the equation of motion:

$$\mathbf{dJ}/\mathbf{dt} = \boldsymbol{\mu} \times \mathbf{H_o} \tag{1.4}$$

$$\mathbf{d\boldsymbol{\mu}}/\mathbf{dt} = \gamma(\boldsymbol{\mu} \times \mathbf{H_o}) \tag{1.5}$$

Classically, the continuous change in the orientation of the vector with time, moving $\boldsymbol{\mu}$ perpendicular to both $\boldsymbol{\mu}$ and $\mathbf{H_o}$, would describe what is called a precessional motion. Quantum mechanics tells us (the solution of the time-dependent Schrödinger equation under the influence of the external field) that the nuclear spin executes a motion around the direction of the magnetic field, and this is called the precessional motion. This motion for spins in the two eigenstates of the $I = 1/2$ system is depicted in Fig. 1.1 and is referred to as the "Larmor precession." If the angular velocity of the vector $\boldsymbol{\mu}$ is ω, then the rate of change of $\boldsymbol{\mu}$ is also described by the equation:

$$\mathbf{d\boldsymbol{\mu}}/\mathbf{dt} = \boldsymbol{\omega} \times \boldsymbol{\mu} \tag{1.6}$$

Equating 1.4 and 1.6, one obtains

$$\omega = -\gamma H_o \tag{1.7}$$

The negative sign in Eq. 1.7 indicates that the precessional motion occurs in a clockwise sense (by convention anticlockwise rotation is taken as positive). The frequency of precession is proportional to the strength of the magnetic field.

Fig. 1.1 Precessional motion for spins in the two eigenstates of the $I = 1/2$ system

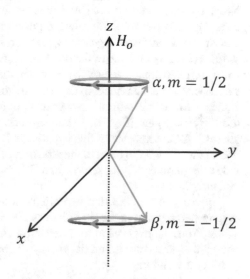

Fig. 1.2 Schematic representation of $(2I + 1)$ eigenstates possible for a nuclear spin I, under the condition of the Zeeman splitting

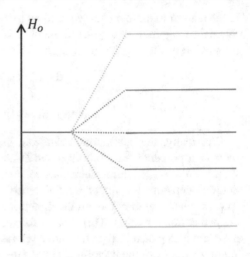

The energy associated with such a motion is the interaction energy between the magnetic moment and the magnetic field and is given by

$$E = -\mu.H_o = -\mu H_o \cos \theta = -\mu_z H_o \tag{1.8}$$

where θ is the angle between the field and the magnetic moment vector. Clearly, the interaction energy depends on the orientation of the magnetic moment with respect to the magnetic field. Thus, for any spin, the different orientations will no longer be of equal energy, and the degeneracy of orientations in space is then said to be lifted. If the nuclear spin is I, the $(2I + 1)$ eigenstates will have different energies as schematically shown in Fig. 1.2. Such a splitting is known as the Zeeman splitting. Now, consider an ensemble of spins, all of which have $I = 1/2$ and therefore constitute a two-level system in the presence of an external magnetic field. Note that since the different eigenstates have now different energies, the superposition states will decay into one or the other eigenstate at thermal equilibrium. The various spins will distribute themselves between the two levels according to Boltzmann statistics. For each orientation, the sense of rotation is the same, and the spins span a conical surface in a uniform manner without any specific phase relationship between the different spins (Fig. 1.3). This is referred to as the "hypothesis of random phases." A consequence of this hypothesis is that the net component of magnetization in the plane orthogonal to the direction of the magnetic field will be zero. This is indeed experimentally observed, and this provides credence to the hypothesis of random phases.

Referring to Fig. 1.3, the two orientations in the top (parallel to the field) and bottom (antiparallel to the field) halves are, respectively, the conventional α and β states. From the expression for interaction energy (Eq. 1.8), the α state has a lower energy than the β state, and the energy difference between them is equal to $2\mu_z H_o$ (see Eq. 1.9 and Fig. 1.4):

Fig. 1.3 Schematic representation of "hypothesis of random phases" shown for a spin ½ system

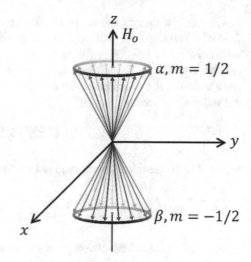

Fig. 1.4 Energy level diagram of a spin ½ system

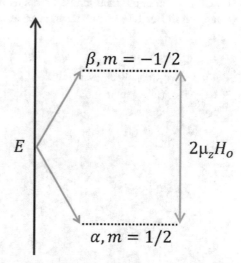

$$\Delta E = 2\mu_z H_o = \gamma H_o \hbar = \omega \hbar \tag{1.9}$$

In general, for a spin I, the $(2I + 1)$ states will have the energy values

$$E_i = -\mu_{zi} H_o = -(m_i/I)\mu H_o; m_I = -I, -(I-1), -(I-2), \ldots +I \tag{1.10}$$

$$E_i = -\gamma m_i H_o \hbar \tag{1.11}$$

$$E = -\mu H_o, \quad -\frac{I-1}{I}\mu H_o, \quad -\frac{I-2}{I}\mu H_o, \ldots +\mu H_o \tag{1.12}$$

Clearly, the energy levels are equally spaced, the separation being $\mu H_o / I$. At a field of 1 Tesla, and for protons, this energy difference is 42.577 MHz (in frequency units) or 2.82×10^{-13} ergs (in energy units). At ordinary temperatures (even 10 K can be considered as ordinary temperature), the various energy levels will be populated as per Boltzmann statistics, and the fractional populations of the individual states are given by

$$p_i = \frac{\exp\left(-E_i/kT\right)}{Z} \tag{1.13}$$

where Z is the partition function, given by the expression

$$Z = \sum_i \exp\left(-\frac{E_i}{kT}\right) = \sum_i \left[1 - \frac{E_i}{kT} + \frac{1}{2!}\left(E_i/kT\right)^2 - \frac{1}{3!}\left(E_i/kT\right)^3 + \ldots\right] \tag{1.14}$$

A simple back of the envelope calculation shows that E/kT is extremely small (Box 1.1) even at as small temperatures as few degrees Kelvin, and thus the series expansion in Eq. 1.14 can be terminated at the first power of energy, E.

Box 1.1: Calculation of $\frac{E}{kT}$

For a spectrometer frequency of 100 MHz, the energy separation for protons is 100 MHz (in frequency units), and the individual energy values will be ± 50 MHz.

(i) At 300K

$$\frac{E}{kT} = \frac{h\nu}{kT} = \frac{6.626 \times 10^{-34} \times 50 \times 10^6}{1.38 \times 10^{-23} \times 300} = 8.0 \times 10^{-6}$$

(ii) At 10K

$$\frac{E}{kT} = \frac{h\nu}{kT} = \frac{6.626 \times 10^{-34} \times 50 \times 10^6}{1.38 \times 10^{-23} \times 10} = 2.4 \times 10^{-4}$$

For other nuclei, say X, these numbers will be multiplied by $\frac{\gamma_X}{\gamma_H}$ and hence will be even smaller.

Thus,

$$p_i = \frac{(1 - E_i/kT)}{\sum_i \left(1 - \frac{E_i}{kT}\right)} \tag{1.15}$$

This approximation is termed as the "high temperature approximation." Equation 1.15 gets further simplified since in the denominator $\sum_i E_i = 0$. Consequently,

$$p_i = \frac{(1 - E_i/kT)}{\sum_i (1)} \tag{1.16}$$

$$= (2I + 1)^{-1}\left(1 + \frac{m\mu H_o}{IkT}\right) \tag{1.17}$$

For the two-level case ($I = 1/2$),

$$p_\alpha = \frac{(1 + \mu H_o/kT)}{2}$$

$$p_\beta = \frac{(1 - \mu H_o/kT)}{2} \tag{1.18}$$

Since the populations are different, with the α state being more populated than the β state, the net magnetization of the ensemble lies parallel to the direction of the magnetic field. The mean value of the magnetization can be computed as

$$M = (p_\alpha - p_\beta)\mu \tag{1.19}$$

$$= \mu^2 H_o/kT \tag{1.20}$$

If there are N nuclei per unit volume of the sample, the total magnetization per unit volume will be NM, and the volume susceptibility χ, which is defined as the ratio of total magnetization per unit volume to applied field, will be

$$\chi = NM/H_o = N\mu^2/kT \tag{1.21}$$

A similar calculation for the general case of the nuclei with spin I yields

$$\chi = N/H_o \sum (2I + 1)^{-1}(1 + m\mu H_o/IkT)\, m\mu/I \tag{1.22}$$

$$\chi = N/H_o(I\,(2I + 1))^{-1}\left[\mu \sum m + (\mu^2 H_o/IkT) \sum m^2\right] \tag{1.23}$$

$$= N/H_o(I(2I + 1))^{-1}(\mu^2 H_o/IkT)\, I(I + 1)(2I + 1)/3 \tag{1.24}$$

$$= N\mu^2(I + 1)/3IkT \tag{1.25}$$

Now, the magnetic susceptibility of a given sample is its characteristic property and represents the extent to which magnetization can be induced in the sample by an externally applied magnetic field. The larger the susceptibility of a system, the higher is the inducible magnetization and vice versa. Equation 1.25 indicates also that the susceptibility is a function of temperature. As the temperature of the sample is increased, the magnetization gets reduced, indicating that the order in the magnetic orientations in the ensemble also gets reduced.

1.3   Spin-Lattice Relaxation

We have seen that the application of a magnetic field to an ensemble of nuclear spins causes splitting of the energy levels and also causes a redistribution of the populations of the individual levels as per Boltzmann statistics. Any change in the strength of the magnetic field changes the spacing between the energy levels, and again there must be redistribution with some levels losing populations and the others gaining populations. What brings about such a redistribution? Clearly, there must be transitions occurring between the levels, and there must be forces within the system itself which cause these transitions. We shall return to the mechanistic aspects later, and at the moment we shall focus on the kinetics of these transitions with some assumed transition probabilities.

Let us consider a system of spin 1/2 nuclei, so that there are only two energy levels and the system is simple enough for an illustrative analysis (Fig. 1.5).

Let n_α and n_β be the populations of the two states α and β (note that these two will be equal in the absence of the field), and let $W_{\alpha\beta}$ and $W_{\beta\alpha}$ be the transition probabilities for α to β and β to α transitions, respectively. At equilibrium, the number of upward transitions equals the number of downward transitions per second. Thus,

$$n_\alpha^o W_{\alpha\beta} = n_\beta^o W_{\beta\alpha} \tag{1.26}$$

where n_α^o and n_β^o are the equilibrium populations at any particular field. From Boltzmann statistics,

$$\frac{n_\beta^o}{n_\alpha^o} = \frac{W_{\alpha\beta}}{W_{\beta\alpha}} = \exp\left(\frac{-\Delta E}{kT}\right) \tag{1.27}$$

Fig. 1.5 Possible transitions between two energy levels in a spin ½ system

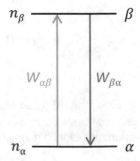

$$= 1 - 2\mu_z H_o/kT \ (\text{under high} - \text{temperature approximation}) \qquad (1.28)$$

If the populations are disturbed from equilibrium values to n_α and n_β, they will return to equilibrium values according to the following rate equations:

$$\frac{dn_\alpha}{dt} = W_{\beta\alpha}n_\beta - W_{\alpha\beta}n_\alpha \qquad (1.29)$$

$$\frac{dn_\beta}{dt} = W_{\alpha\beta}n_\alpha - W_{\beta\alpha}n_\beta \qquad (1.30)$$

So,

$$\frac{d(n_\alpha - n_\beta)}{dt} = 2(W_{\beta\alpha}n_\beta - W_{\alpha\beta}n_\alpha) \qquad (1.31)$$

Defining the variables $N = n_\alpha + n_\beta = n_\alpha^o + n_\beta^o$ and $n = n_\alpha - n_\beta$, one obtains

$$n_\alpha = \frac{(N+n)}{2}, n_\beta = \frac{(N-n)}{2} \qquad (1.32)$$

and

$$\frac{dn}{dt} = W_{\beta\alpha}(N - n) - W_{\alpha\beta}(N + n) \qquad (1.33)$$

$$= N(W_{\beta\alpha} - W_{\alpha\beta}) - n(W_{\beta\alpha} + W_{\alpha\beta}) \qquad (1.34)$$

Further from Eq. 1.27, we find

$$\frac{(W_{\beta\alpha} - W_{\alpha\beta})}{(W_{\beta\alpha} + W_{\alpha\beta})} = \frac{(n_\alpha^o - n_\beta^o)}{(n_\alpha^o + n_\beta^o)} = \frac{n^o}{N} \qquad (1.35)$$

or

$$(W_{\beta\alpha} - W_{\alpha\beta}) = (W_{\beta\alpha} + W_{\alpha\beta})\frac{n^o}{N} \qquad (1.36)$$

Combining Eqs. 1.33 and 1.36, one obtains

$$\frac{dn}{dt} = (W_{\beta\alpha} + W_{\alpha\beta})(n^o - n) \qquad (1.37)$$

Defining an average transition probability,

$$W = \frac{(W_{\beta\alpha} + W_{\alpha\beta})}{2} \qquad (1.38)$$

Equation 1.37 reduces to

$$\frac{dn}{dt} = 2W(n^o - n) \tag{1.39}$$

$$= \frac{(n^o - n)}{T_1} \tag{1.40}$$

where

$$T_1 = \frac{1}{2W} \tag{1.41}$$

The solution of Eq. 1.40 is easily written as

$$(n - n^o) = (n - n^o)_{t=0} \exp(-t/T_1) \tag{1.42}$$

If at time $t = 0$ the system is unmagnetised, then $n = 0$ and

$$n = n^o[1 - \exp(-t/T_1)] \tag{1.43}$$

Equations 1.42 and 1.43 indicate that the population difference approaches the equilibrium value n^o with a characteristic time T_1 which is called the *spin lattice relaxation time*. The inverse of T_1 measures the rate at which the system returns to equilibrium from a perturbed state, after the force is removed.

Now, why is T_1 called the spin-lattice relaxation time? What is a lattice and what is its role in the relaxation process? To answer these questions, we ask another question: Where does the energy released by the spin system in the process of adjustment of populations go? Or if the spin system has to absorb energy, where does it come from? These questions suggest that there must be a sink or a reservoir which we call the lattice and there must be coupling between the spin system under consideration and the reservoir. What this means is that if the spin system loses energy, the reservoir gains energy and vice versa. This is schematically indicated in Fig. 1.6. It implies that the lattice should have energy levels identical to those of the spin system and there must be magnetic interaction between the lattice and the spin system to cause the energy transfer. Thus, the lattice plays a dominant role in bringing about relaxation and helps to maintain a population difference which is so crucial for the absorption of energy from an external source and hence for the observation of the NMR signal. In fact, some of the very first experiments on the NMR phenomenon were unsuccessful, because the system chosen had such a long relaxation time that the continuous absorption of energy from the external source did not occur and the signal could not be detected. The relaxation time is a characteristic property of the sample and depends on a variety of experimental conditions such as temperature, viscosity of the solution, concentration of the material, size of the molecule, etc. The relaxation time also depends upon the efficiency of coupling between the spin system and the lattice, and this in turn depends upon the nature of the interactions between the spins in the system and the spins in the lattice. This

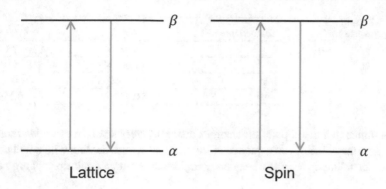

Fig. 1.6 Schematic representation of spin and lattice systems with comparable energy levels

subject has been investigated in detail by several authors and described in several books. A complete analysis of the relaxation phenomenon is beyond the scope of this book, although a qualitative description will be provided in later sections.

1.4 Spin Temperature

We have said before that the distribution of spins between various energy levels of a spin system is dictated by Boltzmann statistics. Naturally, temperature has a definitive role in determining this population distribution. If the system is at thermal equilibrium with the surroundings, the populations will be as per the temperature of the environment. However, if the system is not at equilibrium with the environment, the populations will not correspond to equilibrium values, and under those conditions, the distribution may be described by a virtual temperature which is referred to as the *spin temperature* of the system. Explicitly, if n_α and n_β are the populations of α and β states, respectively, at any nonequilibrium condition, then a temperature T_s can be defined which satisfies the equation

$$\left(\frac{n_\alpha}{n_\beta}\right) = \exp\left(2\mu_z H_o / k T_s\right) \tag{1.44}$$

T_s is called the spin temperature of the system. When the system reaches equilibrium, T_s becomes equal to the thermal temperature T. Spin temperature is a measure of how far the system is away from equilibrium. If somehow the populations of the two states α and β are made equal, the spin temperature becomes infinity and this is called *saturation*. If the upper state has a higher population than the lower state, i.e., if there is a population inversion, then the spin temperature is said to be negative. Table 1.2 lists a few temperature values and their correlation with the populations ($N = n_\alpha - n_\beta$).

Table 1.2 Spin
temperatures and
populations

Temperature (K)	$\frac{n_\alpha}{n_\beta}$	$\frac{(n_\alpha - n_\beta)}{N} \times 10^4$
1	1.0206	101.95
10	1.00204	18.961
300	1.000068	3.400
−10	0.99796	−10.1896

We notice that the population changes are really very small over a wide temperature range, and this is a consequence of the fact that we are dealing with small energy
differences falling in the radio frequency regime or less than a calorie in energy units.

1.5 Resonance Absorption of Energy and the NMR Experiment

The behavior of nuclear spins in the presence of a magnetic field indicates that if the
precessional frequencies can be measured in some way, they provide valuable
information about the magnetic moments of the nuclei. The question therefore is
how to measure these frequencies. Let us consider an experiment schematically
described in Fig. 1.7.

For simplicity, we consider a spin 1/2 system; however, the principles are valid
for other spin systems as well. First a static field H_o is applied to create precessional
motions in the spin system. Then another rotating field of amplitude H_1 is applied in
a plane orthogonal to the direction of the H_o field. Let ω_i and ω_o be the rotational
frequencies of the nucleus and the H_1 field, respectively. If one sits on the rotating
field—this is called transforming into the rotating frame—and looks at the nucleus,
the precessional frequency appears to be $(\omega_i - \omega_o)$ if the two motions are in the same
sense and to be $(\omega_i + \omega_o)$ if they have opposite senses. We consider the former
situation for our discussion here. If we can change the value of ω_o continuously, we
will reach a situation when $\omega_i = \omega_o$, and this is termed as "resonance." Under this
condition, the nucleus appears to be stationary to the observer. The magnetic
moment and the field H_1 interact strongly causing some changes in the energy of
the spin system. In other words, the nuclear spin system would absorb energy, and
this is monitored in the NMR experiment in an indirect manner. A further consequence of the resonance condition is that the nucleus does not see the H_o field at all in
the rotating frame of reference. The only visible field then is the H_1 field, and the
nuclear spins will tend to tilt towards this field. If the H_1 field is applied for a long
time, the nuclear spins will get completely realigned with respect to the H_1 field, and
the time taken for such a realignment will be determined by the relaxation times.
This is often termed as *locking* the spins along the direction of the H_1 field.

The resonance absorption of energy described above can also be described in an
alternative manner with the help of energy level diagrams. This description is in fact
easier to visualize and better for quantifications, particularly in more complex
systems with higher spin values or in an ensemble of different spins resulting in a
multilevel system. Referring to Fig. 1.4 of a simple two-level system of spin 1/2,
resonance condition means the application of an electromagnetic radiation with the

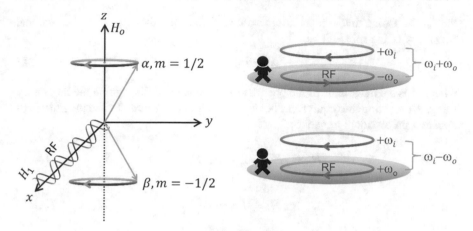

Fig. 1.7 Schematic representation of resonance condition. Left: Spins in the α and β states precess with a frequency ω_i in the anticlockwise direction. The RF with frequency ω_o and amplitude H_1 is applied along the x-axis. Right: Transforming into then rotating frame. The RF is broken into two rotating fields going in clockwise (top) and anticlockwise (bottom) directions; the observer is considered to be rotating with RF. The precessional frequency of the spin seen by the observer will be ($\omega_i + \omega_o$) in the top picture and with frequency ($\omega_i - \omega_o$) in the bottom picture. If $\omega_i = \omega_o$, then it appears to be static in the bottom picture, and this is called resonance condition

frequency of ω_o (radians/sec) or $\omega_o/2\pi$ (cycles/ sec or Hertz) whose energy corresponds to the separation between the energy levels. If the spin system has many energy levels with different separations, then the resonance condition for absorption of energy will be satisfied for more than one frequency values.

We notice from the previous descriptions (Eqs. 1.7 and 1.9) that the precessional frequency in a field H_o is equal to the separation between the two levels caused by the field. Thus, the resonance conditions in the two treatments are the same. The former treatment is termed as classical description and the latter as quantum mechanical description. An apparent discrepancy arises in the two treatments of resonance absorption, and this must be clarified. In the classical treatment, we considered a rotating field which can be represented as $H_1 \exp(i\omega_o t)$, whereas in the other treatment, we considered an electromagnetic radiation whose magnetic vector is represented as $H_1 \cos(\omega_o t)$. A careful inspection reveals that both of them are essentially equivalent. The oscillating field $H_1 \cos(\omega_o t)$ can be conceived of as a combination of two rotating fields going in opposite directions.

$$2\,H_1 \cos(\omega_o t) = H_1 \exp(i\omega_o t) + H_1 \exp(-i\omega_o t) \qquad (1.45)$$

The particular component which goes in the same direction as nuclear precession contributes to resonance absorption of energy, while the other does not. Thus, in all future considerations, we shall consider the oscillating field which is what is generated by the electronics in the NMR experiment.

The application of the radio frequency (RF) field to the system of spins precessing in a static field has another effect, namely, the creation of a phase coherence between the spins. Going into the rotating frame of the RF, the apparent precessional frequency of the spin will be

$$\omega_r = (\omega_i - \omega_o) \qquad (1.46)$$

where ω_i is the magnitude of the precessional frequency of the spin in the laboratory frame. Converting into equivalent fields for ω_r and ω_i, we find, for the magnitude of apparent precessional frequency,

$$\gamma H_r = \gamma H_i - \omega_o = \gamma \left(H_i - \frac{\omega_o}{\gamma} \right) \qquad (1.47)$$

$$\text{or } H_r = \left(H_i - \frac{\omega_o}{\gamma} \right) \qquad (1.48)$$

In this frame, the interaction energy between the spin and the field H_r is given by

$$E_r = -\mu_z H_r \qquad (1.49)$$

For a state with magnetic quantum number m,

$$E_r^m = -\gamma m \hbar H_r \qquad (1.50)$$

$$= -\gamma m \hbar \left(H_i - \frac{\omega_o}{\gamma} \right) \qquad (1.51)$$

$$= E_L^m + m \hbar \omega_o \qquad (1.52)$$

where E_L^m is the energy in the laboratory frame. Now remember that we are dealing with magnitudes for RF frequencies as well as precession frequencies. The energy level diagram in the rotating frame would look as shown in Fig. 1.8.

At the resonance condition, $\omega_i = \omega_o$, and then the two energy levels merge as if there were no magnetic field along the z-axis at all. This is the same conclusion as we noted earlier in the classical description. If we now include the small RF field, the total field in the rotating frame will be along this RF field. Now the question will be: Will the populations readjust to H_1 field? As this energy difference will be almost zero, this relaxation process will take extremely long time, and one does not apply the RF field for so long. However, the spins will interact with this field, and the total magnetization vector tips away from the z-axis, creating in the process a magnetization component in the x-y plane (Fig. 1.9).

Comparing this with the equilibrium situation, a finite x-y magnetization implies the creation of a phase coherence between the spins. This treatment can be easily generalized to off-resonance conditions as well by explicitly calculating the effective field in the rotating frame. Since the effective field is always away from the z-axis, it is clear that the application of an RF always creates a phase coherence between the spins. As soon as the RF is removed, spins start dephasing, resulting in the loss of the phase coherence, i.e., the magnetization or the coherence decays.

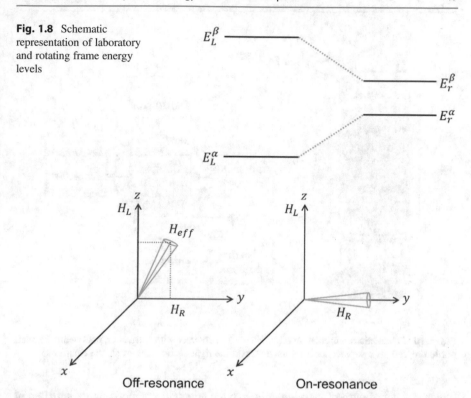

Fig. 1.8 Schematic representation of laboratory and rotating frame energy levels

Fig. 1.9 The effect of external RF field on flipping of the spins. H_L and H_R refer to the amplitudes of the static field and the RF field in the rotating frame. The H_{eff} is the vector addition of these two fields. The spins tend to align with the effective field. Under resonance condition, H_L will be zero

1.5.1   The Basic NMR Spectrometer

Figure 1.10 shows the essential components of the basic NMR spectrometer. The magnet provides the main magnetic field H_o. The sample tube sits vertically in the magnet with the transmitter/receiver RF coil surrounding the tube. There is an RF source which produces the required RF field at a desired frequency. The receiver detects the absorption of energy at resonance in the form of an induced voltage due to the creation of transverse magnetization when the resonance condition is satisfied. The detected signal (voltage) is amplified by an amplifier and fed to a recorder for display. The magnet system also has a *field-frequency lock* which helps to correct for variations in the field due to instabilities in the generated magnetic field. This allows to maintain a constant magnetic field throughout the NMR experiment. Technologically, it is very difficult to construct magnets which produce highly homogeneous fields over wide volume ranges. Inhomogeneities in the field cause variations in the resonance absorption frequencies for the same species and produce broad lines—a feature which is undesirable for high-resolution NMR spectroscopy. Thus, in order to correct for such inhomogeneities, there is a so-called shim system

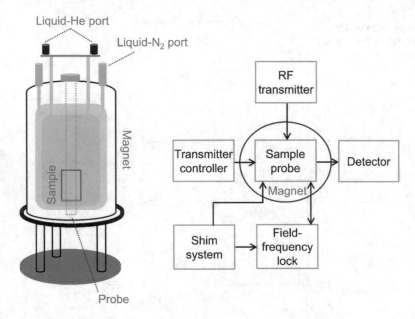

Fig. 1.10 Schematic representation of an NMR spectrometer with different components. The left portion represents a superconducting magnet, and the right portion represents the electronics

which is a set of current carrying coils producing small magnetic fields in different directions. The magnetic fields produced by these coils can be adjusted so as to effectively cancel the main field inhomogeneities. Thus, adjusting the shim system is an important and crucial step for obtaining a good NMR spectrum.

1.6   Kinetics of Resonance Absorption

As in the case of lattice-induced transitions, one can write rate equations for stimulated transitions caused by the radio frequency (RF) field:

$$\frac{dn}{dt} = P(n_\beta - n_\alpha) \tag{1.53}$$

where P is the RF-induced transition probability. By explicit quantum mechanical calculations, P can be shown to be the same for both upward and downward transitions and is given by

$$P = (1/4)\,\gamma^2 H_1^2 \tag{1.54}$$

Using the definitions in Eq. 1.32,

$$\frac{dn_\alpha}{dt} = \left(\frac{1}{2}\right)\frac{dn}{dt} = -Pn \tag{1.55}$$

or,

$$\frac{dn}{dt} = -2Pn \tag{1.56}$$

The solution of Eq. 1.56 can be readily written as

$$n(t) = n(0)\, e^{(-2Pt)} \tag{1.57}$$

where $n(0)$ is the population difference at time $t = 0$. Since every stimulated transition results in either absorption of energy ($\alpha \rightarrow \beta$) or emission of energy ($\beta \rightarrow \alpha$), the net rate of upward transition determines the net rate of energy absorption from the RF field. This is given by

$$\frac{dE}{dt} = n_\alpha P\left(E_\beta - E_\alpha\right) - n_\beta P\left(E_\beta - E_\alpha\right) \tag{1.58}$$

$$= nP\Delta E \tag{1.59}$$

$$= P\Delta En(0)e^{(-2Pt)} \tag{1.60}$$

Equation 1.60 indicates that if the RF is applied for a long time, the rate of energy absorption goes to zero, and then no more energy will be absorbed by the system from the RF. Explicitly, integrating Eq. 1.60,

$$\int dE = \int P\Delta En(0)e^{(-2Pt)}dt \tag{1.61}$$

$$E = P\Delta En(0)(-1/2P)\left[e^{(-2Pt)}\right]_0^t \tag{1.62}$$

$$E = \frac{\Delta En(0)}{2}\left[1 - e^{(-2Pt)}\right] \tag{1.63}$$

Figure 1.11a, b shows plots of dE/dt and E vs t as per Eqs. 1.60 and 1.63, respectively. It is seen from Fig. 1.11b that after some time, there is no more absorption of energy, and this means the NMR signal will disappear.

In reality however, we continuously observe the signal. There must be therefore some other factors which are continuously restoring the population difference between the two levels. This is not hard to see; it is the spin-lattice relaxation which we have discussed a little earlier in the previous sections. The effects of relaxation have not been explicitly included in the above calculations. On including these effects, Eq. 1.56 gets modified to

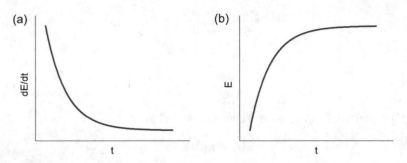

Fig. 1.11 Dependence of (**a**) dE/dt and (**b**) E on t

$$\frac{dn}{dt} = -2Pn - \frac{(n - n^o)}{T_1} \tag{1.64}$$

The solution of this equation can be easily obtained and is given by

$$n(t) = \left(\frac{n}{s}\right)\left[(s - 1)e^{(-st/T_1)} + 1\right] \tag{1.65}$$

where

$$s = (1 + 2PT_1) \tag{1.66}$$

At equilibrium or steady state,

$$\frac{dn}{dt} = 0 \tag{1.67}$$

and

$$2Pn' = \frac{(n^o - n')}{T_1} \tag{1.68}$$

or

$$n' = \frac{n^o}{(1 + 2PT_1)} \tag{1.69}$$

where n' is the steady-state population difference in the presence of the RF field. The rate of energy absorption is then

$$\frac{dE}{dt} = n'P\Delta E = \frac{n^o \Delta E P}{(1 + 2PT_1)} \tag{1.70}$$

If $2PT_1 \ll 1$, the rate of energy absorption will not go to zero, and there will be sustained signal in the experiment. On the other hand, if $2PT_1 \gg 1$, the rate dE/dt can become very small, and the signal intensity will be very low. Thus, the denominator $(1 + 2PT_1)$ in Eq. 1.70 plays a very important role in signal observation and is termed the *saturation factor*. This depends both on the strength of the RF field and the relaxation time T_1 of the system. If T_1 is long, the system will get saturated even at low RF fields causing problems in signal detection. Conversely, for short T_1 systems, high powers can be used safely without causing significant saturation.

1.7 Selection Rules

Until now we considered resonance absorption of energy by a simple ensemble of spins with two energy levels. Although the general discussion applies to more complex spin systems as well, some additional comments need to be made explicitly. It turns out that the stimulated transitions obey a certain selection rule and this follows from the explicit formula for the transition probability, which is given as follows:

$$P = \gamma^2 H_1^2 |< m' |\widehat{I_x}| m >|^2 \qquad (1.71)$$

where m and m' are the m values of Eq. 1.11 for the final and initial states of the transition, respectively. Here the RF is assumed to be applied along the $x-$direction, and hence the $\widehat{I_x}$ operator appears in Eq. 1.71. The effects of the operator $\widehat{I_x}$ or $\widehat{I_y}$ on the spin states can be calculated as indicated in Box 1.2. Here, I^+ and I^- are called the raising and lowering operators. The I^+ operator operating on the state $|I, m>$ produces the spin state $|I, m + 1>$. Therefore, it is called raising operator. Likewise, the lowering operator, I^- operating on the spin state $|I, m>$ produces the state $|I, m - 1>$.

Box 1.2: The Relationship Between $(\widehat{I_x}, \widehat{I_y})$ and (I^+, I^-) Operators and the Effect of I^+, I^- Operators on Spin-State $|I, m>$

$$\widehat{I_x} = \frac{I^+ + I^-}{2}, \widehat{I_y} = \frac{I^+ - I^-}{2i}$$

$$I^+ = |I, m > \hbar\sqrt{(I - m)(I + m + 1)}|I, m + 1 >$$

$$I^- = |I, m > \hbar\sqrt{(I + m)(I - m + 1)}|I, m - 1 >$$

A more general situation would be to consider the RF to be applied anywhere in the transverse plane (with respect to the external field); however, the conclusions regarding the selection rule will not be affected by such a consideration. Equation 1.71 is symmetrical with regard to interchange of m and m'. The quantum mechanical matrix element $< m' |\widehat{I_x}| m >$ vanishes unless

Fig. 1.12 Zeeman splitting
in a spin system of $I = 1$

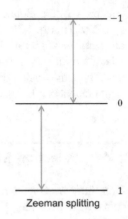

Zeeman splitting

$$| m - m' |= 1 \qquad (1.72)$$

and this puts a condition on the allowed transitions. Thus, for a $I = 1$ system, with three equally spaced energy levels, there will be only two transitions which will be degenerate as shown in Fig. 1.12. The transition from the $m = 1$ state to the $m = -1$ state is a disallowed transition. The symmetry of Eq. 1.71 indicates that the transition probability will be the same for upward and downward transitions.

Individual $(2I+1)$ eigenstates are orthogonal to each other.

1.8   Line Widths

According to Eq. 1.71, resonance absorption should occur at a single sharp frequency given by

$$v = \frac{| E_m - E_{m'} |}{h} \qquad (1.73)$$

However, this never happens in practice, and there is always a width for the resonance line; that is, resonance absorption occurs over a certain range of frequencies around the central frequency. The intensity of the signal is of course maximum at the central frequency and decreases as one moves away from it. The width of the resonance line at half of its maximum height is called its *line width*. Now, what is the origin of this line width? The fundamental reason for the width of a resonance line is the finite lifetime of a spin in the upper state of the spin system, and this follows from the quantum mechanical *uncertainty principle*:

$$\Delta E \Delta t \approx h \qquad (1.74)$$

where h is the Planck's constant. Δt is the uncertainty in time (can be chosen to be as much as the lifetime of the state), and this defines the uncertainty in the energy of the state. This leads to an uncertainty in the energy of absorption and thus to the width of a line. There are several factors contributing to the width of a line, and these will be discussed in brief in the following paragraphs.

(a) Spontaneous emission: Every transition has an intrinsic width due to a phenom-
    enon called spontaneous emission. This is a phenomenon which arises due to
    interaction of electromagnetic radiation with matter, in general, and depends on
    a variety of factors such as the energy of the transition, intensity of the radiation,
    etc. In the case of NMR, however, spontaneous emission is weak (transition
    probability $= \frac{3}{\lambda^3} \left(8\pi^2\gamma^2\right) 10^{\tilde{\,}-25} \sec^{-1}$) and thus does not contribute significantly to
    the line widths.
(b) Width due to spin-lattice relaxation: The transitions induced by the lattice also
    limit the lifetimes of both the upper and the lower states of the two-level system.
    The contribution of this to the width is then dependent on the efficiency of
    interaction between the spin system and the lattice in causing the transitions.
    This, in turn, depends on the molecular motions and the magnetic field
    fluctuations caused by the tumbling of the spins in the lattice. The longer the
    spin-lattice relaxation time, the shorter the line width and vice versa.
(c) Spin-spin interactions or magnetic dipole broadening: The spins in each state
    interact with each other. If the lifetime of the state is long, then such a dipole-
    dipole interaction causes further splitting of the closely spaced levels, and
    transitions can occur from any of these levels. This results in many transitions
    with slightly different frequencies which are not resolved and contribute to an
    overall broadening of the resonance line. This mechanism contributes the
    maximum to the line widths and is particularly dominating in solids and viscous
    liquids. Different frequencies would lead to dephasing of the transverse magne-
    tization or in other words, loss of phase coherence created by the application of
    radio frequency (RF).
(d) Magnetic field inhomogeneity effects: If the field over the volume of the sample
    is not homogeneous, different spins in different volume regions absorb energy at
    different frequencies, resulting in an overall broadening of the line.
(e) Electric quadrupole effects: The nuclei with spin >1/2 have quadrupole
    moments which is a consequence of the nonspherical shape of the nucleus.
    These quadrupole moments interact with the electric field gradients at the
    nucleus and cause spin-lattice relaxation. Thus, the lifetimes get altered and
    eventually lead to the broadening of the lines.

   The net effect of line broadening is that the spins have different precessional
frequencies. This manifests in the decay of x-y magnetization whenever it is created
by the application of the RF as described earlier. Taking into account all the above
factors leading to broad lines, the definition of transition probability has to be
modified so as to show dependence on frequency and is given as

$$P = (1/4)\,\gamma^2 H_1{}^2 g(v) \tag{1.75}$$

where $g(v)$ is referred to as the *line shape function*.

1.9   Bloch Equations

Considering the effects of spin-lattice and spin-spin relaxations in causing line widths, Felix Bloch obtained explicit expressions for the line shape function. These are described below in a qualitative fashion.

Felix Bloch wrote a set of phenomenological equations with macroscopic magnetization M to describe the NMR phenomenon:

$$\frac{\mathrm{d}M}{\mathrm{d}t} = \gamma(M \times H_{\mathrm{eff}}) \tag{1.76}$$

where **M** is a vector representing the total magnetization

$$M = (M_x, M_y, M_z) \tag{1.77}$$

M_x, M_y, and M_z are the components of **M** along x-, y-, and z-axes, respectively, and H is the total magnetic field. It consists of the static field H_o and a rotating field $H_1 \exp(-i\omega t)$ (without any loss of generalizations, the oscillating field is conveniently replaced by a rotating field for easier explanation of the behavior of the spin system and the NMR phenomenon). In the absence of the RF field, the macroscopic magnetization vector M is aligned along the z-axis. When RF is applied, the vector M slowly tilts towards the new effective field direction and eventually aligns along its direction. Now, since the RF is a rotating field, the effective field also rotates and consequently the magnetization vector also rotates with its x and y components oscillating in time with the frequency of the RF (Fig. 1.13).

The explicit behavior of the various components can be described by expanding the vector Eq. 1.76 as follows:

$$\frac{d\left(\vec{i}\,M_x + \vec{j}\,M_y + \vec{k}\,M_z\right)}{dt} = \gamma\left(\vec{i}\,M_x + \vec{j}\,M_y + \vec{k}\,M_z\right)$$
$$\times \left(\vec{i}\,H_x + \vec{j}\,H_y + \vec{k}\,H_z\right) \tag{1.78}$$

where $\vec{i}$, $\vec{j}$, and $\vec{k}$ are the unit vectors along the x, y, and z directions, respectively. Substituting $H_x = H_1 \cos \omega t$ and $H_y = -H_1 \sin \omega t$, one obtains

$$\frac{\mathrm{d}M_x}{\mathrm{d}t} = \gamma(M_y H_o + M_z H_1 \sin \omega t) \tag{1.79}$$

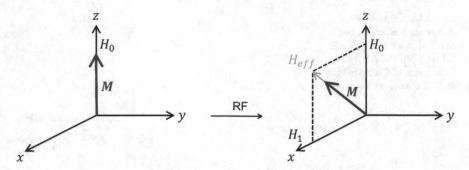

Fig. 1.13 Tilting of the magnetization (M) towards the effective magnetic field, H_{eff}

$$\frac{\mathrm{d}M_y}{\mathrm{d}t} = \gamma(M_zH_1\cos\ \omega t - M_xH_o) \tag{1.80}$$

$$\frac{\mathrm{d}M_z}{\mathrm{d}t} = \gamma\big(-M_xH_1\sin\ \omega t - M_yH_1\cos\omega t\big) \tag{1.81}$$

These do not include effects of relaxation, which we know also affect the rates of changes of net magnetization of the system and consequently attainment of the steady-state condition. Since the effect of relaxation is to reduce the excess magnetization above equilibrium values, Bloch added two empirical terms considering relaxations as first-order processes. Thus, the modified equations including the effects of relaxations were written as

$$\frac{\mathrm{d}M_x}{\mathrm{d}t} = \gamma\big(M_yH_o + M_zH_1\sin\ \omega t\big) - \frac{M_x}{T_2} \tag{1.82}$$

$$\frac{\mathrm{d}M_y}{\mathrm{d}t} = \gamma(M_zH_1\cos\ \omega t - M_xH_o) - \frac{M_y}{T_2} \tag{1.83}$$

$$\frac{\mathrm{d}M_z}{\mathrm{d}t} = \gamma\big(-M_xH_1\sin\ \omega t - M_yH_1\cos\ \omega t\big) - \frac{(M_z - M_o)}{T_1} \tag{1.84}$$

Here T_2 is a new relaxation time constant introduced to describe the decay of the transverse magnetization in analogy with the time constant T_1 which characterizes the decay of the excess z- magnetization. Here, M_o is the equilibrium z-magnetization. The solutions of Eqs. 1.82–1.84 are best obtained by going into the rotating frame of the RF. In this frame, the effective field is stationary, and consequently the magnetization vector is also stationary. Thus, explicit time dependence will be eliminated from the equations, and the equations assume simpler forms. If $\vec{u}$ and $\vec{v}$ are the components of the $\vec{M}$ vector along and orthogonal to the RF direction (Fig. 1.14), then

Fig. 1.14 Transverse components of the magnetic moment in laboratory frame (M_x and M_y) and RF frame (u and v) are depicted in black and blue lines, respectively

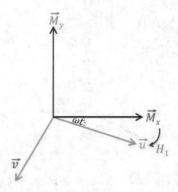

$$M_x = u \cos \omega t - v \sin \omega t; M_y = -u \sin \omega t - v \cos \omega t \qquad (1.85)$$

Substituting these in Eqs. 1.82–1.84, one obtains

$$\frac{du}{dt} + \frac{u}{T_2} + (\omega_o - \omega)v = 0 \qquad (1.86)$$

$$\frac{dv}{dt} + \frac{v}{T_2} - (\omega_o - \omega)u + \gamma H_1 M_z = 0 \qquad (1.87)$$

$$\frac{dM_z}{dt} + \frac{(M_z - M_o)}{T_1} - \gamma H_1 v = 0 \qquad (1.88)$$

At steady state, which is what we need to consider for resonance absorption of energy, all the time derivatives in Eqs. 1.86–1.88 must vanish. This leads to the following solutions:

$$u = M_o \frac{\gamma H_1 T_2^2 (\omega_o - \omega)}{1 + T_2^2 (\omega_o - \omega)^2 + \gamma^2 H_1^2 T_1 T_2} \qquad (1.89)$$

$$v = -M_o \frac{\gamma H_1 T_2}{1 + T_2^2 (\omega_o - \omega)^2 + \gamma^2 H_1^2 T_1 T_2} \qquad (1.90)$$

$$M_z = M_o \frac{1 + T_2^2 (\omega_o - \omega)^2}{1 + T_2^2 (\omega_o - \omega)^2 + \gamma^2 H_1^2 T_1 T_2} \qquad (1.91)$$

The experimental setup can be designed to observe either the u component or the v component of the magnetization in the rotating frame. Under resonance conditions, i.e., $\omega = \omega_o$, Eq. 1.91 reduces to

$$M_z = M_o \frac{1}{1 + \gamma^2 H_1^2 T_1 T_2} \tag{1.92}$$

Comparing this with Eq. 1.69, we recognize that the denominator in Eq. 1.92 is the saturation factor, and one obtains for P:

$$P = \gamma^2 H_1^2 \frac{T_2}{2} \tag{1.93}$$

Thus, at resonance, comparing with Eq. 1.75,

$$[g(v)]_{\omega_o = \omega} = 2T_2 \tag{1.94}$$

Under conditions of no saturation $\gamma^2 H_1^2 T_1 T_2 \ll 1$, the observed signals ($u$ and v components) will have the line shapes as shown in Fig. 1.15, and the functional forms are

$$v \approx \frac{2T_2}{1 + T_2^2 (\omega_o - \omega)^2} \tag{1.95}$$

$$u \approx \frac{(\omega_o - \omega)}{1 + T_2^2 (\omega_o - \omega)^2} \tag{1.96}$$

In Eq. 1.95, the term $2T_2$ is kept for consistency with Eq. 1.94.

These line shapes are referred to as absorptive and dispersive line shapes, respectively. In the absorptive line shape, the signal is maximum at the resonance frequency and dies off at the ends. In contrast, in the dispersive line shape, the signal has zero intensity at the resonance position and shows a maximum and a minimum at symmetrical positions away from the resonance. The maximum and minimum occur at frequencies at which the signal intensity is half of its maximum in the absorptive line shape. It is also seen that the dispersive line shape has longer tails on both sides, in the sense that the signal is still present at frequencies at which the absorptive

Fig. 1.15 ω-dependent line-shapes of $[g(v)]_v$ and $[g(v)]_u$

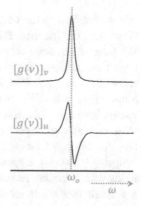

$[g(v)]_v$

$[g(v)]_u$

ω_o

ω

signal has already vanished. The total area under a dispersive line shape is zero in contrast to the maximum value for an absorptive line shape. It is a common practice to represent NMR spectra as consisting of absorptive lines; however, under special circumstances such as for "field frequency locks," dispersive line shapes are used. In practice, the observed line shapes deviate slightly from the two line shapes shown in Fig. 1.15, and this is a consequence of the phenomenological nature of the Bloch equations.

The intensity of the absorption signal can be obtained by integrating Eq. 1.90. One finds that the area under the curve (A) which represents the total signal intensity is given by

$$A \propto \frac{\chi_o H_1}{\left(1 + \gamma^2 H_1{}^2 T_1 T_2\right)^{1/2}} \tag{1.97}$$

and the height of the peak (h) can be derived by putting $\omega_o = \omega$ in Eq. 1.90:

$$h \propto \frac{\chi_o H_1 T_2}{\left(1 + \gamma^2 H_1{}^2 T_1 T_2\right)}$$

$$\approx \chi_o H_1 T_2 \text{ (under conditions of low RF power)} \tag{1.98}$$

where χ_o is the equilibrium magnetic susceptibility of the system. This indicates that peak heights of two lines in a given spectrum can be compared only if their T_2's (or line widths) are identical. The dependence of the intensity on the susceptibility indicates that the intensity is proportional to the number of nuclei per unit volume of the sample which absorb energy at that frequency and also on the square of the magnetic moment of the nucleus. These observations lead to the fact that NMR signal intensities can be used to measure the number of nuclei of any particular type in a given sample.

1.10   More About Relaxation

We now return to the question: What brings about the phenomenon of relaxation? How does the magnetization in a perturbed system return to its equilibrium value? We have to consider here two types of relaxations introduced before, namely, spin-lattice relaxation, also referred to as longitudinal relaxation (T_1), and spin-spin relaxation, also referred to as transverse relaxation (T_2). It is easily realized that longitudinal relaxation has to do with the populations of the spins in the different energy levels, and the transverse relaxation dealing with x-y magnetization has to do with the phase coherence of the precessing magnetization components in the ensemble. As we have already seen before, spin-lattice relaxation is brought by tight coupling between the spin system and the lattice, and the latter must produce rotating fields with appropriate frequencies. Transverse relaxation representing the decay of x-y magnetization must somehow come from random time-dependent changes in frequencies of precessing spins resulting in the loss of phase coherence between them. Notice however that T_2 relaxation does not change the populations of the energy levels and is thus energy conserving. Therefore, this phenomenon can be considered as an entropy effect since the loss of phase coherence between the spins amounts actually to the loss of magnetic order in the spin system. Now, random fluctuations in precession frequencies must imply random fluctuations in the energy values of the energy levels. How do these changes come about? Note that there will be no change in the average energy of a given energy level. Extending the logic further, any change in the value of the energy of a level must come from magnetic interactions with other spins in the system, and these interactions must be fluctuating in time so as to cause fluctuations in the energy values. Thus, it boils down to the fact that T_2 relaxation is caused by fluctuating magnetic fields in the spin system. The same kind of fluctuations are responsible for coupling between the spin system and the lattice and lead to T_1 relaxation as well. Thus, the fundamental mechanisms of both T_1 and T_2 relaxations are essentially the same.

Fluctuating magnetic fields are caused by motions of spins, both in the lattice and in the spin system itself. Every magnetic dipole creates magnetic flux lines around it, and the field varies in both magnitude and direction along these flux lines. Therefore, these fields influence other dipoles in the vicinity, and this varies with distance and relative orientations. The maximum field created by a magnetic dipole with a magnetic moment μ, at a distance R, is $2\mu/R^3$, and this happens to be along the direction of the dipole itself. For a proton, this is 57 Gauss for a distance of 1Å, which is quite a large field indeed. The field at any point due to a dipole keeps fluctuating, both due to translational motions in the solution and also due to precessional motion of the spin. The latter however may not cause too large a change in the distance R if we consider the dipole to be too tiny. If the fluctuations are slow, compared to the lifetime of the states, there will be no averaging effect, and then instead of all nuclei experiencing the same field H, different nuclei in a sample will experience different fields and absorb energy at different frequencies within the above range. This leads to the broadening of the lines. On the other hand, if the fluctuations are faster, they behave like oscillating fields and induce transitions. This phenomenon is pictorially represented in Fig. 1.16.

Every nucleus has magnetic field flux lines around it, and as the nucleus precesses around the field H_o, the flux lines also move with it in a conical fashion. Note that reorientational or tumbling motion of the nucleus in a liquid does not change the orientation of the spin since this is dictated by the main magnetic field orientation and the value of the spin I of the nucleus. Thus, every precessing spin produces a rotating field which has components along the H_o direction and also in the transverse plane. Translational motions of the spins cause variations in the strength of the field at the site of any spin. Thus, there will be fluctuations in both z- and x-y-components of the fields at any particular nucleus. The fluctuations in the z-components are governed by the rates of translational motions within the liquid, and those in the x-y plane are governed by both precessional motions and translational motions of the

Fig. 1.16 Schematic representation of fluctuating magnetic fields due to the motion of dipoles. The flux lines created by one dipole at the site of another dipole are shown. R is the average distance between two precessing dipoles, for the same physical locations of the two. This keeps fluctuating as the dipole move in solution, and accordingly the magnetic fields also keep fluctuating

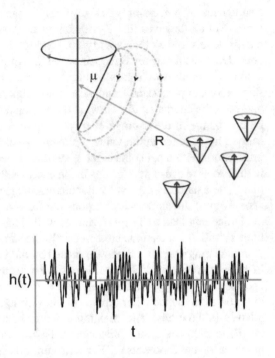

Fig. 1.17 Pictorial representation of the fluctuating magnetic field in the x-y plane as a function of time

spins. The former cause fluctuations in energy levels and thus contribute to line widths as discussed before. The fluctuations in the x-y plane behave like fluctuating *RF* fields with different frequencies and thus cause transitions between the nuclear energy levels.

A better and more rigorous insight into the transitions caused by the fluctuating fields is obtained by Fourier analyzing the fluctuations into its component frequencies. Figure 1.17 shows a pictorial representation of the fluctuating magnetic field in the x-y plane as a function of time.

If $h(t)$ is the fluctuating field about a mean value, zero, due to random motions of the spins, the frequency distribution of the fluctuations in a time interval $(-T, T)$ is given by the Fourier transform:

$$H_T(\omega) = \int_{-T}^{T} h(t)\, e^{-i\omega t} \mathrm{d}t \qquad (1.99)$$

$H_T(\omega)$ is itself a random quantity and if one takes an ensemble average, it vanishes. However, its square does not vanish and reflects the power distribution in the fluctuations. This is referred to as the *spectral density function J (ω)* and is defined mathematically as follows:

$$J(\omega) = \lim_{T \to \infty} \frac{1}{2T} \overline{H_T^*(\omega) H_T(\omega)} \qquad (1.100)$$

The horizontal bar indicates an ensemble average. The $*$ indicates the complex conjugate. It follows that if there is some power $J(\omega)$ at the resonance frequency of a nucleus, then the fluctuations will induce those transitions. Now, it is logical to expect that the spectral density function $J(\omega)$ would be related to the nature of the fluctuations in $h(t)$; in other words, it would depend upon whether the fluctuations have any correlations in time. This is also referred to as memory and will influence the frequency distributions and thus the power distribution in the fluctuations. The memory is described in terms of a correlation function $G(\tau)$ defined as an ensemble average as follows:

$$G(\tau) = \overline{h(t+\tau)^* h(t)} \qquad (1.101)$$

where the average is over time t. The $*$ represents the complex conjugate. If the memory is short, $G(\tau)$ dies away quickly and vice versa, implying that $G(\tau)$ has to be a monotonically decaying function. Considering memory as some sort of a coherence in the motions, its decay must come from interactions between the coherent field components and thus can be considered to follow a first-order process. In other words, $G(\tau)$ can be described as an exponentially decaying function:

$$G(\tau) = \overline{h^*(t)h(t)} e^{(-|\tau|/\tau_c)} \qquad (1.102)$$

where τ_c is a characteristic time called the *correlation time*. It is a measure of the rapidity of the fluctuations or of the extent of correlation in the reorientational motions in the ensemble. Now, since $J(\omega)$ is related to the fluctuations $h(t)$ by a Fourier relation and $G(\tau)$ reflects the memory in the fluctuations $h(t)$, it follows intuitively that both these functions are also related, and it has been shown that they are indeed Fourier transforms of each other; they form a Fourier pair.

$$J(\omega) = \int G(\tau) e^{(-i\omega\tau)} d\tau \qquad (1.103)$$

For an exponentially decaying $G(\tau)$, $J(\omega)$ turns out to be

$$J(\omega) = \overline{h^*(t)h(t)} \frac{2\tau_c}{1 + \omega^2 \tau_c^2} = k \frac{\tau_c}{1 + \omega^2 \tau_c^2} \qquad (1.104)$$

Figure 1.18 shows graphically the nature of $J(\omega)$ as a function of ω.

It is seen that the plot of $J(\omega)$ vs ω resembles a Lorentzian absorptive line, and depending upon the correlation time, the power distribution will be flat over different frequency ranges. In liquids, τ_cs are of the order of 10^{-10} to 10^{-13}, indicating that over a wide range of frequencies, namely, 10^{10} to 10^{13}, the power is uniform and thus can cause transitions for all such resonance frequencies. This covers the frequency ranges encountered in nuclear magnetic resonance and electron spin resonance techniques; in the latter case, one is talking about the transitions between

Fig. 1.18 Lorentzian line shape of $J(\omega)$

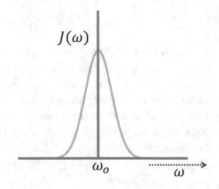

Fig. 1.19 $J(\omega)$ profiles for molecules with different τ_c. ω_o refers to the spectrometer frequency

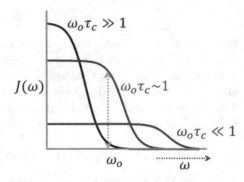

the electron spin levels. In Fig. 1.19, $J(\omega)$ is plotted for three different values of $\omega_o\tau_c$ to bring out these points in greater detail.

$\omega_o\tau_c \ll 1$ corresponds to fast motions, $\omega_o\tau_c \gg 1$ corresponds to slow motion, and $\omega_o\tau_c = 1$ corresponds to intermediate molecular motion. Now, we have said that $J(\omega)$ represents the spectral density distribution, and therefore its integral represents the total molecular power in the system. Integrating Eq. 1.104, writing K for the ensemble average in the expression, we find

$$\int J(\omega)\mathrm{d}\omega = K\frac{\pi}{2} \tag{1.105}$$

It is clear that the total molecular power in a system is constant and is independent of the correlation time. In other words, experimental conditions such as temperature, viscosity, pressure, etc. do not influence the total molecular power in the spin system. We notice from Fig. 1.19 that for fast motions, the power is distributed over a wide range of frequencies and likewise for slow motions, the power is concentrated in low-frequency motions. Since the total power is constant, it follows, as is evident in the figure, that the available molecular power at lower frequencies is much higher in a system with slow molecular motions than in a system with fast molecular motions. Since the rate of transitions between the energy levels is propor-tional to the molecular power available at the resonance frequency, it is obvious from

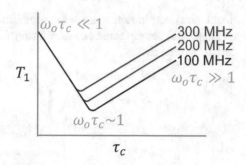

the figure that the transition rate is highest for intermediate motions. In other words, T_1 relaxation will be most efficient for the intermediate motion situation $\omega_o\tau_c = 1$ and thus will have the shortest T_1 value. On eithcr sides of this condition, T_1 would increase, and thus a plot of the relaxation time vs τ_c would look as shown in Fig. 1.20 with a minimum at the condition of intermediate motion.

Note that these motional conditions described above are relative to the frequency of the spectrometer. A particular fast motion condition, on a particular spectrometer, may turn out to be a slow motion condition at a higher-spectrometer frequency. For example, if, for a spectrometer, a particular sample has $\omega_o\tau_c = 0.5$, then the same sample would have $\omega_o\tau_c = 5.0$ at a spectrometer frequency which is ten times the previous one. While the former condition can be considered as fast motion situation, the latter condition belongs to the slow motion regime. Because of this fact, the T_1 values become spectrometer frequency dependent.

Until now, we have discussed the fluctuations in the magnetic fields due to motions and seen how they can cause transitions between the levels. Now the question arises, whether these transitions are also subject to the same kind of selection rules as in the case of induced transition by an applied *RF*. This however is not the case; the reason being that two nuclear magnetic dipoles are involved in these interactions as against a single nuclear magnetic dipole in the former case. Therefore, the perturbation to be used in the equation for Einstein transition probability is the "dipolar Hamiltonian." This perturbation has operator components which can cause transitions between levels with $\Delta m = 0, 1, 2$. That is, the spins of both the magnetic dipoles can flip in these interactions, and therefore, the selection rules will now become

$$\Delta m = 0, 1, 2 \tag{1.106}$$

$\Delta m = 0$ and 2 are called as zero-quantum and double-quantum transitions, respectively. Note that although in a spin system with several spins, there can be higher Δm values, dipole-dipole transitions will be restricted only to the three cases (Eq. 1.106), and thus there is also a selection rule here but different from the one for *RF*-induced transitions. As is evident from the definitions, a double-quantum transition will occur at $2\omega_o$ frequency, and a zero-quantum transition will occur at zero frequency. We notice from the spectral power distribution discussed above that there is indeed uniform power in the whole range for short correlation times, and thus, the fluctuations can indeed cause these transitions with high efficiency. Explicitly, the three spectral density functions causing zero, single, and double-quantum transitions can be written as

$$J_o(\omega) = 2\tau_c \overline{h^*(t)h(t)} \tag{1.107}$$

$$J_1(\omega) = \frac{2\tau_c}{1 + \omega^2\tau_c^2} \overline{h^*(t)h(t)} \tag{1.108}$$

$$J_2(\omega) = \frac{2\tau_c}{1 + 4\omega^2\tau_c^2} \overline{h^*(t)h(t)} \tag{1.109}$$

The above discussion of fluctuation-induced transitions explains how the spin-lattice relaxation can occur in a spin system. However, it is not yet clear how these transitions or the fluctuations can cause the energy conserving spin-spin or the transverse relaxation in a spin system. A little thought indicates that this can occur in three ways. Firstly, the fluctuations in the energy levels due to fluctuations in the $z-$component of the local fields cause fluctuations in precessional frequencies, thus resulting in the loss of phase coherence between the spins and consequently in the decay of net x-y magnetization. This will be a monotonous process. Secondly, if the upward and downward transitions between two spins can occur in correlated manner, the phase memory of the spins can be lost without changing the net energy of the spin system. This process is schematically shown in Fig. 1.21.

The spins k and l are identical nuclei, and their simultaneous transitions amount to spin exchange. The rate of such a spin exchange determines the extent of lifetime changes of the two states and thus determines the extent of line broadening by T_2 relaxation. Thirdly, if the motions are slow, changes in energy levels contribute to multiple frequencies of absorptions, which not only lead to the rapid loss of phase coherence, and the superposition of frequencies leads to substantial line broadening effects. Thus, T_2 decreases monotonously with increasing τ_c unlike the behavior of T_1 described before.

Further insight into the relaxation caused by the fluctuating fields can be obtained by Bloch equations for the interaction of the magnetic moments with the local field h_{loc}. Consider a component of the h_{loc} which rotates like a precessing nucleus

Fig. 1.21 Simultaneous upward and downward transitions between two spins in a correlated manner

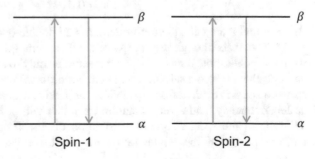

Fig. 1.22 Components of a rotating magnetic field, h, due to nuclear precession. These components will vary for every location on the flux lines. These fluctuations will cause fluctuating magnetic fields at another dipole in the vicinity

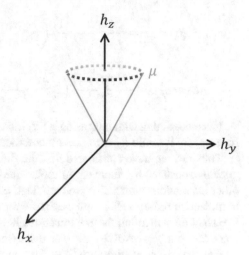

(Fig. 1.22). If h_x, h_y, and h_z are the components of h_{loc} along the x-, y-, and z-axes, respectively, Bloch equations can be written for these components as follows:

$$\frac{dM}{dt} = \gamma(M \times h_{loc})$$

$$h_{loc} = h_x + h_y + h_z$$

$$\frac{dM_x}{dt} = \gamma(h_y M_z - h_z M_y) \tag{1.110}$$

$$\frac{dM_y}{dt} = \gamma(h_z M_x - h_x M_z) \tag{1.111}$$

$$\frac{dM_z}{dt} = \gamma(h_x M_y - h_y M_x) \tag{1.112}$$

Remembering that h_x and h_y are high-frequency time-dependent quantities, while h_z is a slowly varying quantity, it can be observed that the rapidly varying components of the local field contribute to relaxation of all the magnetization components and that the slowly varying component contributes only to the relaxation of transverse magnetization components M_x and M_y. In other words, the rapidly varying component contributes to both T_1 and T_2 relaxation, and the slowly varying component contributes to T_2 relaxation alone. The explicit calculation of relaxation rates for the dipole-dipole interaction leads to the following expressions for two spin $\frac{1}{2}$ nuclei of the same type:

$$R_1 = \frac{1}{T_1} = \frac{2\gamma^4 h^2 I(I+1)}{5r^6} \left(\frac{\tau_c}{1 + \omega_o^2 \tau_c^2} + \frac{4\tau_c}{1 + 4\omega_o^2 \tau_c^2} \right) \tag{1.113}$$

$$R_2 = \frac{1}{T_2} = \frac{\gamma^4 h^2 I(I+1)}{5r^6} \left(3\tau_c + \frac{5\tau_c}{1 + \omega_o^2 \tau_c^2} + \frac{2\tau_c}{1 + 4\omega_o^2 \tau_c^2} \right) \tag{1.114}$$

It is evident that with increasing τ_c, R_1 increases initially and then approaches zero for $\omega_o \tau_c \gg 1$, while R_2 increases monotonically.

Thus far, we have considered that the fluctuating magnetic field at the site of a nucleus is caused by fluctuating dipolar field of another nucleus. There are in fact other interactions which also produce fluctuating magnetic field at a nuclear site due to molecular motions, although the dipole-dipole interaction is the major contributor. We shall only mention these interactions here without going into any great details, since the concepts involved are to be discussed explicitly in later chapters. The other mechanisms of relaxations are (i) scalar coupling-mediated relaxation, (ii) chemical shift anisotropy-mediated relaxation, (iii) chemical exchange-mediated relaxation, (iv) spin rotation relaxation, and (v) paramagnetic relaxation.

1.11   Sensitivity

One of the main concerns in any spectroscopic technique is the sensitivity of the technique. Sensitivity is defined as obtainable signal-to-noise ratio per unit measuring time in the experiment. In an NMR experiment, the signal intensity is determined by the voltage induced by the precessing magnetization in the receiver coil. A detailed calculation shows that the intensity is affected by a number of factors, some of which depend on the sample itself, and some others depend on the experimental arrangements such as the strength of the field, power of RF, properties of the probe, etc. When the experiment is optimized and there is no saturation, the maximum obtainable signal intensity is proportional to

$$N(I+1)\mu\omega_o^2 \tag{1.115}$$

at constant temperature under the condition that spin-lattice and spin-spin relaxation times are equal. In terms of field strength, the intensity is clearly proportional to

$$\frac{N(I+1)\mu^3 H_o^2}{I^2} \tag{1.116}$$

Several facts emerge from Eq. 1.116. First of all, the intensity depends upon the number of nuclei per unit volume (N) and thus is dependent on the concentration of the species in the sample. Second, the intensity depends strongly on the strength of the applied field. This follows from the fact that the strength of the field determines the separation between the nuclear energy levels, which in turn determines

Table 1.3 Relative sensitivities of some of the most common nuclei

Nucleus	Relative resonance frequency (MHz)	Relative sensitivity
^{1}H	100.000	1.0000
^{2}H	15.351	0.0036
^{13}C	25.144	0.0159
^{15}N	10.133	0.0010
^{19}F	94.077	0.8326
^{31}P	40.481	0.0663

the population difference between the levels. The larger the population difference, the higher the intensity and vice versa. Third, the intensity is inherently dependent on the type of the nucleus since the magnetic moment μ and spin I are inherent properties of the nuclei. The third power dependence on the magnitude of the magnetic moment indicates that it is clearly the dominating factor. In other words, the magnetogyric ratio plays a very dominant role, and the nuclei with higher γ values will have higher sensitivities. Among all the nuclei, proton has the highest γ value and therefore is the most sensitive nucleus. Using Eqs. 1.115 and 1.116 and the properties of different nuclei, it is possible to calculate the relative sensitivities of different nuclei. Taking the sensitivity of proton as 1.0, the sensitivities for all the NMR active nuclei have been calculated, and these are listed in tabular forms in several books. The relevant values for some of the most common nuclei are listed in Table 1.3. Relative sensitivities are calculated for the equal number of nuclei, and the numbers represent $(\gamma_X/\gamma_H)^3$; H refers to proton and X refers to any other nucleus. In the Table 1.3, the resonance frequency for ^{1}H is taken to be 100 MHz which corresponds to a magnetic field of 2.34 T. Absolute sensitivities will be the products of natural abundance and relative sensitivity for the individual nuclei.

It is clear from the Eq. (1.116) for the signal intensity that the sensitivity in a particular spectrum can be increased by increasing the magnetic field strength or by increasing the concentration of the nuclear species. In practice, however, the increase with respect to the field will be somewhat lower because of the noise component which has not been considered so far. It has been shown that the signal-to-noise ratio actually has a 3/2 power dependence on magnetic field and 5/2 power dependence on the magnetogyric ratio, as against the second power and third power dependences indicated by the equation, respectively. Different strategies have been used in the literature to enhance the sensitivity in an NMR experiment. The simplest thing to do is to signal average. This means the data must be collected several times and coadded. If n is the number of such coadditions, the signal increases proportionately to n, but the noise adds in proportion to square root of n. Thus, there is a net enhancement in the signal-to-noise (S/N) by a factor of the square root of n. But the price one pays for this is the measuring time in the experiment. The measuring time goes as the square of the desired enhancement in the S/N ratio in the spectrum. Alternatively, the effective volume of the sample in the magnet can be increased so that the total number of nuclei per unit volume in the magnet can be increased. This would involve better designing of the magnets. Finally, the magnetic field strength can be increased. This is of course limited by the magnet technology. The highest magnetic fields that are available today are of the order of 28.2 Tesla which corresponds to a frequency of 1200 MHz for proton. With any increase in magnetic field strength, the entire architecture of electronics associated with resonance absorption also changes, and these also put limitations on the developments.

1.12 Summary

- Every nucleus has a spin angular momentum which is characterized by a quantum number **I**. The angular momentum is a vector quantity. The quantum number of the ground state (lowest energy state) is referred to as the "nuclear spin," I. The magnitude of angular momentum is $h\sqrt{I(I+1)}$.
- A nucleus with a nonzero nuclear spin has a magnetic moment associated with it.
- The z-component of the angular momentum and hence of the magnetic moment can have only discrete values. For a nuclear spin I, there will be $2I + 1$ values of the z-component. These are represented by another quantum number m which is called the magnetic quantum number and may be taken to represent different orientations of the magnetic moment in space. All these have equal energy.
- In the presence of an external magnetic field, the different orientations have different energies.
- When electromagnetic radiation is applied whose energy is equal to the separation between two energy levels, whose m values differ by unity (selection rules), there will be absorption of energy. This is referred to as "resonance absorption." For all nuclei, this energy is in the radio frequency region of the electromagnetic spectrum.
- Bloch equations which provide a classical physics description of the NMR phenomenon are described.
- Concepts of spin-lattice relaxation time (T_1) and spin-spin relaxation time (T_2) are described, and their relation to molecular motion in solution media is explained.

1.13 Further Reading

1. Spin Dynamics by M. H Levitt, 2nd ed. Wiley 2008.
2. High Resolution NMR by J. A. Pople, W. G. Schnieder, H. J. Bernstein, McGraw Hill 1959.
3. Principles of Magnetic Resonance by C. P. Slichter, 3rd ed. Springer 1990.

1.14 Exercises

1.1. Which of the nuclei given below has no magnetic moment?
(a) ^1H (b) ^2D (c) ^{12}C (d) ^{15}N

1.2. If the proton absorbs energy at 500 MHz frequency, then at which frequency nitrogen absorbs?
(a) 10 MHz   (b) 20 MHz   (c) 30 MHz   (d) 50 MHz

1.3. NMR uses the following frequency electromagnetic radiations.
(a) UV   (b) Visible   (c) Radio frequency   (d) Microwave

1.4. The resonance frequency of a proton in a magnetic field strength of 10 Tesla equals to
(a) 42 MHz   (b) 0.42 MHz   (c) 420 MHz   (d) 210 MHz

1.5. Mark the more sensitive nucleus among the following.
(a) ^{31}P (b) ^2D (c) ^{13}C (d) ^{15}N

1.6. The magnetic moment μ of electron is
(a) greater than the magnetic moment of proton
(b) equal to the magnetic moment of proton
(c) less than the magnetic moment of proton
(d) equal to the magnetic moment of carbon

1.7. The magnetic moment of the nucleus is related to
(a) the charge of the nucleus
(b) the angular momentum of the nucleus
(c) the orbital motion of the electron
(d) the mass of the electron

1.8. In a rotating frame, the field along the z-axis is for a particular nucleus is 2 kHz in frequency units. The amplitude of the radiofrequency (RF) is 1 kHz in frequency units. The magnitude of the effective field in frequency units is
(a) 3 kHz   (b) 1 kHz   (c) $\sqrt{5}$ kHz   (d) 5 kHz

1.9. In an NMR experiment, the radiofrequency is oriented
(a) parallel to the main field
(b) at 90° to the main field
(c) antiparallel to the main field
(d) at 45° to the main field

1.10. Among the following techniques, which is the least sensitive?
(a) UV-visible spectroscopy
(b) Infrared spectroscopy
(c) NMR spectroscopy
(d) Microwave spectroscopy

1.11. For a nucleus with spin 3/2, the number of allowed transition is
(a) 3   (b) 6   (c) 12   (d) 2

1.12. In a NMR spectrum, which factor does not contribute to the line width?
(a) Spin-lattice relaxation
(b) Spin-spin relaxation
(c) Magnetic field inhomogeneity
(d) Spontaneous emission

1.13. Radiofrequency-induced transition probability is proportional to
    (a) the spin of the nucleus
    (b) the amplitude of the radiofrequency
    (c) the square of the amplitude of the radiofrequency
    (d) the angular momentum of the nucleus

1.14. The width of the line at half maxima in the NMR spectrum at low RF power is inversely proportional to
    (a) spin-spin relaxation (T_2)
    (b) spin-lattice relaxation (T_1)
    (c) radio-frequency amplitude
    (d) strength of the magnetic field

1.15. A particular proton has a spin-lattice relaxation time (T_1) of 2 s on a 100 MHz spectrometer. Its value on a 300 MHz spectrometer under the same solution conditions and $\omega_0 \tau_c \gg 1$ will be
    (a) 2 s   (b) >2 s   (c) <2 s   (d) none of the above

1.16. A nucleus with a spin value of 5/2 has the following number of degenerate states:
    (a) 10   (b) 4   (c) 5   (d) 6

1.17. Show that the RF-induced transition probability is identical for upward and downward transitions in a spin ½ system.

1.18. Plot the approach to new equilibrium populations (Eq. 1.43) when an ensemble of proton spins is taken from zero field to a given field. Assume T_1 of 0.1 s, 1 s, and 10 s.

1.19. Plot the changes in the effective field when the RF is swept from -100 Hz to $+100$ Hz through resonance, assuming the RF field amplitude in frequency units to be 25 Hz.

1.20. Derive the Bloch equations in the rotating frame.

1.21. For a Lorentzian absorptive line shape, show that the width LW at a height $h = \left(\frac{1}{n}\right) h^{\max}$ where is $h^{\max}$ and the maximum height is given by $LW = \frac{2}{T_2} \times \sqrt{n-1}$.

1.22. What happens if the RF is applied along the z-axis?

High-Resolution NMR Spectra of Molecules

Contents

2.1   Introduction 42
2.2   Chemical Shift 43
      2.2.1   Anisotropy of Chemical Shifts 46
      2.2.2   Factors Influencing Isotropic Chemical Shifts 46
2.3   Spin-Spin Coupling 50
2.4   Analysis of NMR Spectra of Molecules 54
      2.4.1   First-Order Analysis 55
      2.4.2   Quantum Mechanical Analysis 57
2.5   Dynamic Effects in the NMR Spectra 76
      2.5.1   Two-Site Chemical Exchange 77
      2.5.2   The Collapse of Spin Multiplets 80
      2.5.3   Conformational Averaging of J-values 85
2.6   Summary 85
2.7   Further Reading 86
2.8   Exercises 86

Learning Objectives
- Basic NMR parameters of molecules: chemical shifts and scalar coupling constants
- Spectral features of molecules
- Analysis of NMR spectra to extract the NMR parameters
- Dynamic aspects of NMR spectra

© The Author(s), under exclusive license to Springer Nature Switzerland AG 2022
R. V. Hosur, V. M. R. Kakita, *A Graduate Course in NMR Spectroscopy*,
https://doi.org/10.1007/978-3-030-88769-8_2

2.1   Introduction

An NMR spectrum is a display of the resonance frequencies of a particular type of nucleus in a given sample. The sample can be either in the solid state or in the liquid state or in the gaseous state. Accordingly, the structure and appearance of the spectrum will be different. Solid-state NMR spectra of powder samples are actually a superposition of spectra of molecules in different orientations and contain broad lines without structure. The gaseous-state spectra, in contrast, contain sharp lines, but the signal intensities are low due to a small number of molecules present in a given volume of the sample. For chemists and the biologists, the most important are the spectra in the liquid state. In solution media, the molecules tumble around randomly and rapidly, and as a consequence, the dipole-dipole intermolecular interactions which result in substantial complications in the spectra average out to zero. Therefore, liquid-state spectra also contain sharp lines and are termed as high-resolution spectra. All molecules in the solution can be considered to behave independently and consequently, the NMR spectrum can be supposed to represent an individual molecule, and the observed spectrum is a sum of identical contributions from all the molecules in the ensemble. We will be concerned only with this branch of NMR spectroscopy, in the entire course of this book.

Molecules are made up of atoms and atoms contain nuclei. The examination of the periodic table little carefully reveals that almost every element has an isotope which has a magnetic moment. That means it should be possible to obtain an NMR spectrum of every nucleus type in a molecule. However, some nuclei have a quadrupole moment which causes extensive line broadening, and the spectra of such nuclei are not amenable to analysis by the standard procedures described in this chapter.

We have already seen in the previous chapter that the different nucleus types absorb energy at different nonoverlapping frequency regions because of their different gyromagnetic ratios. Therefore, in a system consisting of different types of nuclei, it is possible to selectively observe the resonances of a particular nucleus type and derive specific information. Thus, for a given molecule, it is possible to think of a ^{1}H-spectrum, a ^{13}C-spectrum, a ^{15}N spectrum, etc. This is what makes NMR one of the most powerful spectroscopic techniques having applications in several branches of science, namely, physics, chemistry, biology, and medicine.

How does an NMR spectrum of a molecule look like? If we are observing the protons, do all the protons in the molecule absorb energy at the same frequency? If it were so, the information derivable from the spectra would be very limited. Fortunately, this is not so. The molecules contain nuclei distributed in a well-defined electron cloud, and various nuclei and electrons interact with each other. This must in some way influence the NMR spectrum. These interactions cause differences in the linewidths as we already discussed before, and they also cause differences in resonance absorption frequencies for various proton types in different local surroundings. Indeed, the NMR spectra contain an enormous amount of information regarding the structure and dynamics in the molecules. This information is described in terms of a set of NMR parameters, whose extraction and interpretation are the ultimate objective of a chemist or a biologist or an NMR spectroscopist in general. These parameters are (1) chemical shifts, (2) spin-spin coupling constants, (3) intensities of resonance lines, and (4) relaxation times and linewidths. While the first three parameters are intimately connected with the three-dimensional structures of molecules, the last parameter describes the dynamism in the molecules. The focus in this chapter is on the structural parameters with which the chemists and the biologists are concerned most of the time.

2.2   Chemical Shift

Discovered in 1951, this is the most important and fundamental parameter which must be determined for any kind of analysis and interpretation of NMR spectra. In a molecule containing several nuclei of a particular type, say ^1H or ^{13}C, the different nuclei in different parts of the molecule having different environments absorb energy at different frequencies, and the chemical shift of a nucleus quantifies this electronic environmental difference. For example, Fig. 2.1 shows the spectrum of a simple molecule, namely, ethyl alcohol recorded at low resolution by Dharmatti and coworkers, the discoverers of the chemical shift phenomenon in liquids.

The spectrum shows three lines belonging to methyl, methylene, and hydroxyl protons in the molecule. These three groups of protons have different electronic environments and consequently resonate at different frequencies. The intensities of the individual lines are in the same proportion as the number of protons of each type, namely, 3:2:1.

What is the origin of the chemical shift? Detailed theories have been worked out by several authors. A complete description of these theories is beyond the scope of this book; they can be found in the excellent texts by Pople et al., Emsley et al., and

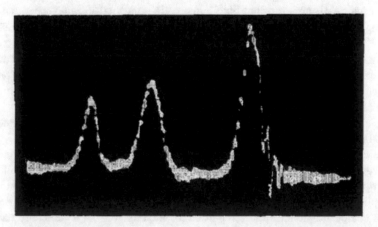

Fig. 2.1 ^1H NMR spectrum of ethanol recorded at low resolution by Dharmatti and co-workers. (Reproduced from J. Chem. Phys. **19**, 507 (1951), with the permission of AIP Publishing)

many others. Briefly, the phenomenon arises due to the fact that the electronic environment around a nucleus screens the applied magnetic field. The applied field induces currents in the electron cloud which produces a magnetic field that mostly opposes the externally applied magnetic field. As a result, the magnetic field appears altered at the site of the nucleus. This is termed as screening by the electron cloud. Because of the asymmetry of the electron cloud around a nucleus in a given molecule, which has a definite shape, the induced field will have different orientations, depending on the orientation of the molecule with respect to the applied field. However, it is the z-component of this field which is of importance, and the ensemble average of this component determines the extent of screening. It is important to mention here that electronic currents in any given orbital can change their sense of rotation as the molecule tumbles in solution, so that the z-component of the field that is produced would oppose the external field, in a diamagnetic situation. Because of this, the average of the induced field would not vanish. Depending on the extent of screening, the resonance frequency will be obviously different. The effective field at the site of a nucleus is given by the equation

$$H_{loc} = H_o(1 - \sigma_{loc})$$ (2.1)

In general, σ_{loc} can be positive or negative; a positive value implies shielding, and a negative value implies deshielding. A detailed calculation indicates that the screening constant has actually two components:

$$\sigma_{loc} = \sigma_d + \sigma_p$$ (2.2)

where σ_d is called the diamagnetic contribution and σ_p is called the paramagnetic contribution. Strong paramagnetic contributions occur only for heavier nuclei where energetically low-lying atomic orbitals are available and the applied field causes mixing of the wave functions of these excited states with the ground state wave functions. Therefore, for protons, it is mostly the diamagnetic contribution which is of importance. It turns out that the diamagnetic contribution is positive, while the paramagnetic contribution is negative. Thus, the sign of the screening constant in general is dependent on which is the dominant contributor. Further, as will be discussed later, the remote electronic currents in complex molecules can also influence the screening constant at a given nucleus. These effects can be both negative and positive depending upon the relative orientation of the remote group, which can alter the field at the site of the nucleus under question.

The changes in the energy levels due to screening for a single spin 1/2 system are shown in Fig. 2.2.

The change in the field at the site of the nucleus due to the screening is called chemical shift, and σ is called the screening constant or the shielding constant which is a property of the electronic environment. Clearly, the chemical shift depends on the applied field H_o. The higher the field, the higher will be the separation between the resonance frequencies of any particular nucleus type in a molecule and vice versa. But a consequence of this is that, in a complex system consisting of several absorption frequencies, chemical shifts will be different at different magnetic fields,

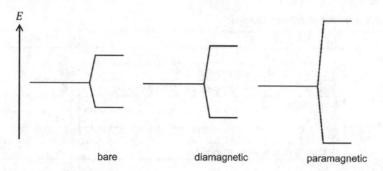

Fig. 2.2 Diamagnetic shielding contribution decreases the separation between the energy levels, whereas paramagnetic shielding increases the separation

and then it would not be possible to characterize the electronic environments uniquely. In order to circumvent this problem, a field independent definition is given by expressing the chemical shift as a ratio:

$$\delta_i = \frac{(H_r - H_i)}{H_r} \times 10^6 \tag{2.3}$$

where H_r is the field experienced by a reference nucleus, H_i is the field at the nucleus of interest i, and δ_i is the chemical shift of the nucleus i. In terms of frequencies, this is given as

$$\delta_i = \frac{(v_i - v_r)}{v_o} \times 10^6 \tag{2.4}$$

Note that here the denominator is v_o instead of v_r, and this is permissible since v_o is much larger than the difference $(v_o - v_r)$ and thus v_r can be approximated to v_o. δ_i is expressed in parts per million (ppm), and we now see that it is independent of the applied field but depends only on the screening constants. On this scale, chemical shift is a dimensionless quantity. The reference frequency is always taken to be at 0 ppm. A higher value of chemical shift implies low shielding, and conversely, a lower value of chemical shift implies higher shielding. Different types of reference compounds have been used, but the most convenient ones are those for which the resonance frequency occurs at the highest field. Thus, with respect to such a reference compound, the chemical shifts of nuclei in other compounds would have positive values. For protons, tetramethylsilane (TMS) has been the preferred reference compound, and Table 2.1 lists the chemical shifts of other reference compounds. The reference compound is either added to the sample, in which case it is called an internal reference, or can be kept separately in a capillary in the sample tube, in which case it is called an external reference. In the former situation, the reference has to be soluble in the solvent used. TMS is the common standard for all nonpolar solvents. For aqueous solutions in which TMS is insoluble, TSP and DSS (see Table 2.1) are the common standards.

Table 2.1 δ values of reference compounds

Nucleus	Reference compound	Chemical shift (ppm)*
^{1}H	TMS	0.0
	Tetramethylsilyl propionate (TSP)	0.0
	Sodium 4,4-dimethyl-4-silapentane sulfonate (DSS)	0.0
	Acetonotrile	2.0
	Dioxane	3.64
	t-butanol	1.27
	Tetramethylammonium chloride	3.10
^{13}C	TMS	0.0
	TSP	0.0
	DSS	0.0
^{15}N	Ammonium chloride	0.0
	Ammonia	0.0
^{31}P	85% phosphoric acid	0.0

* a non-zero number indicates the chemical shift of the reference line with respect to TMS

In some of these reference compounds, there are more than one line in the spectrum, but the most upfield line belonging to the methyls is taken as the reference. For simplifications, it is customary to use reference compounds in which all the unnecessary protons are substituted by deuterons.

If in a particular spectrum two nuclei resonate at the same frequency, they are said to be "chemically equivalent." Thus, for example, in the spectrum of ethyl alcohol shown in Fig. 2.1, three methyl protons are equivalent, and the two methylene protons are equivalent.

2.2.1 Anisotropy of Chemical Shifts

As mentioned earlier, the shielding generated by the electronic environments around the nuclei in a molecule is dependent on the molecular structure itself and can depend on the orientation of the molecule with respect to the direction of the magnetic field. This is termed as chemical shift anisotropy. Then, one gets broad lines because of the superposition of resonances corresponding to different orientations, as it happens in the case of oriented systems or in solids. This is schematically shown in Fig. 2.3. In solutions, because of the rapid tumbling motion, the anisotropy gets averaged out, and one gets sharp lines representing the isotropic chemical shifts.

2.2.2 Factors Influencing Isotropic Chemical Shifts

From the discussion on the origin of chemical shifts, it is clear that anything that affects the electron density around a nucleus affects its chemical shift. The higher the electron density, the higher the shielding and vice versa. Several intra- and

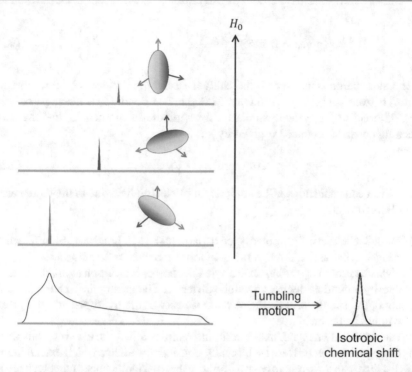

Fig. 2.3 The molecular orientation with respect to the field influences the chemical shifts. In a powder sample, for example, where all orientations are possible, the superposition of chemical shifts originating from the different orientations results in a very broad NMR line. A typical appearance is shown in the bottom picture. See Chap. 7, Appendix 2 for more theoretical details on chemical shift anisotropy. The effect of averaging of anisotropy due to tumbling motions in solutions is indicated. This results in sharp lines representing the isotropic chemical shifts

intermolecular factors can cause these changes, and we will count here the most important ones commonly encountered in the spectra of molecules. In a molecule, the electron density around a ^{1}H in a C-H bond, for example, is directly dependent on the polarity of the C-H bond. This in turn is influenced by the following factors:

(i) Electronegativities of the substituents: An electron-withdrawing group reduces the electron density around the ^{1}H and thus causes deshielding or downfield shifts.

(ii) Direct electrostatic effects of charges and dipoles: Charges and dipoles produce electric fields at the site of the nucleus and thus polarize the electron distributions (Equation 2.5a–b). The change in screening constant caused by a charge distribution around a C-H bond is given by

$$\Delta\sigma\ (\text{ppm}) = 0.125\sum_j \frac{q_j\ \cos\varnothing_j}{R_j^2} - 0.17\left(\sum_j \frac{q_j}{R_j^2}\right)^2 \tag{2.5a}$$

The summation runs over all the charges in the distribution. R_j is the distance (in nm) between the charge q_j and the ^{1}H in the C-H bond. $\varnothing_j$ is the angle between the C-H bond and q_j vector. Similarly, for a dipole of moment μ, the change in screening constant produced is given by

$$\Delta\sigma\ (\text{ppm}) = -k\,\mu\,e_z \tag{2.5b}$$

where k is a constant and e_z is the component of electric field due to the dipole along the C-H vector.

(iii) Inductive effects: The polarity of a particular C-H bond can be affected by relayed polarization through the neighboring bonds in the molecule.
(iv) Hybridization: Depending on the hybridization of the carbon atom, the electron density around the hydrogen will be different. The greater the s-character of the bond, the smaller will be the electron density at the hydrogen and the greater will be the deshielding.
(v) van der Waal effects: Direct stearic interactions affect the electron densities.
(vi) Long-range effects: Here we basically consider the effects which extend to long distances. In this category, two effects are of significance, namely, (a) ring current effects of aromatic rings and (b) "contact shifts" due to unpaired electrons of paramagnetic species. The π electron clouds in aromatic rings generate currents in the presence of a magnetic field, and these currents produce secondary magnetic fields which extend to long distances and produce chemical shift changes at nuclei coming under their influence (Eq. 2.6a–c). Loci of such shifts around the aromatic rings have been calculated, and these help in predicting the chemical shift changes that would occur for different positions of a ^{1}H around the ring. The contributions to the shielding constant ($\Delta\sigma$) of a nucleus by an aromatic ring are given by

$$\Delta\sigma(\text{ppm}) = \frac{\mu_0}{4\pi}\ \frac{e^2 r^2}{2m_e}\left(\frac{1}{R^3}\right) \tag{2.6a}$$

where r is the radius of the aromatic ring and R is the distance from the center of the ring to the proton. For standard values of μ_0(permeability of free space), charge of electron (e), mass of electron (m_e), this turns out to be

$$\Delta\sigma(\text{ppm}) = -\frac{0.0276}{R^3} \tag{2.6b}$$

If there are multiple rings contributing to the shielding at a particular nucleus, the net effect will be an addition as given in Eq. 2.6c:

$$\Delta\sigma(\text{ppm}) = -0.0276\sum_j \frac{1}{R_j^3} \tag{2.6c}$$

Figure 2.4 shows the results of such calculations for a benzene ring. The contact shift, on the other hand, arises due to coupling between an unpaired electron and the nucleus under observation. The electron need not be in the same molecule, and thus this interaction can also be intermolecular. Consequently, this interaction can really extend to long distances. From detailed quantum mechanical calculations of these interactions, it is known that the contact coupling, also termed the "Fermi interaction" arises due to a finite probability of the electron occupying the same space as the nucleus. This interaction causes changes in the energy levels of the nucleus. The absorption of energy by the nucleus will then depend upon the orientation of the electron spin, and consequently, the resonance frequencies will be different for the two different orientations with respect to the external field. However, if the electronic transitions between the two orientations are too rapid, then, the nucleus will not be able to see the two orientations distinctly and would absorb energy at a frequency which corresponds to the weighted average of the two individual

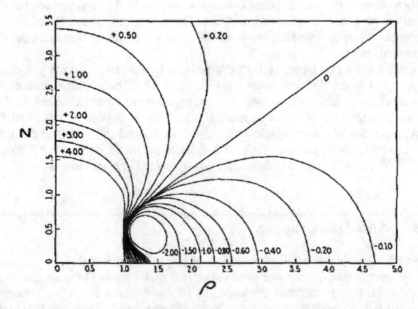

Fig. 2.4 "Isoshielding" lines (in ppm) in the neighborhood of a benzene ring. The plot represents one quadrant of a plane passing normally through the center of the ring. The lines represent the shift in the NMR shielding value which will be experienced by protons as a result of the magnetic field of the benzene ring. Here ρ and z are the cylindrical coordinates in units of 1.39 Å. (Reproduced from J. Chem. Phys. **29**, 1012 (1958), with the permission of AIP Publishing)

Table 2.2 ^{1}H chemical shift ranges with reference to TMS

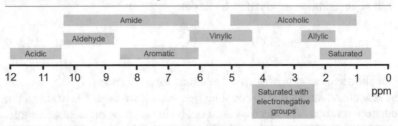

frequencies. The shift caused by such effects can sometimes be very large, of the order of 50–100 ppm.

(vii) H-bond: This is a weak interaction between two electronegative atoms sharing a H atom, e.g., N-H → O, N-H → N, O-H →N, etc. The H atom is covalently linked to one electronegative atom, and the other interacts with the partial positive charge on the H through its lone pair of electrons. The two heavy atoms cannot come closer than about 2.8–3.1 Å. The formation of H-bond causes considerable deshielding of the ^{1}H nucleus and causes downfield shifts. The effects of H-bonds will also get relayed to the neighboring bonds.

In addition to the specific interactions discussed above, the solvents can also and do affect the chemical shifts of the nuclei. The polarity of the solvent molecules influences the polarities of the bonds in the solute molecules and thus changes the electron densities.

From a study of a large number of organic molecules, some general guidelines with regard to the chemical shifts have emerged which help a great deal in the identification of functional groups and carbon skeletons in unknown molecules. These have rendered NMR as an analytical tool for an organic chemist, and the characterization of a new organic compound is considered incomplete without the analysis of its NMR spectrum. Tables 2.2, 2.3, and 2.4 list the chemical shift ranges for ^{1}H, ^{13}C, and ^{15}N nuclei in different types of molecular environments.

2.3   Spin-Spin Coupling

Figure 2.5 shows the ^{1}H NMR spectrum of ethyl alcohol recorded on a modern NMR spectrometer. In contrast to the spectrum of the same molecule shown in Fig. 2.1, the present spectrum shows higher resolution and exhibits fine structure. This is due to a phenomenon known as "spin-spin coupling." Discovered also in the years 1950–1951 by Proctor and Wu, this is the next most important NMR parameter which allows unambiguous characterization of molecules. Spin-spin coupling reflects an interaction between spins which leads to the splitting of resonance lines, and the extent and pattern of splitting is characteristic of the strength of the

Table 2.3 ^{13}C chemical shift ranges with reference to TMS

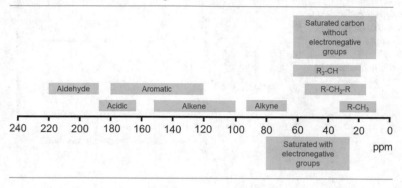

Table 2.4 ^{15}N chemical shift ranges with reference to liquid NH$_3$ (0 ppm)

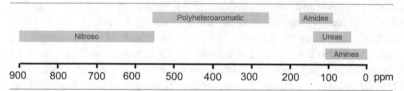

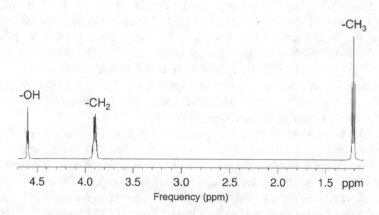

Fig. 2.5 ^{1}H NMR spectrum of ethanol on a high-resolution NMR spectrometer

interaction and the number of interacting spins. For example, in Fig. 2.5 the methyl resonances are split as a triplet with the three peaks having an intensity distribution of 1:2:1. Likewise, the methylene proton resonances appear as a quartet with the four peaks having an intensity distribution of 1:3:3:1. Both these fine structures are a consequence of spin-spin coupling between methyl and methylene protons.

What is the mechanism of spin-spin coupling and how do splittings occur? Here again, detailed theories have been worked out which are beyond the scope of this

Fig. 2.6 Concerted polarization of nuclear and electron spins. The upper and the bottom pictures indicate two possible relative orientations of the two nuclear spins, and they will have different energies

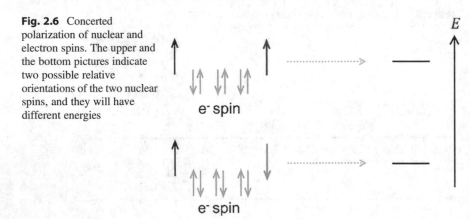

book. Qualitatively, spin-spin coupling, also referred to as J-coupling, occurs due to an interaction mediated by the electrons in the intervening bonds. A nuclear spin polarizes an electron spin in the adjacent bond orbital, which in turn polarizes the spin of the partner electron in the same bond orbital. This electron then polarizes the spin of the neighboring nucleus, and this process continues as far as the spin order generated by such an interaction can be maintained. Depending on the nature of the bonds (single bonds, double bonds, etc.), the nature of the intervening atoms in the bonding sequence, and the relative configurations of the atoms, the spin-spin coupling can extend up to 4–5 bonds. The strength of the coupling is called coupling constant and is denoted by the symbol ^{n}J for two spins separated by n-bonds. Figure 2.6 shows pictorially the concerted polarization of the nuclear and electron spins, bringing about such a nuclear spin-spin coupling interaction.

This coupling interaction between two spins I_1 and I_2 is quantitatively given as $J_{12} I_1 \cdot I_2$, and J_{12} is called the coupling constant. A simple explanation of how the J-coupling leads to splitting of lines is depicted in Fig. 2.7.

Consider a system of two $I = 1/2$ spins, say A and X. That there is a J-coupling between A and X implies that the energy of spin A in any of its two Zeeman states α and β depends upon whether the spin X is in the α or the β state. Thus, the spin configurations $\alpha\alpha$, $\alpha\beta$, $\beta\alpha$, and $\beta\beta$ will have different energies. The coupling is said to be **positive** if the energy of the α state of A spin increases for the X spin in the α state and decreases for the X spin in the β state. Likewise, the energy of the β state of A spin will increase if the X spin is in the β state and will decrease if the X spin is in the α state, i.e., the parallel orientation of spins increases the energy, and the antiparallel orientation decreases the energy of the state. As a result, the resonance frequency of absorption of A spin for the transition from its α state to its β state will be dependent upon whether the X spin is in the α or the β state. The transition frequency will be higher for the β orientation and lower for the α orientation of the X spin than the resonance frequency in the absence of the coupling. In a similar manner, there will be two transitions for the X spin around the no-coupling resonance frequency. The two frequencies will be equally spaced from the no-coupling

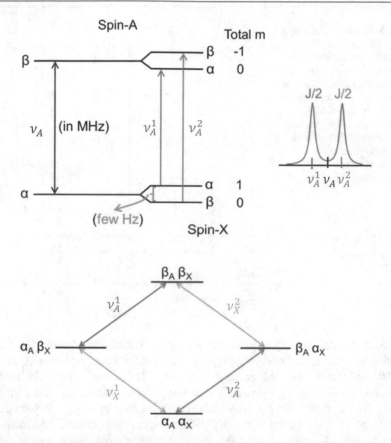

Fig. 2.7 A schematic energy level representation of a two-spin (A and X) ½ system. Here, for the isolated A-spin, only one transition is possible; hence, that results in a single transition. On the other hand, when A is coupled with X, two transitions, $\alpha_A\alpha_X \rightarrow \beta_A\alpha_X$ and $\alpha_A\beta_X \rightarrow \beta_A\beta_X$, are possible. This situation eventually results in two spectral lines, and that is known as spin-spin coupling

frequency, and the separation between the two frequencies represents the coupling constant. The situation with regard to the relative energies of the states will be exactly opposite for a negative value of the coupling constant between the two spins.

Referring to Fig. 2.8, it is clear that for one bond coupling, the antiparallel orientation of nuclear spins leads to lower energy, and therefore this coupling is positive. For nuclear spins separated by two bonds, the parallel orientation of the spins leads to lower energy, and thus this coupling is negative. Likewise, three-bond couplings are positive. For longer separations, these simple arguments are not adequate, and detailed calculations will be necessary to find out the relative energies of the states. In any case it is important to note that whatever be the sign of the coupling constant, the appearance of the NMR spectrum remains the same.

The magnitude of a coupling constant is dependent on factors such as the product of the magnetogyric ratios of the interacting nuclear spins, electron distributions in

Fig. 2.8 In the case of one-bond-separated spins, the antiparallel orientation of spins leads to lower energy; thus, coupling is positive ($J >$ 0, upper part). On the other hand, for the two-bond-separated spins, the parallel orientation of spins leads to lower energy; therefore, coupling is negative ($J <$ 0, lower part)

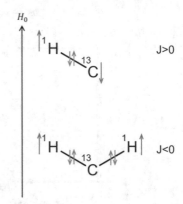

Table 2.5 Ranges of important coupling constants

Spin pair	J (Hz)-range
^2J: ^1H-^1H	5–20
^1J: ^1H-^{13}C	100–250
^1J: ^1H-^{15}N	80–100
^1J: ^1H-^{31}P	A few hundred Hz

the bonds, stereochemistry around the interacting nuclei, etc. Among the various coupling constants, one-bond coupling constants are the strongest being of the order of 100–200 Hz for the heteronuclear ^{13}C-^1H, ^{15}N-^1H, and ^{31}P-^1H interactions; one-bond ^1H-^1H coupling constant is of course higher, but this is hardly of much interest in chemistry since it can only occur in the hydrogen molecule. The magnitudes of the coupling constants provide valuable structural information on molecules and also play crucial roles in the design of many modern experiments. These aspects will be discussed in later chapters. Table 2.5 lists the ranges of some of the most important homo- and heteronuclear coupling constants derived from the study of a large number of organic molecules.

2.4   Analysis of NMR Spectra of Molecules

The basic step in deriving useful information from NMR spectra of a given molecule is to extract the chemical shift and coupling constant information for all the nuclei in the molecule. In other words, in a ^1H NMR spectrum, for example, the various resonances must be assigned as belonging to specific protons, and then the fine structures of the individual groups of lines belonging to them must be quantitatively explained. This yields the chemical shifts (δ-values) and coupling constants (J-values) for all the protons in the molecule. Many times, such an analysis can be performed by simple inspection following certain basic rules; then it is termed as "first-order analysis." In more complex systems, a more rigorous treatment based on quantum mechanical calculations is warranted. We shall describe these two treatments in that order.

In all these analyses, a certain convention is followed for the nomenclature of the nuclear spins. The spins are labeled by upper-case alphabets, A, B, C, etc., and in a spin system consisting of more than one spin; the choice of alphabets is based upon the difference in their resonance frequencies vis-a-vis the coupling constant between them. For example, an AX system would indicate that the two spins A and X have resonance frequencies widely separated compared to their J-coupling constant. Likewise, in an ABX system, spin X is widely separated from both A and B, and spins A and B are themselves fairly close in relation to the coupling constant J_{AB}. Spin systems such as AX, AMX, AMQX, etc. are said to be weakly coupled, while AB, ABC, etc. are said to be strongly coupled. It is important to note here that these terminologies "strongly coupled," "weakly coupled," etc. are not a reflection on the magnitude of the coupling constant alone, but is rather a reflection on the strength of J in comparison to the chemical shift difference between the coupled nuclei. Thus, it is clear that heteronuclear spin systems always belong to the weak coupling category, and strong coupling situations occur in homonuclear spectra. The first-order analysis always deals with weakly coupled systems, and strongly coupled situations will have to be necessarily dealt with by quantum mechanical calculations. A group of nuclei are said to be "magnetically equivalent" if (i) they all have the same chemical shifts and (ii) each of them has the same coupling constant to every other spin outside the group.

2.4.1   First-Order Analysis

As the name suggests, this is the first level of analysis one would do with a given spectrum and is satisfactory when the resonance lines belonging to individual nuclei (protons for example) are well separated from each other in the spectrum. The important observations one makes in such an analysis are (i) the multiplet patterns in different groups of lines, (ii) relative intensities of the individual groups, and (iii) the measurement of coupling constants from the splittings in the individual groups and identifications of groups having common coupling constants. Following a set of rules are generally applicable in such an analysis.

(i) The centers of the multiplets represent the chemical shifts. The relative chemical shifts of different multiplets must correspond to the order expected from the chemical environments of the individual nuclei.

(ii) In a group of magnetically equivalent nuclei, A_2, B_2, A_3, etc., the J-coupling between themselves does not lead to the splitting of lines. A rationale for this comes from detailed quantum mechanical calculations which will be discussed in the later section.

(iii) The integral of any multiplet is proportional to the total number of equivalent nuclei in that group. For example, a CH_2 group may have two equivalent protons; a CH_3 group may have three equivalent protons; etc. Thus, if a multiplet which belongs to a single proton can be identified, then the intensities of other groups can be appropriately scaled to determine the number of protons in the individual groups.

(iv) The resonance line of a nucleus J-coupled to n equivalent nuclei of another type gets split into $(2nI + 1)$ lines where I is the spin of the individual nucleus in the equivalent group. It follows that if the nucleus is J-coupled to two groups of equivalent nuclei having n and m nuclei and if their spins are I_1 and I_2, respectively, then the total number of lines in the multiplet will be $(2nI_1 + 1)$. $(2mI_2 + 1)$. Thus, for spin 1/2 nuclei (say different groups of protons, fluorines, phosphorous, etc.), the multiplet for a particular nucleus would have $(n + 1)$. $(m + 1)$ lines.

(v) For a single group of n-equivalent nuclei of spin $\frac{1}{2}$ coupled to a particular nucleus, say A, the different lines in the A multiplet will have the intensities in the proportion of binomial coefficients in the expansion of $(a + b)^n$. For example, as seen in the spectrum of ethyl alcohol, the CH_2 protons coupled to the three equivalent protons in the CH_3 group are split into four lines having intensities in the proportion $1 : 3 : 3 : 1$. Similarly, the CH_3 protons are split into three lines with intensities in the ratio $1 : 2 : 1$, due to the coupling to the two equivalent protons in the CH_2 group. Extending in this manner, the expected intensity pattern for any multiplet can be calculated, and for a single group of spin $\frac{1}{2}$ nuclei, the expected patterns for the different number of equivalent nuclei are shown in Fig. 2.9. This pattern of splittings is referred to as Pascal's triangle.

Various splitting patterns in a first-order spectrum can be understood with the help of simple energy level diagrams based on α and β orientations of the different spins as discussed before. Energy level diagrams for two (a) and three (b) nonequivalent spins are shown as illustrations in Fig. 2.10.

The various transitions belonging to different spins have been labeled, and the splitting trees leading to the expected patterns in spectra are also indicated. These explain the origin of Pascal's triangle for equivalent spin ½ nuclei.

Fig. 2.9 A schematic representation of transition intensities in a multiplet with the aid of Pascal's triangle

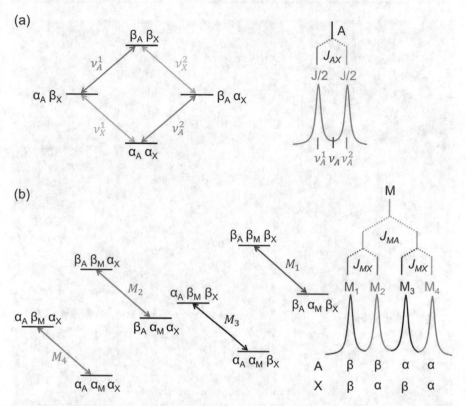

Fig. 2.10 Energy level diagrams and scalar coupling splitting trees for two and three spin ½ systems are shown in (**a**) and (**b**), respectively

2.4.2   Quantum Mechanical Analysis

This is the most rigorous analysis of NMR spectra and is applicable under all conditions of chemical shifts and coupling constants in the spectra. Here we have to assume for the reader a basic knowledge of the principles of quantum mechanics such as operators, Hamiltonian, eigenvalues, eigenfunctions and their properties, etc. Particularly important are the angular momentum operators and their properties. Some of the important properties are listed in the Box 2.1 for ready use.

> **Box 2.1: Fundamental Properties of Angular Momentum Operators**
> (a) I is the operator for the total angular momentum of the spin system. I_x, I_y, and I_z are the operators for the components of angular momentum along x, y, and z directions, respectively.

(continued)

Box 2.1 (continued)

(b) For a single spin $\frac{1}{2}$ system, the two orientations of the spin with respect to the magnetic field are represented by the states α and β, and these are eigenfunctions of the I_z operator, i.e.,

$$I_z |\alpha> = \frac{1}{2} |\alpha>$$

$$I_z |\beta> = \frac{-1}{2} |\beta>$$

The functions α and β are orthonormal, i.e.,

$$<\alpha \mid \alpha> = <\beta \mid \beta> = 1$$

$$<\beta \mid \alpha> = <\alpha \mid \beta> = 0$$

(c) The operators I_x, I_y, and I_z for any spin do not commute and obey a cyclic relation:

$$[I_p, I_q] = iI_r$$

where indices p, q, and r are cyclic permutations of x, y, and z.

(d) Two commuting operators with nondegenerate eigenvalues have common eigenfunctions. For example, if $\mathbf{A}$ and $\mathbf{B}$ are commuting operators and $\varnothing$ is the eigenfunction of $\mathbf{B}$ with non-degenerate eigenvalue b, then

$$\mathbf{A}(\mathbf{B}\varnothing) = \mathbf{A}(b\varnothing) = b(\mathbf{A}\varnothing) = \mathbf{B}(\mathbf{A}\varnothing)$$

Since b is a nondegenerate eigenvalue of $\mathbf{B}$, $\mathbf{A}\varnothing$ must be a scalar multiple of $\varnothing$.

$$\mathbf{A}\varnothing = a\varnothing$$

This means $\varnothing$ is an eigenfunction of $\mathbf{A}$ and a is the eigenvalue of $\mathbf{A}$.

If b is a degenerate eigenvalue of B, then $A\varnothing$ can be a different function, say $\varnothing'$.

$$\mathbf{A}\varnothing = \varnothing'$$

Then $\varnothing$ is not an eigenfunction of A.

(continued)

Box 2.1   (continued)
(e) Raising (I^+) and lowering (I^-) operators

$$I^+ = I_x + iI_y; I^- = I_x - iI_y$$

$$I^+ \mid \alpha >= 0; I^- \mid \alpha >= \mid \beta >; I^+ \mid \beta >= \mid \alpha >; I^- \mid \beta >= 0$$

$$I_x \mid \alpha >= \frac{1}{2}(I^+ + I^-) \mid \alpha >= \frac{1}{2} \mid \beta >$$

$$I_x \mid \beta >= \frac{1}{2}(I^+ + I^-) \mid \beta >= \frac{1}{2} \mid \alpha >$$

$$I_y \mid \alpha >= \frac{1}{2i}(I^+ - I^-) \mid \alpha >= -\frac{1}{2i} \mid \beta >$$

$$I_y \mid \beta >= \frac{1}{2i}(I^+ - I^-) \mid \beta >= \frac{1}{2i} \mid \alpha >$$

The general approach in any quantum mechanical solution of a problem is to set up a proper Hamiltonian taking into account all the possible interactions, solve the Schrödinger equation, obtain the eigenvalues and the eigenfunctions, and then calculate the specific observables of interest, being the transition frequencies and their intensities in the present case.

In high-resolution NMR, the energy of the nuclear spin system will have two contributions, namely, the interaction with the external magnetic field—termed the Zeeman interaction—and the J-coupling interaction between the spins. The direct dipole-dipole interaction which depends on the angle between the internuclear vector and the magnetic field (see Chap. 7, Appendix 1) averages out to zero due to rapid tumbling motions of the molecules. For the same reason, J is also an exclusively intramolecular property. Thus, the general isotropic nuclear spin Hamiltonian is given by

$$\mathcal{H} = \mathcal{H}_o + \mathcal{H}_1 \tag{2.7}$$

with

$$\mathcal{H}_o = \sum_i \gamma_i H_i I_{iz} \tag{2.8}$$

and

$$\mathcal{H}_1 = 2\pi \sum_{i<j} J_{ij} I_i . I_j = 2\pi \sum_{i<j} J_{ij} \{ I_{ix} I_{jx} + I_{iy} I_{jy} + I_{iz} I_{jz} \} \tag{2.9}$$

The external field H_o is taken in the negative z direction. $\mathcal{H}_o$ and $\mathcal{H}_1$ are termed as the Zeeman and the J-coupling parts of the Hamiltonian, respectively. Indices i and

j run over all the spins in the spin system. I_i and I_j are the angular momentum operators of the spins i and j, respectively. I_{iz} is the operator for the z-component of the angular momentum for spin i. In high-resolution NMR, in chemistry and biology, one is concerned mainly with spin ½ nuclei such as ^{1}H, ^{13}C, ^{15}N, ^{31}P, etc., and therefore we shall focus our attention with such nuclei in the ensuing discussion. We have to now obtain the eigenvalues and the eigenfunctions for the Hamiltonian in Eq. 2.9.

Let us consider a system with p spin $\frac{1}{2}$ nuclei. Each of the spins can exist in an α state or a β state. Then the state of the spin system as a whole can be any combination of these two states of the individual spins. Let us denote a particular state by a function Φ_n defined as

$$\Phi_n = \alpha_1 \beta_2 \alpha_3 \beta_4 \alpha_5 \alpha_6 \ldots \ldots \ldots \ldots \beta_p \qquad (2.10)$$

There will be 2^p such combinations possible, and consequently there will be that many state functions for the system. Now, since α and β are eigenfunctions of the Iz operator for each nucleus, it is clear that these product functions will be eigenfunctions of the Zeeman Hamiltonian $\mathcal{H}_o$, i.e.,

$$\mathcal{H}_o \Phi_n = E_n \Phi_n \qquad (2.11)$$

where E_n for different Φ_n are the corresponding eigenvalues. However, the J-coupling Hamiltonian $\mathcal{H}_1$ causes mixing of these states, and the total Hamiltonian $\mathcal{H}$ will not be diagonal in the product basis. The correct eigenfunctions will have to be determined as linear combinations of the product functions.

$$\Psi_k = \sum_n c_{nk} \Phi_n \qquad (2.12)$$

The eigenvalues of the Hamiltonian will be determined from the solutions of the secular equation:

$$|\mathcal{H}_{mn} - E\delta_{mn}| = 0 \qquad (2.13)$$

$\mathcal{H}_{mn}$ are the matrix elements of the Hamiltonian defined as

$$\mathcal{H}_{mn} = <\Phi_m \mid \mathcal{H} \mid \Phi_n> \qquad (2.14)$$

The secular Eq. 2.13 is of the order 2^p. However, it can be factorized into a number of equations of lower order by adopting a trick, namely, grouping the basis functions Φ_n according to total spin component along the z-axis. Now let us define an operator:

$$F_z = \sum_i I_{iz} \qquad (2.15)$$

where the summation runs over the p spins in the system. The product functions Φ_n are eigenfunctions of F_z with eigenvalues corresponding to the total magnetic quantum number of the system in the particular state. It can be easily shown that the operator F_z commutes with the Hamiltonian $\mathcal{H}$, i.e.,

$$[\mathcal{H}, F_z] = [\mathcal{H}F_z - F_z\mathcal{H}] = 0 \qquad (2.16)$$

(**Note:** Since F_z has degenerate eigenvalues, the commutation given in Eq. 2.16 does not imply that eigenfunctions of F_z will be eigenfunctions of $\mathcal{H}$.)

Now, we will show that there will be no matrix elements of Hamiltonian between product functions which have different eigenvalues for the F_z operator.

Let

$$F_z\Phi_n = f_n\Phi_n \qquad (2.17)$$

$$F_z\Phi_m = f_m\Phi_m \qquad (2.18)$$

f_n and f_m are the eigenvalues corresponding to the functions Φ_n and Φ_m, respectively. From the commutation relationship in Eq. 2.16, it follows that

$$< \Phi_m|\mathcal{H}F_z - F_z\mathcal{H}|\Phi_n >= 0 \qquad (2.19)$$

$$(f_n - f_m) < \Phi_m|\mathcal{H}|\Phi_n >= 0 \qquad (2.20)$$

If f_n is not equal to f_m, then the matrix element in Eq. 2.20 will have to vanish. Thus, the matrix representation of the Hamiltonian in the basis of $\{\Phi_n\}$ will be block-diagonalized as shown in Fig. 2.11.

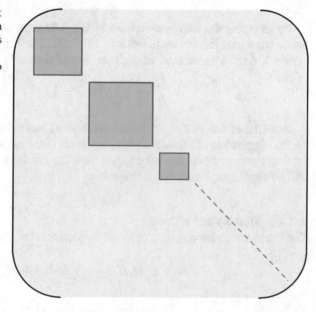

Fig. 2.11 A schematic block diagonal matrix representation of the Hamiltonian in the basis of $\{\Phi_n\}$. Matrix elements are nonzero when $f_n = f_m$ and zero when $f_n \neq f_m$

Consequently, the secular determinant gets block-diagonalized, and the problem of determining the eigenvalues will be reduced in dimension. The individual blocks belonging to specific f values can be treated separately.

The eigenfunctions for each of the eigenvalues can then be determined as follows:

$$\mathcal{H}\Psi_k = \varepsilon_k \Psi_k \tag{2.21}$$

where ε_k are the eigenvalues of the total Hamiltonian and Ψ_k are the corresponding eigenfunctions. Substituting for Ψ_k from Eq. 2.12, one gets

$$\mathcal{H}\left[\sum_n c_{nk}\Phi_n\right] = \varepsilon_k \left[\sum_n c_{nk}\Phi_n\right] \tag{2.22}$$

Multiplying from the left by Φ_m and calculating the matrix element, we get

$$\sum_n c_{nk} <\Phi_m|\mathcal{H}|\Phi_n> = \varepsilon_k \sum_n c_{nk} <\Phi_m \mid \Phi_n > \tag{2.23}$$

Since the basis functions $\{\Phi_n\}$ are orthogonal, Eq. 2.23 simplifies to

$$\sum_n c_{nk} <\Phi_m|\mathcal{H}|\Phi_n> = \varepsilon_k c_{mk} \tag{2.24}$$

In short notation,

$$\sum_n c_{nk}\mathcal{H}_{mn} = \varepsilon_k c_{mk} \tag{2.25}$$

By changing Φ_m for multiplication in Eq. 2.23, one can get a set of 2^p homogeneous equations for the coefficients in Eq. 2.25. Further, the state function of the system has to be normalized which leads to the condition

$$\sum_n |c_{nk}|^2 = 1 \tag{2.26}$$

From these, the coefficients can be determined and thereby the eigenfunctions for all the eigenvalues. In the following, we shall carry out explicit calculations for a two-spin system as an illustration, and then the results for more complex systems will be simply presented and discussed.

2.4.2.1 Two-Spin AB Case

The Hamiltonian for a system of two $I = \frac{1}{2}$ spins is written as

$$\mathcal{H} = \frac{1}{2\pi}[\gamma_1 H_1 I_{z1} + \gamma_2 H_2 I_{z2}] + J I_1.I_2 \tag{2.27}$$

$$= \nu_1 I_{z1} + \nu_2 I_{z2} + J I_1.I_2 \tag{2.28}$$

We then construct a basis set of four product functions $\Phi_1 = \alpha\alpha$, $\Phi_2 = \alpha\beta$, $\Phi_3 = \beta\alpha$, and $\Phi_4 = \beta\beta$. All these functions are orthogonal by virtue of the fact that α and β are orthogonal. That means, for example,

$$< \alpha\alpha|\alpha\beta >=< \alpha|\alpha >< \alpha \mid \beta >= 0 \tag{2.29}$$

Similar equations hold for the other functions as well. Using these properties and also the property that the α and β functions are eigenfunctions of the I_z operators, we can calculate the matrix elements of the Hamilto/nian.

$$\mathcal{H}_{11} =< \alpha\alpha|H|\alpha\alpha > = v_1 < \alpha|I_{z1}|\alpha >< \alpha|\alpha > +v_2 < \alpha|I_{z2}|\alpha >< \alpha|\alpha > \\ +J < \alpha\alpha|I_{x1}I_{x2} + I_{y1}I_{y2} + I_{z1}I_{z2}|\alpha\alpha > \tag{2.30}$$

$$= \frac{(v_1 + v_2)}{2} + J\{< \alpha\alpha|I_{x1}I_{x2}|\alpha\alpha > + < \alpha\alpha|I_{y1}I_{y2}|\alpha\alpha > + < \alpha\alpha|I_{z1}I_{z2}|\alpha\alpha >\} \tag{2.31}$$

$$= \frac{(v_1 + v_2)}{2} + J\{ < \alpha|I_{x1}|\alpha >< \alpha|I_{x2}|\alpha > + < \alpha|I_{y1}|\alpha >< \alpha|I_{y2}|\alpha > + \\ < \alpha|I_{z1}|\alpha >< \alpha|I_{z2}|\alpha >\} \tag{2.32}$$

Matrix elements of I_x and I_y can be calculated by making use of the "raising and lowering" operators I^+ and I^- as follows:

$$I_x = \frac{I^+ + I^-}{2} ; I_y = \frac{I^+ - I^-}{2i} \tag{2.33}$$

These operators have the following properties:

$$I^+ \mid \alpha > = 0 \qquad\qquad I^+ \mid \beta > = \mid \alpha > \tag{2.34}$$

$$I^- \mid \alpha >= \mid \beta > \qquad\qquad I^- \mid \beta >= 0 \tag{2.35}$$

Thus, Eq. 2.32 simplifies to

$$\mathcal{H}_{11} = \frac{(v_1 + v_2)}{2} + J(0 + 0 + \frac{1}{4}) = \frac{(v_1 + v_2)}{2} + \frac{J}{4} \tag{2.36}$$

Similarly, the other matrix elements can be calculated, and these are

$$\mathcal{H}_{12} = 0; \mathcal{H}_{13} = 0; \mathcal{H}_{14} = 0$$

$$\mathcal{H}_{21} = 0; \mathcal{H}_{22} = \frac{(v_1 - v_2)}{2} - \frac{J}{4}; \mathcal{H}_{23} = \frac{J}{2}; \mathcal{H}_{24} = 0$$

$$\mathcal{H}_{31} = 0; \mathcal{H}_{32} = \frac{J}{2}; \mathcal{H}_{33} = \frac{(v_2 - v_1)}{2} - \frac{J}{4}; \ \mathcal{H}_{34} = 0$$

$$\mathcal{H}_{41} = 0; \mathcal{H}_{42} = 0; \mathcal{H}_{43} = 0; \mathcal{H}_{44} = -\frac{(v_1 + v_2)}{2} + \frac{J}{4} \qquad (2.37)$$

Thus, the matrix representation of the Hamiltonian in the product basis is

$$\mathcal{H} = \begin{bmatrix} \frac{(v_1 + v_2)}{2} + \frac{J}{4} & 0 & 0 & 0 \\ 0 & \frac{(v_1 - v_2)}{2} - \frac{J}{4} & \frac{J}{2} & 0 \\ 0 & \frac{J}{2} & \frac{(v_2 - v_1)}{2} - \frac{J}{4} & 0 \\ 0 & 0 & 0 & -\frac{(v_1 + v_2)}{2} + \frac{J}{4} \end{bmatrix} \qquad (2.38)$$

The matrix in Eq. 2.38 must be diagonalized to get the eigenvalues of the Hamiltonian. These yield the energy levels of the system. It is clear that $\mathcal{H}_{11}$ and $\mathcal{H}_{44}$ are themselves two of the eigenvalues and yield two energy values E_1 and E_4. The central 2×2 matrix yields the remaining two energy values E_2 and E_3. Defining $(v_1 - v_2) = \delta$, the central 2×2 matrix can be written as

$$\begin{bmatrix} \frac{\delta}{2} - \frac{J}{4} & \frac{J}{2} \\ \frac{J}{2} & -\left(\frac{\delta}{2} + \frac{J}{4}\right) \end{bmatrix}$$

The eigenvalues of this matrix are the solutions of the determinantal equation:

$$\begin{vmatrix} \frac{\delta}{2} - \frac{J}{4} - E & \frac{J}{2} \\ \frac{J}{2} & -\left(\frac{\delta}{2} + \frac{J}{4}\right) - E \end{vmatrix} = 0 \qquad (2.39)$$

This leads to the two eigenvalues:

$$E_2 = -\frac{J}{4} + \frac{1}{2}\sqrt{\delta^2 + J^2}; E_3 = -\frac{J}{4} - \frac{1}{2}\sqrt{\delta^2 + J^2} \qquad (2.40)$$

Defining

$$C \cos 2\theta = \frac{\delta}{2} \text{ and } C \sin 2\theta = \frac{J}{2} \qquad (2.41)$$

$$C = \frac{1}{2}\sqrt{\delta^2 + J^2} \qquad (2.42)$$

The four energy values are thus

$$E_1 = \frac{(v_1 + v_2)}{2} + \frac{J}{4};$$

$$E_2 = -\frac{J}{4} + C;$$

$$E_3 = -\frac{J}{4} - C; \tag{2.43}$$

$$E_4 = -\frac{(v_1 + v_2)}{2} + \frac{J}{4}$$

(Note: The $\alpha\alpha$ state has the highest energy since the magnetic field has been taken to be along the negative z-axis.)

We will now illustrate the determination of the eigenfunctions corresponding to the eigenvalues. First of all, the fact that the matrix representation of the Hamiltonian in Eq. 2.38 has only diagonal elements in the first and the last rows implies that the Hamiltonian operator leaves the $\alpha\alpha$ and the $\beta\beta$ states unaltered; in other words, it does not mix these states. Thus

$$\mathcal{H}|\alpha\alpha> = E_1|\alpha\alpha> \qquad\qquad \mathcal{H}\,|\,\beta\beta> = E_4\,|\,\beta\beta> \tag{2.44}$$

That is, $\alpha\alpha$ and $\beta\beta$ states are two of the eigenfunctions of the Hamiltonian with the eigenvalues E_1 and E_4, respectively.

$$\Psi_1 = |\alpha\alpha>$$
$$\Psi_4 = |\beta\beta> \tag{2.45}$$

For the other two eigenfunctions corresponding to E_2 and E_3, we follow the procedure described before. However, we show here an explicit calculation only for E_2 as an illustration. Referring to Eq. 2.24, one can write the linear equations for the coefficients using the elements of the 2×2 matrix portion of the Hamiltonian.

$$(H_{22} - E_2)C_{22} + H_{23}C_{32} = 0$$
$$H_{32}C_{22} + (H_{33} - E_2)C_{32} = 0 \tag{2.46}$$

Substituting for the different matrix elements and the eigenvalue, we obtain

$$\left(\frac{\delta}{2} - C\right)C_{22} + \frac{J}{2}C_{32} = 0$$
$$\frac{J}{2}C_{22} + \left(-\frac{\delta}{2} - C\right)C_{32} = 0 \tag{2.47}$$

Substituting for C from Eq. 2.41, we get

$$C_{22}(\cos 2\theta - 1) + C_{32}(\sin 2\theta) = 0 \tag{2.48}$$

Using the normalization condition

$$C_{22}{}^2 + C_{32}{}^2 = 1 \tag{2.49}$$

one obtains after a little bit of algebra

$$C_{22} = \cos \theta \qquad\qquad C_{32} = \sin \theta \qquad\qquad (2.50)$$

Thus, the eigenfunction corresponding to E_2 is given by

$$\Psi_2 = \cos \theta |\alpha\beta > + \sin \theta |\beta\alpha > \qquad\qquad (2.51)$$

Similarly, Ψ_3 corresponding to E_3 can be shown to be

$$\Psi_3 = - \sin \theta |\alpha\beta > + \cos \theta |\beta\alpha > \qquad\qquad (2.52)$$

Equations 2.51 and 2.52 indicate that the product functions $\alpha\beta$ and $\beta\alpha$ get mixed due to the coupling between the spins. The extent of mixing is determined by the coefficients which are dependent on the relative magnitudes of the coupling constant and the chemical shift separation between the spins. Larger the $\frac{J}{\delta}$ ratio stronger will be the mixing and vice versa.

Having obtained the eigenfunctions and the eigenvalues of the Hamiltonian, we will now try to calculate the intensities of the various transitions by calculating the transition probabilities. For the radiofrequency-induced transitions between any two of these states, say m and m', the quantum mechanical transition probability is given by

$$P_{mm'} = \gamma^2 H_1^2 |< m'|I : x|m >|^2 g(\nu) \qquad\qquad (2.53)$$

Here, the RF is assumed to be applied along the x-axis, H_1 is the amplitude of the RF, and $g(\nu)$ is the lineshape function. From the properties of the I_x operator defined in Eqs. 2.33–2.35 and the eigenfunctions derived above (Eqs. 2.45, 2.51, and 2.52), it can be seen that there will be four transitions with nonzero transition probabilities and these are

$$\Psi_2 \rightarrow \Psi_1, \Psi_4 \rightarrow \Psi_2, \Psi_3 \rightarrow \Psi_1, \Psi_4 \rightarrow \Psi_3 \qquad\qquad (2.54)$$

This is the origin of the selection rule for observable transitions in NMR. The relative intensities of the four transitions turn out to be as indicated in Table 2.6.

Now,

$$\sin 2\theta = \frac{J}{2C} = \frac{J}{\sqrt{\delta^2 + J^2}} = \frac{1}{\sqrt{1 + \left(\frac{\delta}{J}\right)^2}} \qquad\qquad (2.55)$$

Table 2.6 Transitions and relative intensities in an AB system

Transition	Frequency	Relative intensity
$3 \rightarrow 1$	$\frac{(\nu_1+\nu_2)}{2} + \frac{J}{2} + C$	$1 - \sin 2\theta$
$4 \rightarrow 2$	$\frac{(\nu_1+\nu_2)}{2} - \frac{J}{2} + C$	$1 + \sin 2\theta$
$2 \rightarrow 1$	$\frac{(\nu_1+\nu_2)}{2} + \frac{J}{2} - C$	$1 + \sin 2\theta$
$4 \rightarrow 3$	$\frac{(\nu_1+\nu_2)}{2} - \frac{J}{2} - C$	$1 - \sin 2\theta$

The midpoint of transitions $3 \to 1$ and $4 \to 2$ is $\frac{(v_1+v_2)}{2} + C$, and that between transitions $2 \to 1$ and $4 \to 3$ is $\frac{(v_1+v_2)}{2} - C$. These do not correspond to the chemical shifts of spins 1 and 2, as in the case of a weakly coupled AX system.

If $\left(\frac{\delta}{J}\right)^2 \gg 1$, i.e., if the spins are weakly coupled, then $\sin 2\theta = 0$. Then all the four transitions listed in Table 2.6 will have identical intensities. It also follows that $\cos\theta = 1$ and $\sin\theta = 0$. Because of this, the mixing of product functions seen in the eigenfunctions vanishes, and the product functions become the eigenfunctions of the Hamiltonian. The four transitions can then be identified as belonging to particular spins. Then we can see that the center of each doublet corresponds to the chemical shift of the respective spin. For example, the midpoint of $4 \to 2$ and $3 \to 1$ transitions in which spin 1 undergoes a transition from α to β state is

$$\frac{1}{2}[v_{4\to 2} + v_{3\to 1}] = \frac{1}{2}(C + v_1 + v_2) = \frac{1}{2}(\delta + v_1 + v_2) = v_1 \qquad (2.56)$$

These are indeed the rules in the first-order analysis of spectra. The parameter θ which manifests in the presence of strong coupling and alters the intensities and frequencies of the transitions is therefore called the strong coupling parameter. We shall examine a few special situations to bring out the effects of strong coupling in the appearance of the spectra. Figure 2.12 shows schematically how the spectrum of a two-spin system changes as the ratio $\frac{J}{\delta}$ changes.

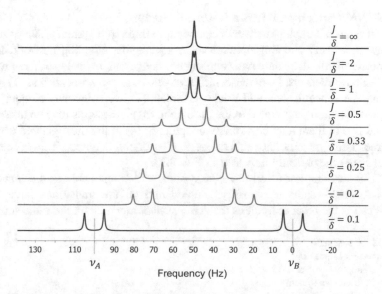

Fig. 2.12 Simulated NMR spectra for an AB spin system with 10 Hz of scalar coupling ($J_{AB} = 10$ Hz). These spectra are generated at different ratios of $\frac{J}{\delta}$

We see that at large values of the ratio, the outer transitions almost vanish, and if their intensities are lower than the S/N ratio in the spectrum, then the spectrum will consist of only two lines. This can be completely misinterpreted as belonging to two uncoupled singlets or a doublet with a wrong coupling constant. In the limit of infinite value of $\frac{J}{\delta}$, i.e., when the two spins are equivalent, the outer two transitions would have identically zero intensity ($\sin 2\theta = 1$), and the inner two transitions would have the same energy $v = v_1 = v_2 = \frac{1}{2}(v_1 + v_2)$. This explains why the coupling between equivalent spins does not lead to the splitting of lines in the NMR spectrum. Under this limit of equivalence of two spins (A_2), $\sin \theta = \cos \theta = \frac{1}{\sqrt{2}}$, the eigenfunctions will become $\Psi_1 = \alpha\alpha$, $\Psi_2 = \frac{1}{\sqrt{2}}(\alpha\beta + \beta\alpha)$, $\Psi_3 = \frac{1}{\sqrt{2}}(\alpha\beta - \beta\alpha)$, and $\Psi_4 = \beta\beta$. Functions Ψ_1, Ψ_2, and Ψ_4 are symmetric with respect to interchange of spin states of the two spins, and Ψ_3 is antisymmetric (the eigenfunction changes sign on interchanging the spin states of the two spins).

In summary, the effects of strong coupling are as follows: (i) the spins lose their identity; the transitions cannot be described as belonging to this spin, that spin, etc.; (ii) the transitions will have different intensities; (iii) the product functions corresponding to the same total z-component of spin get mixed; and (iv) the chemical shifts of the spins cannot be determined by simple inspection of the spectra; the coupling constant can however still be measured from the separation of the lines as in the first-order analysis.

2.4.2.2 NMR Spectra of Three Coupled Nuclei

Different types of coupling networks can occur in three spin systems. The spins can be either in a linear coupling network or in a triangular coupling network. In the former case, there will be only two coupling constants, and, in the latter, there will be three coupling constants. In all these, different situations can arise; such as (1) all the spins may be nonequivalent and weakly coupled; (2) two spins may be equivalent and weakly or strongly coupled to the third spin; (3) all the spins may be nonequivalent and two of them may be strongly coupled; or (4) in the most general case, all the three spins may be nonequivalent and strongly coupled. These coupling networks are schematically indicated in Fig. 2.13.

AMX is the simplest of all the spin systems and is amenable to the first-order analysis. All the spins are weakly coupled and all the transitions have equal intensities. At the other extreme is the ABC system in which all the three spins are

Fig. 2.13 A schematic representation of linear and triangular coupling networks possible in three spin systems

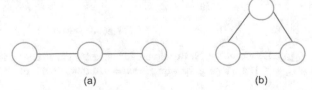

(a)   (b)

Table 2.7 Eigenvalues and eigenfunctions in an AB_2 system

S. no.	Symmetry label[a]	Eigenfunction	Energy
1	$s_{3/2}$	$\alpha\alpha\alpha$	$\frac{1}{2}\nu_A + \nu_B + \frac{1}{2}J_{AB} + \frac{1}{4}J_{BB}$
2	$1s_{1/2}$	$\alpha(\alpha\beta + \beta\alpha)/\sqrt{2}$	$\frac{1}{2}\nu_A + \frac{1}{4}J_{BB}$
3	$2s_{1/2}$	$\beta\alpha\alpha$	$-\frac{1}{2}\nu_A + \nu_B - \frac{1}{2}J_{AB} + \frac{1}{4}J_{BB}$
4	$1s_{-1/2}$	$\alpha\beta\beta$	$\frac{1}{2}\nu_A - \nu_B - \frac{1}{2}J_{AB} + \frac{1}{4}J_{BB}$
5	$2s_{-1/2}$	$\beta(\alpha\beta + \beta\alpha)/\sqrt{2}$	$-\frac{1}{2}\nu_A + \frac{1}{4}J_{BB}$
6	$s_{-3/2}$	$\beta\beta\beta$	$-\frac{1}{2}\nu_A - \nu_B + \frac{1}{2}J_{AB} + \frac{1}{4}J_{BB}$
7	$a_{1/2}$	$\alpha(\alpha\beta - \beta\alpha)/\sqrt{2}$	$\frac{1}{2}\nu_A - \frac{3}{4}J_{BB}$
8	$a_{-1/2}$	$\beta(\alpha\beta - \beta\alpha)/\sqrt{2}$	$-\frac{1}{2}\nu_A - \frac{3}{4}J_{BB}$

[a]s and a indicate symmetric and antisymmetric eigenfunctions

Table 2.8 Mixing of states and corresponding energies in the AB_2 system

S. no.	Symmetry label	Eigenfunction	Energy
2'	$1s'_{1/2}$	$\cos\theta_+(1s_{1/2}) + \sin\theta_+(2s_{1/2})$	$\frac{1}{2}\nu_B + \frac{1}{4}(J_{BB} - J_{AB}) + C_+$
3'	$2s'_{1/2}$	$-\sin\theta_+(1s_{1/2}) + \cos\theta_+(2s_{1/2})$	$\frac{1}{2}\nu_B + \frac{1}{4}(J_{BB} - J_{AB}) - C_+$
4'	$1s'_{-1/2}$	$\cos\theta_-(1s_{-1/2}) + \sin\theta_-(2s_{-1/2})$	$-\frac{1}{2}\nu_B + \frac{1}{4}(J_{BB} - J_{AB}) + C_-$
5'	$2s'_{-1/2}$	$-\sin\theta_-(1s_{-1/2}) + \cos\theta_-(2s_{-1/2})$	$-\frac{1}{2}\nu_B + \frac{1}{4}(J_{BB} - J_{AB}) - C_-$

strongly coupled. While this represents the most general case for detailed quantum mechanical analysis, it is also the most difficult to obtain analytical solutions for the eigenvalues and the eigenfunctions. Fortunately, however, strongly coupled systems can be converted to weakly coupled networks by going to higher magnetic fields, whereby the separations between the resonance frequencies increase, while the coupling constants remain unchanged. We do not intend to show the various calculations here but shall present the results for some special cases as illustrations. Table 2.7 gives the eigenvalues and the eigenfunctions for the AB_2 system in the limit of weak coupling between A and B. One can see that these are products of α and β states for spin A and the eigenfunctions of the B_2 system. In the case of strong coupling, there will be further mixing of these functions leading to the following eigenfunctions and energies as indicated in Table 2.8. Tables 2.9 and 2.10 give the transition frequencies and the relative intensities of the AB_2 and AX_2 systems, respectively.

C_+ and C_- and angles θ_+ nad θ_- are positive quantities defined as

$$C_+ \cos 2\theta_+ = \frac{1}{2}(\nu_A - \nu_B) + \frac{1}{4}J_{AB}$$

$$C_+ \sin 2\theta_+ = \frac{1}{\sqrt{2}}J_{AB}$$

Table 2.9 Transitions and relative intensities in an AB_2 system

S. no.	Transition	Nucleus	Transition frequencies	Relative intensity
1	$3' \to 1$	A	$\frac{1}{2}(\nu_A + \nu_B) + \frac{3}{4}J_{AB} + C_+$	$[\sqrt{2}\sin\theta_+ - \cos\theta_+]^2$
2	$5' \to 2'$	A	$\nu_B + C_+ + C_-$	$[\sqrt{2}\sin(\theta_+ - \theta_-) + \cos\theta_+ \cos\theta_-]^2$
3	$8 \to 7$	A	ν_A	1
4	$6 \to 4'$	A	$\frac{1}{2}(\nu_A + \nu_B) - \frac{3}{4}J_{AB} + C_-$	$[\sqrt{2}\sin\theta_- + \cos\theta_-]^2$
5	$4' \to 2'$	B	$\nu_B + C_+ - C_-$	$[\sqrt{2}\cos(\theta_+ - \theta_-) + \cos\theta_+ \sin\theta_-]^2$
6	$2' \to 1$	B	$\frac{1}{2}(\nu_A + \nu_B) + \frac{3}{4}J_{AB} - C_+$	$[\sqrt{2}\cos\theta_+ + \sin\theta_-]^2$
7	$5' \to 3'$	B	$\nu_B - C_+ + C_-$	$[\sqrt{2}\cos(\theta_+ - \theta_-) - \sin\theta_+ \cos\theta_-]^2$
8	$6 \to 5'$	B	$\frac{1}{2}(\nu_A + \nu_B) - \frac{3}{4}J_{AB} - C_-$	$[\sqrt{2}\cos\theta_- - \sin\theta_-]^2$
9	$4' \to 3'$	Comm	$\nu_B - C_+ - C_-$	$[\sqrt{2}\sin(\theta_+ - \theta_-) + \sin\theta_+ \sin\theta_-]^2$

$$C_- \cos 2\theta_- = \frac{1}{2}(v_A - v_B) - \frac{1}{4}J_{AB}$$

$$C_- \sin 2\theta_- = \frac{1}{\sqrt{2}}J_{AB}$$

Defining $\delta = (v_A - v_B)$,

$$C_+ = \frac{1}{2}\left[\delta^2 + \delta J_{AB} + \frac{9}{4}J_{AB}^2\right]^{\frac{1}{2}}$$

$$C_- = \frac{1}{2}\left[\delta^2 - \delta J_{AB} + \frac{9}{4}J_{AB}^2\right]^{\frac{1}{2}}$$

In the limit of weak coupling between A and B, θ_- and 0_+ will tend to be zero, and the primed states reduce to the unprimed states. There will be no transitions between symmetric and antisymmetric eigenfunctions; hence, there will be a total of nine transitions, eight symmetrical, and one antisymmetrical. These are indicated in Table 2.9. Transition, in the limit of weak coupling reduces to $\alpha\beta\beta$ to $\beta\alpha\alpha$, which cannot be ascribed to any spin and is thus referred to as combination line.

In the limit of weak coupling $\theta_+ = \theta_-$, Table 2.9 transforms as Table 2.10.

Table 2.11 gives the eigenvalues and eigenfunctions of the ABX system. Table 2.12 gives the transitions frequencies and their relative intensities.

$$D_+ \cos 2\varphi_+ = \frac{1}{2}(v_A - v_B) + \frac{1}{4}(J_{AX} - J_{BX}); D_- \cos 2\varphi_-$$

$$= \frac{1}{2}(v_A - v_B) - \frac{1}{4}(J_{AX} - J_{BX})$$

$$D_+ \sin 2\varphi_+ = \frac{1}{2}J_{AB}; D_- \sin 2\varphi_- = \frac{1}{2}J_{AB}$$

Table 2.10 Transitions and relative intensities in an AB_2 system in the limit of weak coupling, i.e., AX_2 system

S. no.	Transition	Nucleus	Transition frequency	Relative intensity
1	$3' \to 1$	A	$(v_A) + J_{AB}$	1
2	$5' \to 2'$	A	(v_A)	1
3	$8 \to 7$	A	(v_A)	1
4	$6 \to 4'$	A	$(v_A) - J_{AB}$	1
5	$4' \to 2'$	B	$(v_B) + J_{AB}/2$	2
6	$2' \to 1$	B	$(v_B) + J_{AB}/2$	2
7	$5' \to 3'$	B	$(v_B) - J_{AB}/2$	2
8	$6 \to 5'$	B	$(v_B) - J_{AB}/2$	2
9	$4' \to 3'$	Comm	$2v_B - v_A$	0

Table 2.11 Eigenvalues and eigenfunctions in an ABX system

S. no.	Eigenfunction	Energy
1	$\alpha\alpha\alpha$	$\frac{1}{2}(v_A + v_B + v_X) + \frac{1}{4}(J_{AB} + J_{BX} + J_{AX})$
2	$\alpha\alpha\beta$	$\frac{1}{2}(v_A + v_B - v_X) + \frac{1}{4}(J_{AB} - J_{BX} - J_{AX})$
3	$\cos\varphi_+(\alpha\beta\alpha) + \sin\varphi_+(\beta\alpha\alpha)$	$\frac{1}{2}v_X - \frac{1}{4}J_{AB} + D_+$
4	$-\sin\varphi_+(\alpha\beta\alpha) + \cos\varphi_+(\beta\alpha\alpha)$	$\frac{1}{2}v_X - \frac{1}{4}J_{AB} - D_+$
5	$\cos\varphi_-(\alpha\beta\beta) + \sin\varphi_-(\beta\alpha\beta)$	$-\frac{1}{2}v_X - \frac{1}{4}J_{AB} + D_-$
6	$-\sin\varphi_-(\alpha\beta\beta) + \cos\varphi_-(\beta\alpha\beta)$	$-\frac{1}{2}v_X - \frac{1}{4}J_{AB} - D_-$
7	$\beta\beta\alpha$	$\frac{1}{2}(-v_A - v_B + v_X) + \frac{1}{4}(J_{AB} - J_{BX} - J_{AX})$
8	$\beta\beta\beta$	$\frac{1}{2}(-v_A - v_B - v_X) + \frac{1}{4}(J_{AB} + J_{BX} + J_{AX})$

Table 2.12 Transitions and relative intensities in an ABX system

S. no.	Transition	Nucleus	Energy	Relative intensity
1	$8 \to 6$	B	$v_{AB} + \frac{1}{4}[-2J_{AB} - J_{AX} - J_{BX}] - D_-$	$1 - \sin 2\varphi_-$
2	$7 \to 4$	B	$v_{AB} + \frac{1}{4}[-2J_{AB} + J_{AX} + J_{BX}] - D_+$	$1 - \sin 2\varphi_+$
3	$5 \to 2$	B	$v_{AB} + \frac{1}{4}[2J_{AB} - J_{AX} - J_{BX}] - D_-$	$1 + \sin 2\varphi_-$
4	$3 \to 1$	B	$v_{AB} + \frac{1}{4}[2J_{AB} + J_{AX} + J_{BX}] - D_+$	$1 + \sin 2\varphi_+$
5	$8 \to 5$	A	$v_{AB} + \frac{1}{4}[-2J_{AB} - J_{AX} - J_{BX}] + D_-$	$1 + \sin 2\varphi_-$
6	$7 \to 3$	A	$v_{AB} + \frac{1}{4}[-2J_{AB} + J_{AX} + J_{BX}] + D_+$	$1 + \sin 2\varphi_+$
7	$6 \to 2$	A	$v_{AB} + \frac{1}{4}[2J_{AB} - J_{AX} - J_{BX}] + D_-$	$1 - \sin 2\varphi_-$
8	$4 \to 1$	A	$v_{AB} + \frac{1}{4}[2J_{AB} + J_{AX} + J_{BX}] + D_+$	$1 - \sin 2\varphi_+$
9	$8 \to 7$	X	$v_X - \frac{1}{2}[J_{AX} + J_{BX}]$	1
10	$5 \to 3$	X	$v_X + D_+ - D_-$	$\cos^2(\varphi_+ - \varphi_-)$
11	$6 \to 4$	X	$v_X - D_+ + D_-$	$\cos^2(\varphi_+ - \varphi_-)$
12	$2 \to 1$	X	$v_X + \frac{1}{2}[J_{AX} + J_{BX}]$	1
13	$7 \to 2$	Comb	$2v_{AB} - v_X$	0
14	$5 \to 4$	Comb (X)	$v_X - D_+ - D_-$	$\sin^2(\varphi_+ - \varphi_-)$
15	$6 \to 3$	Comb (X)	$v_X + D_+ + D_-$	$\sin^2(\varphi_+ - \varphi_-)$

$$v_{AB} = \frac{1}{2}(v_A + v_B)$$

Midpoint of group 1: $v_{AB} - \frac{1}{2}(J_{AX} + J_{BX})$. Mid-point of group 2: $v_{AB} + \frac{1}{2} \times (J_{AX} + J_{BX})$.

The separation between midpoints: $J_{AX} + J_{BX} =$ the separation between two transitions of X ($8 \to 7$ and $2 \to 1$). In the limit of weak coupling between A and B, $\varphi_+ = \varphi_- = 0$; all the 12 transitions will have equal intensities.

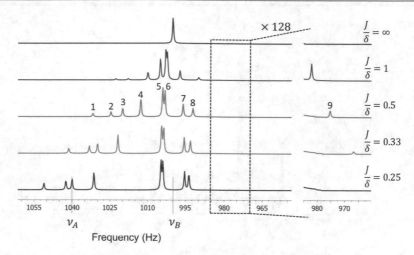

Fig. 2.14 Simulated NMR spectra for an AB_2 spin system with 10 Hz of scalar coupling ($J_{AB} = 10$ Hz). These spectra are generated at different ratios of $\frac{J}{\delta}$. Herein, in order to see the combination line (transition 9), the expanded region is plotted at an increased vertical scale

In Tables 2.7, 2.8, 2.9, 2.10, 2.11, and 2.12, the transitions are assigned to particular spins, but it must be noted that this assignment is in the limit of the spectra going over to first order. Otherwise, the transitions do not have any specific identity. We should also note that there are combination lines which occur due to more than one spins changing their polarizations. Such tables for other complex systems involving four and five spins are available in other texts, and computer programs are available for calculating the spectra of more complex spin systems. However, to be able to appreciate the results of such calculations, we shall discuss here the spectra of AB_2 and ABX systems in some detail. Figures 2.14, 2.15, and 2.16 show schematically the spectral features of these two-spin systems.

The AB_2 system has a total of nine transitions, and their relative positions and intensities are a sensitive function of the strong coupling-related parameter $\frac{J}{\delta}$. At low values of $\frac{J}{\delta}$, only eight transitions are visible, and these are separated in two groups which could be identified as belonging to A and B spins. As $\frac{J}{\delta}$ increases, the transitions start separating nonuniformly, and the intensities also do not show a simple pattern. The combination line (transition 9) makes its appearance. Slowly the outer transitions start diminishing in intensity just like in the two-spin AB case, and eventually a single will be seen in the limit of complete equivalence of all the three spins, that is, an A_3 system. From the transition energies and the relative intensities given in Table 2.11, three useful features may be derived: (i) line 3 directly gives the chemical shift of spin A (ν_A); its position is independent of the AB coupling. (ii) The chemical shift of B (ν_B) is given by the mean of transitions 5 and 7. (iii) Once the chemical shifts of A and B are determined, the coupling constant J_{AB} can be determined by adding transitions 1 and 6.

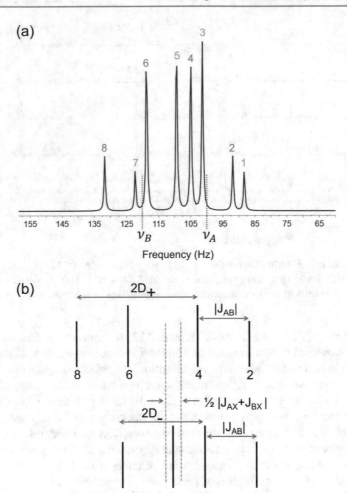

Fig. 2.15 (a) Expanded AB part of the simulated NMR spectra for a ABX spin system. This spectrum is generated with scalar coupling values of $J_{AB} = 13$ Hz, $J_{AX} = 3$ Hz, and $J_{BX} = 10$ Hz. (b) A schematic description of the two AB-type segments and parameters corresponding to the separation between the lines

$$v_1 + v_6 = v_A + v_B + \frac{3J_{AB}}{2} \qquad (2.57)$$

The AB_2 pattern is sensitive to the sign of δ but is insensitive to the sign of J, the pattern shown in Fig. 2.14 being for the positive value of δ. For the negative value of δ, the order of the transitions will have to be reversed (see Table 2.11). However, for a negative value of J, the appearance of the spectrum remains unaltered although the numbering of the transitions in decreasing order of energy will have to be changed.

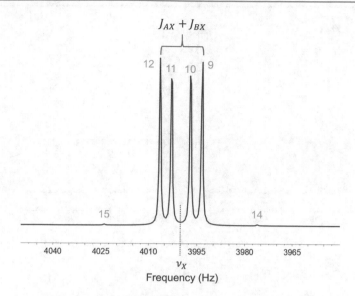

Fig. 2.16 Expanded X part of the simulated NMR spectra for an ABX spin system. This spectrum is generated with scalar coupling values of $J_{AB} = 13$ Hz, $J_{AX} = 3$ Hz, and $J_{BX} = 10$ Hz

The ABX spectrum contains two parts: an AB part (Fig. 2.15) and an X part (Fig. 2.16). The AB part consists of two groups of four lines having the definite pattern of a two-spin AB system. The two groups are interspersed, and often the difficult task is to identify the two groups properly. The X part of the spectrum contains six lines with an intensity pattern as shown. In order to facilitate the analysis of these spectra, the AB and X parts are separately presented in Figs. 2.15 and 2.16, respectively, indicating the separations between the transitions in terms of the different parameters. In Fig. 2.15a the simulations for a particular choice of coupling constants are shown. In Fig. 2.15b, the two AB groups are separately shown schematically to facilitate the analysis of the spectra as indicated.

The two groups of transitions in AB are labeled as 1, 3, 5, and 7 and 2, 4, 6, and 8, and we notice that the easiest parameter to identify is the J_{AB} coupling constant. The midpoints of the two quartets are separated by $\frac{|J_{AX}+J_{BX}|}{2}$. Notice, however, that the signs of the coupling constants cannot be determined from these spectra. The parameters D_+ and D_- which depend on the difference $J_{AX} - J_{BX}$ enable however the determination of the relative signs of the coupling constants. The spectral patterns are sensitive to the relative signs of J_{AX} and J_{BX} couplings as shown in Fig. 2.17.

In the X part (Fig. 2.16), the transitions 9 and 12 have the highest intensity, and there are two weak transitions outside of these, and there are two weak transitions on the interior. The separations also allow determination of $|J_{AX} + J_{BX}|$ and also $|D_+ + D_-|$ and $|D_+ - D_-|$.

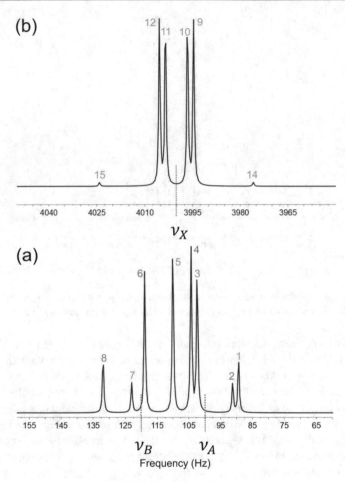

Fig. 2.17 Expanded AB (**a**) and X (**b**) parts of the simulated NMR spectra for an ABX spin system. Herein, in order to see the effect of scalar coupling signs, spectra are generated with scalar coupling values of $J_{AB} = 13$ Hz, $J_{AX} = 3$ Hz, and $J_{BX} = -10$ Hz

---

2.5 Dynamic Effects in the NMR Spectra

Until now we had been concerned with spectra of spin systems not undergoing any time-dependent changes, and the spectral features were calculated accordingly. However, in several situations, this is no longer valid, and the dynamism in the system does reflect in line positions and lineshapes in the NMR spectra. We are now talking about phenomena such as chemical exchange of a nuclear species between two or more sites in a molecule, the sites having different chemical environments, the exchange of a nucleus between two molecules either of the same type or of different types, conformational transitions in a single molecule, etc. All these phenomena constitute the so-called dynamic NMR spectroscopy. In the following, we shall describe in some detail the effects of time-dependent perturbations on the NMR spectra of molecules.

2.5.1 Two-Site Chemical Exchange

Let us consider first a simple case of a spin exchanging between two chemically different sites so that it has different resonance frequencies at the two sites. Let us also assume that the spin does not have any J coupling with any other spin; the conclusions however are also valid for systems with J coupling. A common example would be a hydroxyl proton in an alcohol exchanging with water protons. If the exchange is slow, that is, the hydroxyl proton stays for enough time in each of the positions and undergoes spin flip by absorbing energy at the respective frequencies, then there will be two separate lines in the NMR spectrum. If, on the other hand, the exchange is faster than the time it takes for the spin flip to occur at either of the positions, the proton will only see an average chemical environment during the time of its spin slip, and thus there will be only one line in the spectrum, at a frequency dictated by the weighted average of the two frequencies.

A quantitative analysis of these effects is best obtained by considering Bloch equations modified to include magnetization changes due to exchange phenomena. We have seen earlier in the first chapter that the signal shapes are dependent on the behavior of the magnetization components in the rotating frame of reference of the applied RF field. When the RF is weak—a condition which is easily satisfied—the Bloch equations in the rotating frame are given by

$$\frac{du}{dt} + \frac{u}{T_2} + (\omega_o - \omega)v = 0 \tag{2.58}$$

$$\frac{dv}{dt} + \frac{v}{T_2} + (\omega_o - \omega)u = -\gamma H_1 M_o \tag{2.59}$$

where u and v are the rotating frame magnetization components parallel and orthogonal to the RF field; they are also referred to as the real and imaginary components of the rotating frame magnetization. Taking this definition and defining a complex magnetization

$$G = u + iv \tag{2.60}$$

an equation for rate of change in G can be written as

$$\frac{dG}{dt} + \left[\frac{1}{T_2} - i(\omega_o - \omega)G\right] = -i\gamma H_1 M_o \tag{2.61}$$

Now, for an exchanging system, the complex magnetization at the two sites can be different. Thus, defining these as G_A and G_B for the sites A and B, respectively,

$$\frac{dG_A}{dt} + \alpha_A G_A = -i\gamma H_1 M_{oA} \tag{2.62}$$

$$\frac{dG_B}{dt} + \alpha_B G_B = -i\gamma H_1 M_{oB} \qquad (2.63)$$

where α_A and α_B are defined as

$$\alpha_A = \frac{1}{T_{2A}} - i(\omega_A - \omega) \qquad (2.64)$$

$$\alpha_B = \frac{1}{T_{2B}} - i(\omega_B - \omega) \qquad (2.65)$$

The Bloch equations given in Eqs. 2.62–2.63 have to be now modified to include the effects of exchange, $A \rightarrow B$, which affects the magnetizations at the two sites. Assuming first-order kinetics for the exchange, the two equations can be written as

$$\frac{dG_A}{dt} + \alpha_A G_A = -i\gamma H_1 M_{oA} + \frac{G_B}{\tau_B} - \frac{G_A}{\tau_A} \qquad (2.66)$$

$$\frac{dG_B}{dt} + \alpha_B G_B = -i\gamma H_1 M_{oB} + \frac{G_A}{\tau_A} - \frac{G_B}{\tau_B} \qquad (2.67)$$

where τ_A and τ_B are the lifetimes of the spin in the A and the B sites, respectively; thus their inverses represent the rate constants for the exchange process. The steady-state solutions of these equations can be obtained as before by setting

$$\frac{dG_A}{dt} = \frac{dG_B}{dt} = 0 \qquad (2.68)$$

The total complex moment is given by

$$G = G_A + G_B = -i\gamma H_1 M_o \frac{[\tau_A + \tau_B + \tau_A \tau_B(\alpha_B p_A + \alpha_A p_B)]}{[(1 + \alpha_A \tau_A)(1 + \alpha_B \tau_B) - 1]} \qquad (2.69)$$

where p_A and p_B are the populations of the A and B sites and M_o is the total magnetization; that is,

$$p_A = \frac{\tau_A}{\tau_A + \tau_B}; p_B = \frac{\tau_B}{\tau_A + \tau_B} \qquad (2.70)$$

$$M_{oA} = p_A M_o; M_{oB} = p_B M_o \qquad (2.71)$$

The imaginary part of G determines the absorptive lineshape in the spectrum. It is clear that the lineshape will depend upon the lifetimes of the spin in the two sites, relaxation times of the spin at the two sites, and also the two populations. The equations however simplify under extreme conditions of slow and fast exchanges.

In the limit of slow exchange, meaning, where τ_A and τ_B are much larger than $(\omega_A - \omega_B)^{-1}$, the spectrum will consist of two distinct lines. In this limit, G will be nearly equal to G_A when ω is near ω_A and will be nearly equal to G_B when ω is near ω_B. The expression for G, when ω is near ω_A, for example, will be (from Eq. 2.66 by setting $G_B = 0$)

$$G = G_A = -i\gamma H_1 M_o \frac{p_A \tau_A}{1 + \tau_A \alpha_A} \tag{2.72}$$

The imaginary part of G_A which determines the shape of the absorptive line will be

$$v = -\gamma H_1 M_o \frac{p_A T'_{2A}}{1 + (T'_{2A})^2 (\omega_A - \omega)^2} \tag{2.73}$$

where T'_{2A} is given by

$$(T'_{2A})^{-1} = (T_{2A})^{-1} + (\tau_A)^{-1} \tag{2.74}$$

Thus, we see that the line is broadened by an amount $(\tau_A)^{-1}$. This provides a means of estimating exchange rates by measuring the line broadenings.

In the limit of rapid exchange, the lifetimes are short, and the exchange rates are much larger than the separation between the resonance frequencies of the two sites. One can then neglect all terms involving products of lifetimes in Eq. 2.69, and the expression for G becomes

$$G = -i\gamma H_1 M_o \frac{\tau_A + \tau_B}{\tau_A \alpha_A + \tau_B \alpha_B} = -\frac{i\gamma H_1 M_o}{p_A \alpha_A + p_B \alpha_B} \tag{2.75}$$

The imaginary part of G is given by

$$v = -i H_1 M_o \frac{T'_2}{1 + (T'_2)^2 (p_A \omega_A + p_B \omega_B - \omega)^2} \tag{2.76}$$

We now observe that there will be a single line in the spectrum at a frequency which is the weighted average of the two frequencies ω_A and ω_B. The width of this line will now be related to T'_2 which is given by

$$T'_2{}^{-1} = p_A T_{2A}{}^{-1} + p_B T_{2B}{}^{-1} \tag{2.77}$$

Thus, rapid exchange leads to the collapse of signals to a single mean frequency with a width which is also the weighted mean of the two linewidths in the absence of exchange. The intensity of the line will be twice the intensity of the individual lines in the limit of slow exchange.

For all intermediate exchange rates, the line-shapes will look fairly complex and must be calculated from the complete expression for G. Figure 2.18 shows calculated line-shapes for a few exchange rates starting from slow exchange to fast exchange passing through a set of intermediate values, under the simplifying assumptions

$$p_A = p_B = \frac{1}{2}; \tau_A = \tau_B = 2\tau; T_{2A}{}^{-1} = T_{2B}{}^{-1} = 0 \tag{2.78}$$

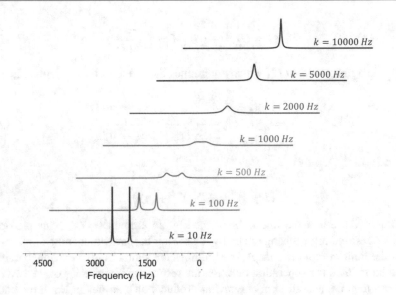

Fig. 2.18 Simulated NMR spectra for a two-site exchange at different exchange rates

We observe that at one stage when $\tau(\omega_A - \omega_B) = 1.414$, a single broad line occurs, and this condition is referred to as the "coalescence condition." Figures 2.19 and 2.20 show experimental illustrations for two uncoupled and two coupled spins, respectively, of the changes in the band shapes as the exchange rates are changed by changing the temperatures. The coalescence condition is a distinctive situation which can be easily detected. This permits the calculation of the exchange rates and consequently the life times under this condition and provides another "NMR timescale" which is applicable to exchanging systems. It is important to note that the coalescence condition is field-dependent since the Larmor frequency difference between the two sites is dependent on the strength of the applied field.

2.5.2   The Collapse of Spin Multiplets

So far, we considered an isolated spin exchanging between two chemically different sites. The theory developed however is applicable to coupled spin systems as well, wherein the resonance frequency of a particular spin changes from one value to another in its multiplet lines due to spin flips of the spins to which it is J coupled. In this situation, the exchange rate will be determined by the relaxation times of the other coupled spins. It is also possible that a spin jumps from one molecule to another, and in so doing, it changes its relative polarization. Because of this, the other spins J-coupled to the hopping spin experience the same kind of exchange within the multiplet lines. The same thing happens with the hopping spin as well.

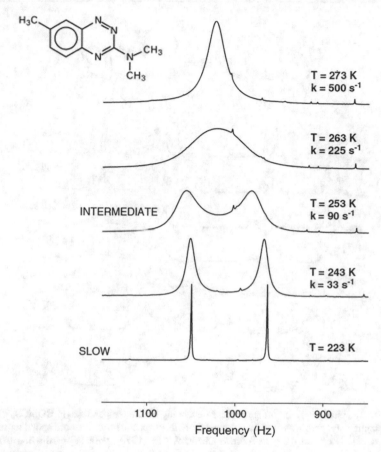

Fig. 2.19 Two-site chemical exchange experimental spectra recorded on 3- dimethylamino-7-methyl-1,2,4-benzotriazine, as a function of temperature. (Reproduced from Annual Reports on NMR Spectroscopy. **63**, 23 (2008), with the permission of Elsevier Publishing)

Depending upon the exchange rates, the multiplets can collapse as in the case of the rapid exchange described above.

A classic example of this is the spectrum of ethyl alcohol. For ease of understanding, we will consider one of the methylene protons (H_A) and the hydroxyl proton (H_B) which are J-coupled (Fig. 2.21a). The methylene proton resonance (H_A) is a doublet because of the two possible polarizations of the hydroxyl proton H_B. Similarly, the H_B proton is also a doublet because of two possible polarizations of H_A. However, in alcohol, the hydroxyl proton of one molecule is hydrogen bonded to the oxygen of another molecule, and there is a continuous exchange between two sites. The hydroxyl protons jump between different alcohol molecules.

Referring to Fig. 2.21b, the H_A proton experiences this jumping of hydroxyl protons between molecules. Therefore, the H_A proton may see different polarizations of the hydroxyl proton (H_B in α state, H_B' in β state). Thus, within the H_A doublet,

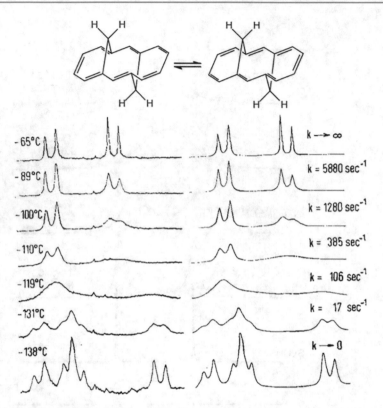

Fig. 2.20 Resonances of the bridge protons depending on the temperature (in COS/CS$_2$ (80:20); 100 MHz); left- (experimental) and right-hand (simulated spectra) with the corresponding reaction rate constants. (Reproduced from Angew. Chem, **9**, 513 (1970), with the permission of Wiley Publishing)

there will be continuous exchange between the two lines; if this exchange is slow compared to the J-coupling constant, the two lines will be seen separately. On the other hand, if the exchange is very rapid compared to the coupling constant, there will be averaging as indicated in the previous section, and there is only one resonance line at v_A. The same argument can be extended to the second magnetically equivalent methylene proton.

Similarly, referring to Fig. 2.21c, the hydroxyl proton (H$_B$) when it jumps between two molecules may see the methylene protons (H$_A$ and H$_A$') in α and β polarization states. Therefore, this exchange depending upon the exchange rate can again result in an averaging of two lines of the H$_B$ doublet producing a single resonance at v_B. The same argument can be extended to the second magnetically equivalent methylene proton (not indicated in the figure). Thus, the hydroxyl proton will appear as a singlet instead of a triplet.

The exchange rates can be influenced by changing the experimental conditions. For example, an addition of a small amount of an acid to pure alcohol increases the exchange rates rapidly and results in the collapse of the multiplets. The coupling of

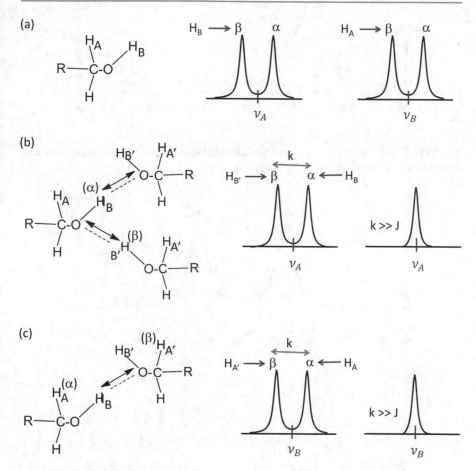

Fig. 2.21 (a) The splitting patterns due to J-coupling with the hydroxyl proton (H_B) for an alcohol considering one of the methylene protons (H_A). v_A represents the chemical shift of H_A, and v_B represents the chemical shift of H_B. The two lines in the doublet H_A arise because of different polarizations of H_B (α and β). Similarly, the two lines in H_B doublet arise because of different polarization of H_A (α and β). (b) Hydrogen bonding network of the hydroxyl protons in alcohol: The hydroxyl protons jump between two sites; therefore, when a molecule loses its hydroxyl proton to a second molecule, it may gain the same from a third molecule. It can happen that the polarization of the lost and gained hydroxyl protons could be different; thus, H_B and H_B' may have α and β polarizations; because this is a dynamic process, the H_A proton will see the hydroxyl protons in either of the orientations. If the exchange process is slow, then the H_A will appear as a doublet, whereas if the exchange rate (k) is much higher than the coupling constant (J), then it will average to a single line (v_A). (c) The hydroxyl proton (H_B) may see the methylene proton (H_A) in different polarizations (α and β) on the two molecules. So, this can result in averaging as in (b) depending upon the magnitude of the exchange rate

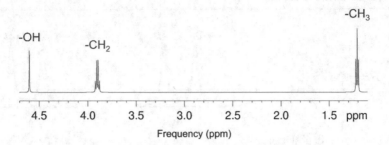

Fig. 2.22 ^{1}H NMR spectrum of ethanol with added acid impurities, which results in the collapse of scalar coupling between CH_2 and -OH groups

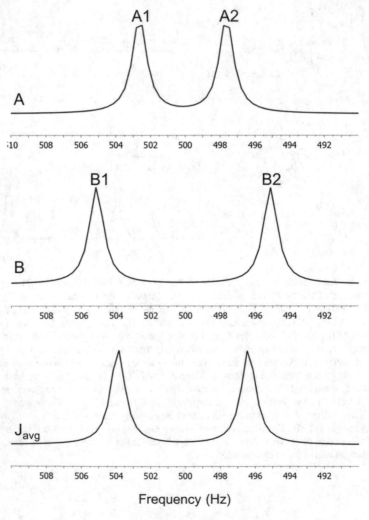

Fig. 2.23 Scalar couplings between two spins at different conformations A (upper trace) and B (middle trace). These scalar couplings average due to the rapid conformational transition, and that results in J_{avg} (lower trace)

hydroxyl proton to methylene protons will be eliminated. Similar arguments can be extended to other alcohols and exchanging systems (Fig. 2.22).

Such effects of multiplet collapse are also encountered when one of the spins coupled to the observed nucleus is an unpaired electron which undergoes rapid transitions between its two Zeeman states. This results in a substantial broadening of the resonance line of the nucleus.

2.5.3 Conformational Averaging of J-values

A consequence of the multiplet collapse due to rapid exchange is the "averaging of J-values" due to conformational transitions in molecules. This phenomenon is schematically shown in Fig. 2.23.

In conformation A, a particular proton has a coupling constant J_A and is accordingly split into two lines A1 and A2 with that separation. In conformation B, the particular proton has coupling constant J_B and would be split into two lines B1 and B2 with this new separation. Transitions A1 and B1 correspond to the same configuration of the spin to which the proton is coupled, as well as the A2 and the B2 transitions but with the different spin configuration. Now if the molecule is undergoing rapid exchange between the two conformations A and B, transitions A1 and B1 collapse to a single line at their weighted mean position, and transitions A2 and B2 collapse to their weighted mean position. Thus the separation between the two transitions will now be different from both J_A and J_B; the new coupling constant will be a weighted average of J_A and J_B.

$$J_{avg} = p_A J_A + p_B J_B \qquad (2.79)$$

This kind of averaging occurs almost invariably in all molecules and complicates the interpretation of the coupling constants in terms of structures of molecules in solution media. These aspects will be discussed in later chapters.

2.6 Summary

- NMR spectral features, namely, chemical shift and coupling constant are described.
- The analysis of the spectra for different most common two- and three-spin systems is described. The concepts of weak and strong coupling and their influence on the spectral appearance are described.
- Spectral alterations due to dynamism in the molecular systems are described.

2.7 Further Reading

1. High Resolution NMR by J. A. Pople, W. G. Schnieder, H. S. Bernstein, McGraw Hill, 1959
2. High Resolution NMR, E. D. Becker, 3rd ed., Elsevier, 2000
3. NMR Spectroscopy: Basic Principles, Concepts and Applications in Chemistry, H. Günther, 3rd ed., Wiley, 2013

2.8 Exercises

2.1. Two protons, A and B, give peaks at 3 ppm and 5 ppm, respectively, on a 500 MHz spectrometer. The separation between A and B on a 300 MHz spectrometer will be
   (a) 2 MHz   (b) 2 kHz   (c) 600 Hz   (d) 200 Hz   (e) 6 kHz

2.2. A molecule with a molecular formula C_3H_6O produces a single line in the proton NMR spectrum. How many peaks would a proton-decoupled ^{13}C spectrum of the molecule produce?
   (a) 1   (b) 2   (c) 3   (d) 5

2.3. How many peaks would a 1D ^{13}C spectrum of 50% deuterated chloroform will have?
   (a) 1   (b) 2   (c) 3   (d) 5

2.4. The proton spectrum of vinyl chloride ($H_2C=CH-Cl$) will consist of
   (a) a doublet and a triplet
   (b) two doublets
   (c) two triplets
   (d) three doublets of doublets

2.5. Which among the following commutation relationships is correct?
   (a) $[I_x, I_y] = iI_z$
   (b) $[I_x, I_y] = -iI_z$
   (c) $[I_x, I_z] = iI_y$
   (d) $[I_y, I_z] = -iI_x$

2.6. The proton spectrum of a 25% ^{13}C-labeled CH_3COOD will have
   (a) one peak
   (b) three peaks of equal intensities
   (c) three peaks of intensities of ratio 1:6:1
   (d) three peaks of intensities of ratio 1:1:6

2.7. Which proton will be most downfield shifted in $CH_3-CH=CH-CHO$ molecule?
   (a) Proton in the methyl group
   (b) Proton attached to the carbon adjacent to the methyl group
   (c) Proton attached to the carbon adjacent to the aldehyde group
   (d) Proton in the aldehyde group

2.8. The chemical shift (ppm) of the proton in the NMR spectrum reflects on

(a) the relative population of the different protons in the sample

(b) the electronic environment of the protons in the molecule

(c) the relative population of the protons with respect to a reference

(d) none of the above

2.9. Hydrogen bond causes

(a) downfield shift of the bonded protons

(b) no change in the chemical shift

(c) upfield shift of the bonded proton

(d) disappearance of the peak

2.10. In an aromatic ring, a proton in the plane of the ring due to ring current effect will be

(a) upfield shifted

(b) downfield shifted

(c) broadened out

(d) not affected

2.11. A molecule with two protons coupled to each other produces a spectrum of four lines. The chemical shifts of the two protons are 2.7 and 3.0 ppm. The coupling constant is 10 Hz. The intensities of four lines in the decreasing order of frequency on a 100MHz spectrometer will be

(a) 0.68, 1.32, 0.68, 1.32

(b) 0.68, 0.68, 1.32, 1.32

(c) 1.32, 0.68, 0.68, 1.32

(d) 0.68, 1.32, 1.32, 0.68

2.12. For a weakly coupled two spin system say (k, l), the high-resolution Hamiltonian is

(a) $\omega_k I_{kz} + \omega_l I_{lz} + 2\pi J_{kl} I_{kz} I_{lz}$

(b) $\omega_k I_{kx} + \omega_l I_{lz} + 2\pi J_{kl} I_{kx} I_{lz}$

(c) $I_{kz} + I_{lz} + 2\pi J_{kl} I_{kz} I_{lz}$

(d) $\omega_k I_{kx} + \omega_l I_{ly} + 2\pi J_{kl} I_{kx} I_{ly}$

2.13. The eigenfunctions for a weakly coupled two-spin ½ system are $\alpha\alpha$, $\alpha\beta$, $\beta\alpha$, and $\beta\beta$. Strong coupling leads to the mixing of

(a) $\alpha\alpha$, $\alpha\beta$

(b) $\alpha\alpha$, $\beta\alpha$

(c) $\alpha\beta$, $\beta\alpha$

(d) $\alpha\alpha$, $\beta\beta$

2.14. Considering a two-spin system, prove $[\mathcal{H}, F_z] = [\mathcal{H}F_z - F_z\mathcal{H}] = 0$.

2.15. Draw the schematic 400 MHz NMR spectra of the AMX spin system assuming the chemicals shifts A=2.0 ppm, M=4.0 ppm, and X=8.0 ppm and the coupling constants J(A-M) = 10 Hz, J(M-X) = 5 Hz, and J(A-X) = 7 Hz.

2.16. Extract the chemical shifts (Hz) and the coupling constants (Hz) of the different protons in the given ^{1}H NMR spectrum.

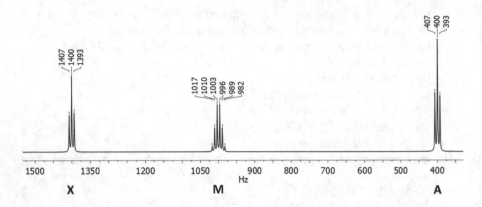

2.17. Extract the chemical shifts (ppm) and the C-H coupling constants (Hz) of the different carbons from the ^{13}C NMR spectrum given.

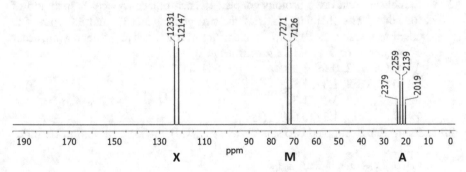

2.18. Show by explicit calculation that for an A2 system the coupling between the two spins does not appear in the spectrum.

2.19. Calculate the expected spectrum for an AB system given that the chemical shift separation on a 100 MHz spectrometer is 50 Hz and the coupling constant is 15 Hz. How would the spectrum look like when it is recorded on a 600 MHz spectrometer?

2.20. Given that $v_A = 200$ Hz, $v_B = 230$ Hz, and $J_{AB} = 15$ Hz, calculate the frequencies and intensities of the AB$_2$ spectrum, and simulate it using a linewidth of 1 Hz.

2.21. Given the $v_A = 150$ Hz, $v_B = 175$ Hz, $v_X = 4000$ Hz, $J_{AB} = 12$ Hz, $J_{AX} = 8$ Hz, and $J_{BX} = 4$ Hz, calculate the frequencies and intensities of the ABX spectrum, and simulate it using a linewidth of 1 Hz.

Fourier Transform NMR

3

Contents

3.1	Introduction	90
3.2	Principles of Fourier Transform NMR	91
3.3	Theorems on Fourier Transforms	97
3.4	The FTNMR Spectrometer	100
3.5	Practical Aspects of Recording FTNMR Spectra	101
	3.5.1 Carrier Frequency and Offset	101
	3.5.2 RF Pulse	101
	3.5.3 Free Induction Decay (FID) and the Spectrum	101
	3.5.4 Single-Channel and Quadrature Detection	103
	3.5.5 Signal Digitization and Sampling	106
	3.5.6 Folding of Signals	107
	3.5.7 Acquisition Time and Resolution	109
	3.5.8 Signal Averaging and Pulse Repetition Rate	110
3.6	Data Processing in FT NMR	112
	3.6.1 Zero Filling	112
	3.6.2 Digital Filtration or Window Multiplication or Apodization	113
3.7	Phase Correction	117
3.8	Dynamic Range in FTNMR	122
3.9	Solvent Suppression	122
3.10	Spin Echo	126
3.11	Measurement of Relaxation Times	129
	3.11.1 Measurement of T_1 Relaxation Time	129
	3.11.2 Measurement of T_2 Relaxation Time	133
3.12	Water Suppression Through the Spin Echo: Watergate	133
3.13	Spin Decoupling	134
3.14	Broadband Decoupling	136
3.15	Bilinear Rotation Decoupling (BIRD)	138
3.16	Summary	139
3.17	Further Reading	139
3.18	Exercises	139

© The Author(s), under exclusive license to Springer Nature Switzerland AG 2022
R. V. Hosur, V. M. R. Kakita, *A Graduate Course in NMR Spectroscopy*,
https://doi.org/10.1007/978-3-030-88769-8_3

Learning Objectives
* Introducing Fourier transform NMR
* NMR data acquisition and processing
* Mathematical aspects of Fourier transforms
* Signal phases in FT NMR spectra
* Dynamic range effects in FT NMR spectra
* Spin echo and its benefits
* Measurement of relaxation times

3.1   Introduction

In the early days of NMR, the spectra were recorded by keeping the frequency of the RF fixed and sweeping the field continuously so as to match the resonance conditions for the various lines sequentially in a spectrum. This was termed as "slow passage" or "continuous wave" (CW) spectroscopy. The field sweep had to be slow so that the spins follow the changes in the field and there is enough time for the populations to readjust as dictated by the changing field. Since this is dependent on the spin-lattice relaxation times of the spins, the sweep rate was dictated by the relaxation times. The longer the relaxation times, the slower the sweep so that all the frequencies were free of disturbances arising due to interferences between spins lagging behind in following the magnetic field. Thus, typically for a ^{1}H high-resolution spectrum spanning about 1000 Hz, about 20–30 min would be required to scan through the spectrum.

Now, in an NMR spectrum, the signal-to-noise (S/N) ratio for any peak is defined as

$$\frac{S}{N} = \frac{\text{peak height above a mean noise level}}{\text{maximum peak} - \text{peak separation in noise}} \times 2.5 \qquad (3.1)$$

Given the fact that NMR is inherently an insensitive spectroscopic technique compared to optical techniques, the $\frac{S}{N}$ ratios are inherently poor, and it is invariably necessary to adopt "signal averaging" techniques to enhance the sensitivities. What this means is that the spectra have to be scanned several times and the data coadded. In such an event, the peak height or the signal intensity increases proportionately to the number of additions, but the maximum peak-peak separation in noise increases as the square root of the number of additions. If n is the number of coadditions, then the net gain in S/N will be a factor of square root of n. Thus, if a S/N enhancement by a factor of p is desired, then the time required to achieve it will increase by a factor p^2. For example, if a scan through a NMR spectrum takes 30 min, then to achieve a S/N enhancement by a factor of 5 would require 25 scans to be coadded, and the time required would be 750 min or 12 h 30 min. This places stringent demands on the spectrometer stability, temperature, etc. Alternatively, one has to work with highly concentrated samples so that excessive signal averaging may not be required to observe the desired signals. Samples with low solubility or nuclei with low natural abundance such as ^{13}C or ^{15}N are almost impossible to study. These turned out to be serious limitations for applications of NMR.

The discovery of Fourier transform NMR in 1966 constituted the greatest revolution in NMR methodology and opened flood gates of applications in chemistry, biology, and medicine. It was a totally new concept of recording NMR spectra and enabled many possibilities for spin manipulations and observation of transient and dynamic effects, hitherto impossible to investigate. In this chapter we shall discuss the basic technique and the practical aspects of Fourier transform NMR.

3.2   Principles of Fourier Transform NMR

The fundamental difference between this new technique and the conventional CW technique is that in FT NMR there is no sweeping of either the magnetic field or the frequency. Information about all the resonances in the spectrum is collected indirectly in a few seconds in the so-called time domain, and the frequency spectrum is obtained by a mathematical transformation, namely, the Fourier transformation of the collected data.

The trick is to simultaneously apply a large number of radiofrequency fields (several thousands) covering a wide range of frequencies at any desired intervals at one time so that there is always a RF frequency to satisfy the resonance condition of every line in the spectrum. This is achieved by the application of a so-called RF pulse. The pulse generates also frequencies which do not have resonance counterparts, but these are automatically filtered out by the detection system as we shall see in a short while.

A RF pulse is a RF applied for a short time τ, typically of the order of few microseconds. This is shown schematically in Fig. 3.1.

Such an electronic switching produces an output which has a frequency distribution around the main RF frequency, say ω_o, as shown in Fig. 3.2.

If τ is of the order of few microseconds, it is clear that a range of a MHz or at least several KHz range of frequencies with similar amplitudes will be generated around the central frequency ω_o. This is certainly more than enough to cover all the resonance frequencies present in any spectrum. The superposition of these different frequency waves produces a wave pattern schematically shown in Fig. 3.3.

Because this pattern looks like a square wave, the RF pulse is often represented as a square barrier. In the schematic Fig. 3.3, all the frequency waves are assumed to have identical amplitudes. This is important because the amplitude of RF determines the RF power and consequently the signal intensity. If the intensities of two transitions have to be compared for some derivable information, it is necessary that they are excited with identical RF powers. Therefore, referring to Fig. 3.2, we see that only a small width around the central frequency should be retained and the others filtered out by suitable electronic devices. However, for 10–20 KHz ranges, this condition is easily satisfied, and this is enough for most situations. In certain

Fig. 3.1 A schematic
representation of a radio
frequency pulse for a duration
of τ

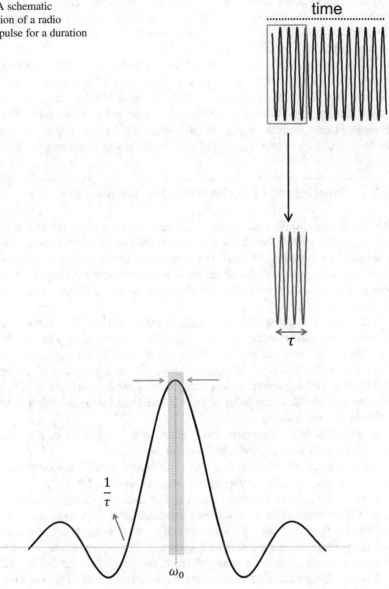

Fig. 3.2 The frequency domain output of RF pulse. Here, the uniform excitation around the main
RF frequency ω_o is highlighted in gray color

situations, such as in ^{13}C where spectral ranges are very large, some difficulties arise,
and then it becomes necessary to record spectra by applying pulses with different
central frequencies to observe different regions in the spectra. It is also obvious in
this context that the shorter the pulse, the better will be the spectral range of

Fig. 3.3 Simulated wave from the superposition of different frequencies. Thick blue wave form represents the result of superposition

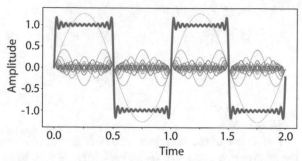

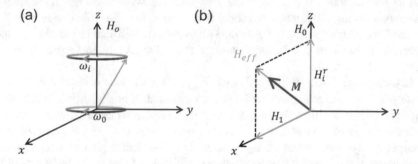

Fig. 3.4 (a) Rotating frame representation of precessional frequency, $\omega_i^r = \omega_i - \omega_o$. (b) Translating ω_i^r into magnetic field along z-axis is equal to H_i^r (not shown in the figure). The effective field (H_{eff}) for nucleus i in the rotating frame is the vector addition of H_1 and H_i^r

observation. We will see later the factors which govern the choice of the pulse widths in the context of optimizations of experimental conditions.

Now, what is the response of the system to the RF pulse, and what is the signal we collect in an FTNMR experiment? This can be appreciated readily by going into the rotating frame of the RF applied along the x-axis and looking at Larmor precession in the classical picture (see Fig. 3.4).

In the rotating frame, the precessional frequency, ω_i^r, of spin i is $(\omega_i - \omega_o)$ as indicated in Fig 3.4a. Translating this into magnetic field, the field along z-axis is given by $\frac{\omega_i - \omega_o}{\gamma}$, which is denoted by H_i^r in Fig 3.4b.

$$H_i^r = \frac{\omega_i - \omega_o}{\gamma} \qquad (3.2)$$

The effective field in the rotating frame will be a vector addition of H_1 and H_i^r which will be in the x-z plane.

$$H_{i,\text{eff}}^r = \frac{\left[(\omega_i - \omega_o)^2 + (\gamma H_1)^2\right]^{\frac{1}{2}}}{\gamma} \qquad (3.3)$$

$$\tan \theta = \frac{H_i^r}{H_1} \tag{3.4}$$

$$\omega_{i,\text{eff}}^r = -\gamma H_{i,\text{eff}}^r$$

If $\gamma H_1 \gg |\omega_i - \omega_o|$
$H_{i,\text{eff}}^r \cong H_1$

It is clear that the magnitude and direction of effective field critically depends on the relative magnitudes of H_1 and H_i^r. If H_1 is very large compared to H_i^r, then $H_{i,\text{eff}}^r$ will be almost along the H_1 axis. If this condition can be satisfied for all the spins in the sample having different precessional frequencies, then the effective field will be along H_1 for all the spins in the system. The magnitude will also be practically equal to H_1, and then one can simply consider the behaviour of the spins under the influence of this field.

In the rotating frame, the direction of $H_{i,\text{eff}}^r$ acts as the axis of quantisation of the spins, and they tend to orient with respect to this field. If this effective field is along H_1, for all the spins, then they have to undergo substantial changes with respect to their energy levels, redistribution among these levels, etc., and the rate of these changes will be governed by the relaxation times. The time requirement for complete realignment would be of the order of seconds. Figure 3.5 shows the trajectory of the magnetization as a function of time.

The frequency of precession will be given by

$$\omega^r = -\gamma H_1 \tag{3.5}$$

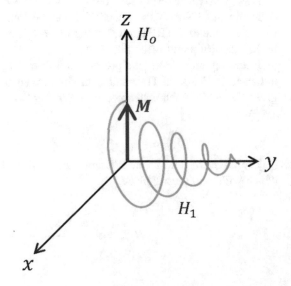

Fig. 3.5 The trajectory of rotation of net magnetization when the effective field is H_1, leading to an eventual alignment along H_1

If, however, the RF is applied for a short time τ as a pulse, the magnetization will simply make a rotation in the y-z plane. The angle of rotation, called the flip angle θ, is given by

$$\theta = -\gamma H_1 \tau \tag{3.6}$$

If τ is adjusted to cause a 90° rotation then, we say, we have 90° pulse; if it is adjusted for a 180° rotation, then we have a 180° pulse; etc. After the RF pulse is over, the effective field returns to H_o, and the magnetization returns to the z-axis. This is schematically shown in Fig. 3.6.

This recovery is also governed by relaxation, and during this time, the various spins precess with their characteristic frequencies. Such precessing magnetization components can induce signal in a detector in the x-y plane, and the total signal detected $g(t)$ will have contributions from all the frequency components. In other words, $g(t)$ is the Fourier transform of the frequency spectrum of the sample.

$$g(t) = \sum_n a_n \cos \omega_n t + \sum_n b_n \sin \omega_n t$$

or

$$g(t) = \frac{1}{2\pi} \int F(\omega) e^{i\omega t} d\omega \tag{3.7}$$

Thus the response of the system to the RF pulse, recorded as a function of time, enables unscrambling of the frequencies present in the spin system. The

Fig. 3.6 After the RF pulse, the return of the magnetization to the z-axis is schematically shown in the black color spiral cone

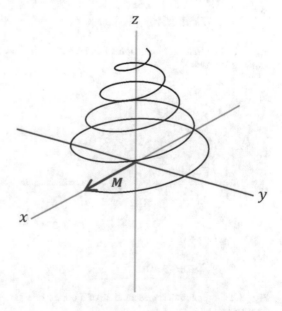

time-dependent function which represents the total magnetization in the transverse plane may be written as a new function $f(t)$, as

$$f(t) = g(t)e^{-t/T_2} \tag{3.8}$$

The function $f(t)$ is called the "free induction decay" (FID), since it is recorded during free precession in the absence of any perturbation, and it also decays due to T_2 relaxation processes. The Fourier relation between FID and the spectrum is pictorially shown in Fig. 3.7.

The above-described method of obtaining an NMR spectrum is the principle of FTNMR technique and is summarized in Fig. 3.8.

The duration of the FID is governed by the T_2 relaxation and thus will be of the order of a few hundreds of milliseconds to seconds. A data collection of one FID corresponds to one scan through the spectrum in the slow passage experiment. Comparing the time factors of the two modes of NMR experiments, it is clear that

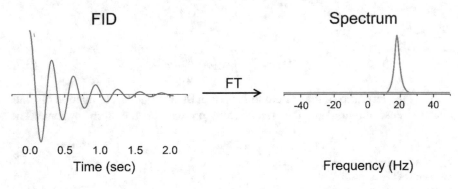

Fig. 3.7 Pictorial representation of Fourier relation between FID and the spectrum. Arbitrary numbers are used

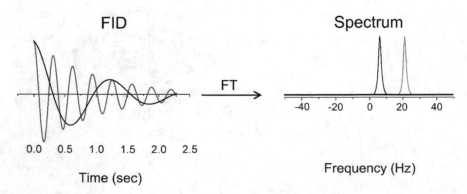

Fig. 3.8 A pictorial representation of Fourier relation between FID and spectrum with two frequencies

the FTNMR experiment results in 2–3 orders of magnitude saving in time. As a consequence, in the time required for one CW scan, several transients can be collected, and this amounts to an enhancement in the S/N ratio.

The advantages of such an enhancement in sensitivity are enormous. (1) Low concentrations of the samples can be used. The concentrations are often limited by solubility, availability, viscosity changes at high concentration, etc. (2) Nuclei with low natural abundance such as ^{13}C and ^{15}N can be studied, providing additional probes for the characterization of molecules. Signal averaging which is a must in these cases can be easily performed. (3) Because of the enhanced speed of data acquisition, short-lived species having half-lives of the order of seconds only can be readily studied, and their kinetics of transformations can be investigated. (4) Dynamical processes can be investigated, and data can be collected as a function of time. These have opened up enormous applications of NMR in various areas of chemistry and biology.

In addition, the fact that, in FTNMR spin system, excitation and detection are separated in time is of great significance for all the modern developments which will be discussed in later chapters.

3.3   Theorems on Fourier Transforms

Since the NMR spectrum is now obtained by a mathematical manipulation, namely, Fourier transformation, of the collected time domain data, the properties of Fourier transforms in general become relevant to the characteristics of the derived NMR spectra. The following are some of the useful theorems in this regard.

We consider, in these theorems, time and frequency as Fourier pairs:

$$f(t) = \frac{1}{2\pi} \int_{-\infty}^{\infty} F(\omega)e^{i\omega t} d\omega$$

$$F(\omega) = \int_{-\infty}^{\infty} f(t) \, e^{-i\omega t} dt \tag{3.9}$$

(a) If $f(t)$ is real (if only M_x or M_y is detected), then $F(\omega)$ is complex and Hermitian:

$$F(\omega) = F^*(-\omega)$$

$$F^*(\omega) = F(-\omega) \tag{3.10}$$

(b) If $f(t)$ is even, e.g., $\cos(\omega t)$, then $F(\omega)$ is also even.

$$f(t) = f(-t)$$

$$F(\omega) = F(-\omega) \tag{3.11}$$

If $f(t)$ is odd, then $F(\omega)$ is also odd.

$$f(-t) = -f(t)$$

$$F(-\omega) = -F(\omega) \tag{3.12}$$

(c) If $f(t)$ is even, then

$$f(t) = \frac{1}{\pi} \int_0^\infty F(\omega) \cos(\omega t) d\omega \tag{3.13}$$

$$F(\omega) = 2F_c(\omega) \tag{3.14}$$

$$F_c(\omega) = \int_0^\infty f(t) \cos(\omega t) dt \tag{3.15}$$

This is called the cosine transform.
Similarly, for odd $f(t)$,

$$(d) \quad f(t) = \frac{1}{\pi} \int_0^\infty F(\omega) \sin(\omega t) d\omega \tag{3.16}$$

$$F(\omega) = -2i\, F_s(\omega) \tag{3.17}$$

$$F_s(\omega) = \int_0^\infty f(t) \sin(\omega t) dt \tag{3.18}$$

This is called the sine transform and

Additive Theorem

$$(e) \quad F^+[f(t) \pm g(t)] = F^+[f(t)] \pm F^+[g(t)] = F(\omega) \pm G(\omega) \tag{3.19}$$

F^+ represents the Fourier transformation operator:

$$(f) \quad F^+[f(at)] = \frac{1}{2\pi\,|a|} F\left(\frac{\omega}{a}\right) \tag{3.20}$$

This theorem has important implications for signal averaging. Several FIDs can be coadded and Fourier transformed at the end, to get the frequency spectrum.

Multiplication of FID

$$(g) \quad F^+[af(t)] = a\,F^+[f(t)] = a\,F(\omega) \tag{3.21}$$

Delayed Acquisition

$$(h) \quad F^+[f(t + \delta t)] = e^{(i\omega\delta t)}F(\omega) \tag{3.22}$$

δt represents the delay in the start of the data acquisition. The frequency domain spectrum $F(\omega)$ is phase modulated by delayed acquisition.

(i) The area under the function $F(\omega)$ is equal to the value of the FID at $t = 0$:

$$f(0) = \frac{1}{2\pi} \int F(\omega)d\omega \tag{3.23}$$

(j) Convolution: Multiplication in the time domain translates into convolution in the frequency domain.

$$F^+[f(t).g(t)] = F(\omega) * G(\omega) \tag{3.24}$$

(k) Digitization: If $f(t)$ is sampled and thus can be considered as a series of Dirac δ-functions, τ seconds apart, then its Fourier transform is also a series of Dirac δ-functions, $\frac{1}{\tau}$ Hz apart.

$$F^+ \sum_{-\infty}^{\infty} \delta(t - n\tau)f(t) = \frac{1}{\tau} \sum_{-\infty}^{\infty} F(\omega) * \delta\left(\omega - \frac{n}{\tau}\right)$$
$$= \frac{1}{\tau} \sum_{-\infty}^{\infty} F\left(\omega - \frac{n}{\tau}\right) \tag{3.25}$$

These theorems have the following implications in FTNMR.

(i) Signal averaging can be done in the time domain, and the Fourier transformation can be done only once at the end. This results in substantial saving in data processing time.

(ii) Signal-to-noise (S/N) ratio or resolution in the spectra can be enhanced, as desired by the situation by suitable data processing techniques such as window multiplication (commonly called as apodization) in the time domain. This is

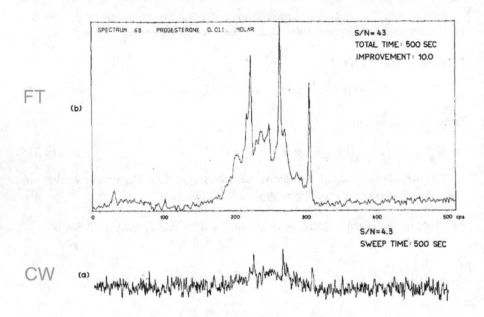

SPECTRUM 68   PROGESTERONE 0.011 MOLAR

S/N= 43
TOTAL TIME: 500 SEC
IMPROVEMENT: 10.0

FT
(b)

0   100   200   300   400   500 cps

S/N=4.3
SWEEP TIME: 500 SEC

CW   (a)

Fig. 3.9 ^{1}H NMR spectra of a 0.011 M solution of progesterone in hexafluorobenzene, both spectra performed in 500 s. (**a**) Single scan, the absorption mode signal is recorded directly by CW NMR. (**b**) Pulse FT NMR method, with 500 scans. (Reproduced from Review of Scientific Instruments **52**, 1876 (1981) with the permission from AIP Publishing)

done after data collection and is thus independent of the spectrometer itself. The improvements in the quality of the spectra are exemplified in Fig. 3.9.

(iii) The time domain signal (FID) can be collected in a digitized manner, providing time gaps in between data points, during which specific manipulations are possible.

3.4   The FTNMR Spectrometer

The completely different method of obtaining an NMR spectrum necessarily implies different basic elements and design of the spectrometer. RF has now been applied as pulse, and it must be possible to control the duration of these pulses very precisely. Note that these are in microsecond ranges, and thus their control is highly demanding. It is also necessary to achieve high RF powers, and the rise times of these pulses should be very short (ns). Since the data is collected in digital form, special devices are needed for the conversion of the analog NMR signal into digital form. For all such precision timing of pulses and precise digitization, computers become essential ingredients of an NMR spectrometer. A computer is also the key data manipulator and carries out all the mathematical manipulations for obtaining good spectra. Figure 3.10 shows schematically the essential elements of a FTNMR spectrometer.

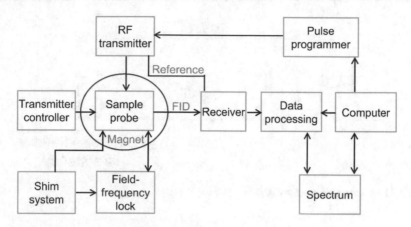

Fig. 3.10 A schematic representation of the essential elements of an FT NMR spectrometer

3.5 Practical Aspects of Recording FTNMR Spectra

3.5.1 Carrier Frequency and Offset

The frequency of the RF pulse is referred to as carrier frequency. In order for the power spectrum of the RF to be flat over the desired spectral region (for uniform excitation), it will be necessary to move the carrier frequency to the region of interest; this shift is called *offset*. Such shift of the main frequency is achieved by different mixing processes electronically. This is schematically indicated in Fig. 3.11.

3.5.2 RF Pulse

The RF can be applied along any axis in the x-y plane; the angle it makes with the x-axis is referred to as the phase of the pulse. Thus a 90° pulse applied along the x-axis denoted as 90_x is said to have 0° phase, a 90_y pulse has a 90° phase, etc.; such a definition is only a convention. The phases strictly speaking determine the phase relation between the transmitter and receiver phases; a zero-phase difference is taken to imply that the RF is applied along the x-axis. The special hardware devices used for bringing about phase changes are called phase shifters.

3.5.3 Free Induction Decay (FID) and the Spectrum

As discussed in Sect. 3.2, free induction decay (FID) is the response of the spin system to a RF pulse. The RF pulse rotates the equilibrium z-magnetization into the x-y plane, for a 90° pulse, and this magnetization then precesses and decays as the

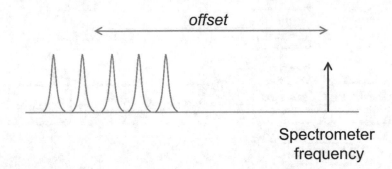

Fig. 3.11   A schematic representation of offset in FT NMR

Fig. 3.12 A schematic representation of the rotation of magnetization in the transverse plane following the application of a 90° pulse along the x-axis and its detection by the receiver along the y-axis

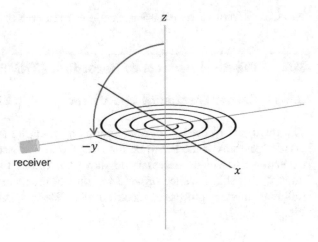

system returns to equilibrium. The precessing magnetization induces a signal in the receiver kept in the *x-y* plane, as indicated in Fig. 3.12.

Depending upon whether the x-component or the y-component of the magnetization is detected following a 90_x pulse, the FID has a sine or cosine functional form for a single frequency as shown in Fig. 3.13.

Mathematically, the two functional forms are

$$x_{\text{component}} = M_o \sum_i \sin \omega_i t$$

$$-y_{\text{component}} = M_o \sum_i \cos \omega_i t$$

The two types of FIDs give rise to different line shapes after real Fourier transformation: the sine function gives the dispersive line shape and the cosine function produces an absorptive line-shape.

Fig. 3.13 Cosine $(-M_y)$ and sine (M_x) magnetizations, after the application 90_x pulse

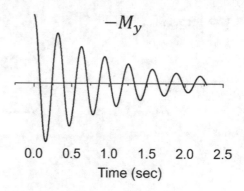

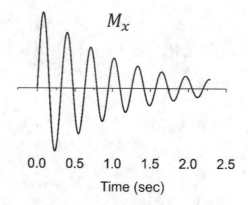

There is a definite relation between the phase of the RF pulse and the phase of a FID or of the spectrum as can be seen from Fig. 3.14.

3.5.4   Single-Channel and Quadrature Detection

This is intimately connected with the location of the offset. Suppose we are detecting the y-component of the magnetization only, following a 90_x pulse (called the "single-channel detection"), the FID generated by a single frequency ω_i will have cosine dependence on time.

$$F(t) = \cos(\omega_i t)e^{(-\lambda_i t)} \tag{3.26}$$

where λ_i is the decay rate constant $(\lambda_i = 1/T_{2i})$.

Complex Fourier transformation of such a signal will however lead to two lines at $\omega = \omega_i$ and $\omega = -\omega_i$; this is clearly an effect of FT only, and it is not reality (Fig. 3.15). If the carrier is located in the center of the spectrum, then the peak at $-\omega_i$ will interfere with another real peak in the spectrum. The real peaks and artifacts of the FT will overlap. Therefore, under these conditions, it becomes necessary to place

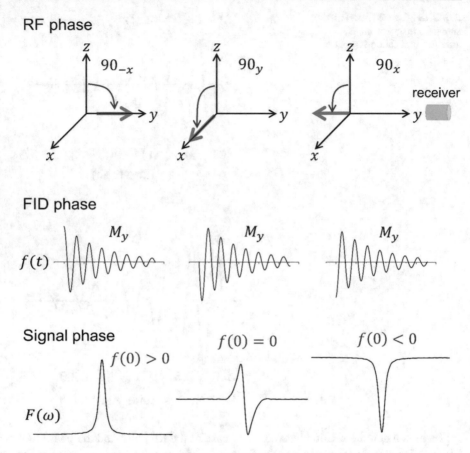

Fig. 3.14 For the RF pulses with different phases, when the receiver is put along the y-axis, the phase relation between FID and signal are shown

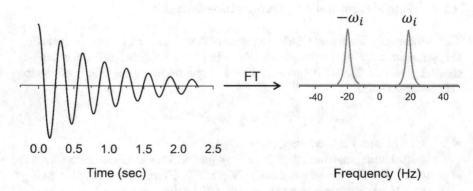

Fig. 3.15 A complex FT of a signal detected in single-channel mode results in two frequencies

the carrier at one end of the spectrum, and then the FT artifacts will be located in a distinctly different region. The same argument holds if only the x-component of the magnetization is detected in the FID.

Now, if both x- and y-components are detected at the same time by separate detectors, then the combined FID is a complex function $e^{(-i\omega_i t)}$, and the complex FT will produce a single peak at ω_i (Fig. 3.16); then it is also possible to distinguish positive and negative frequencies, and consequently the carrier can be placed anywhere in the spectral region. This is commonly known as "quadrature detection."

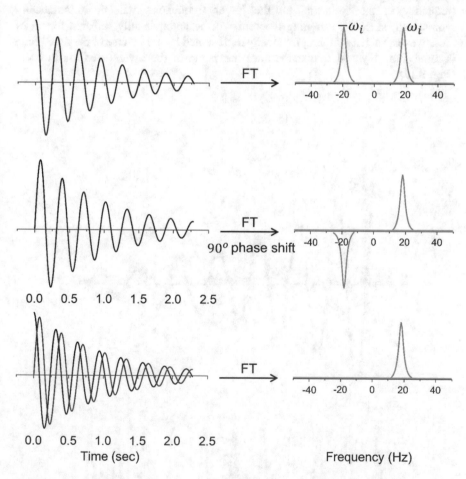

Fig. 3.16 The simultaneous detection of signal along both x- and y-axes (quadrature detection), subsequent FT of the individual FIDs separately, and coaddition after suitable phase correction as indicated in the middle spectrum yields a single frequency (see Sect. 3.7 for phase correction)

3.5.5 Signal Digitization and Sampling

As mentioned earlier, the FID is digitized; the time interval between two consecutive points is a constant for the whole length of the FID and is called the dwell time. How is this time determined? This is determined by the sampling theorem, which states that to represent a sine/cosine wave precisely by digital points, there must be at least two points per cycle of the wave. Since the FID is a superposition of all the frequencies in the spectrum, every data point in the FID has contributions, from all the frequencies. Thus, if the spacing of the data points is selected to suite the largest frequency in the spectrum, all the lower frequencies will be automatically represented, since the sampling theorem will be automatically satisfied for all of these waves; such a sampling frequency represented by the inverse of the dwell time is termed the Nyquist frequency, after the name of the inventor of the theorem (Fig. 3.17).

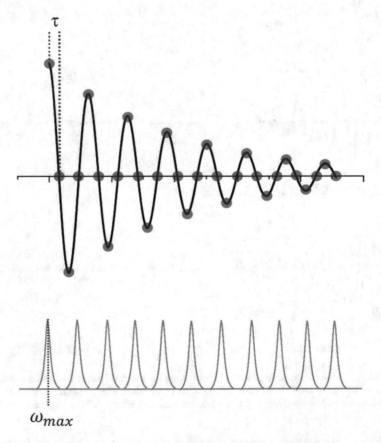

Fig. 3.17 Nyquist relation between data sampling and maximum representative frequency. τ is the dwell time and ω_{max} is the maximum frequency in the spectrum; $\tau = (1/2\omega_{max})$

3.5.6   Folding of Signals

The property of the sampling theorem poses a difficulty that one has to know the frequency range in the spectrum, even before collecting the data. However, this problem can be generally circumvented fairly easily by choosing initially an arbitrarily large spectral range to locate the relative positions of the signals. And then the sampling frequency can be progressively optimized to suite the desired spectral range. This optimization involves also proper positioning of the carrier. If the spectral range determined by the sampling rate and offset is not appropriately selected, the signals presented outside the spectral region fold into the selected region with a distorted phase (Fig. 3.18). This permits the detection of folded signals and corrections can be applied.

According to the digitization theorem, the FT of a digitized FID generates a series of spectra $F(\omega)$, displaced $\frac{1}{\tau}$ Hz apart, where τ is the dwell time (Fig. 3.19).

The largest frequency represented by the dwell time τ is $\omega_s = \frac{1}{2\tau}$. This is equivalent to having a carrier in the middle of $\frac{1}{\tau}$ range of $F(\omega)$; then the set of points represents equally well equidistant frequencies on both sides of this carrier. Now suppose dwell time selection is wrongly made and there is a frequency ω', $\Delta\omega$ rad/sec outside the largest frequency, ω_s, i.e, $\omega' = \omega_s + \Delta\omega$. This ω' will be present in each of the $F(\omega)$ spectral repetitions, $\frac{1}{\tau}$ Hz apart. The central spectrum of $F(\omega)$ $(n = 0)$ gets into it ω' from the spectrum on its left $(n = 1)$, and likewise the ω' of the central spectrum appears in the $F(\omega)$ of the next one $(n = -1)$ and so on. In effect, the central section got a frequency at $-\omega_s + \Delta\omega = -(\omega_s - \Delta\omega)$. Thus, it appears as though ω' is reflected around ω_s frequency represented by the selected dwell time. This represents the case when a quadrature detection is employed, where the carrier is placed in the middle (center of $n = 0$ block, Fig. 3.19I) and positive and negative frequencies can be distinguished.

Fig. 3.18 A schematic representation of spectral folding in NMR. (I) In the case of a quadrature detection, the peak (at position "b") which is outside the selected spectral region folds into the spectrum at position "a" with distorted phase. (II) In the case of a single-channel detection, the position of "a" will be near the other end of the spectrum as indicated

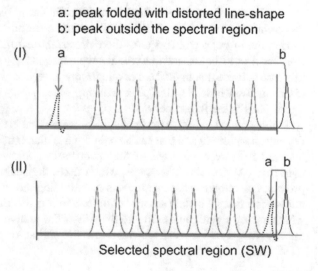

a: peak folded with distorted line-shape
b: peak outside the spectral region

(I)

(II)

Selected spectral region (SW)

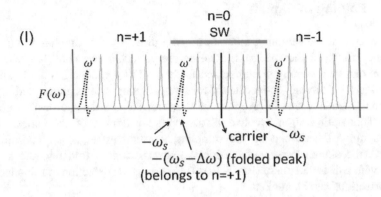

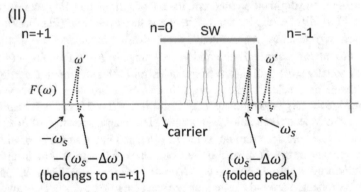

Fig. 3.19 A schematic representation of a series of spectra that are generated by the FT of FID, as per the digitization theorem in the case of a quadrature detection (I) and single-channel detection (II). The peak belonging to $n = +1$ appears in the region of $n = 0$

In the case of a single-channel detection, the carrier is placed at one end of the spectrum as shown in Fig. 3.19II. If the actual spectrum is on the right side of the carrier, the ω' peak will appear as the extra peak near the left blue line that comes from the $n = +1$ block, and this would have a frequency $-(\omega_s - \Delta\omega)$; note that ω_s in this case is twice that in the case of quadrature detection. Now, since there is no sign discrimination possible, this peak will appear at $+(\omega_s - \Delta\omega)$ as indicated in the figure. This is the folded peak.

In this context, it is also important to consider what is the largest spectral range or the largest sampling rate one can have in a particular experiment. This is determined by the hardware component of the spectrometer known as the analog-to-digital converter (ADC). This is a device, which takes the analog signals as the input and produces an output as binary numbers representing the strength of the signal, such a conversion takes a certain amount of time, and it is this time which limits the rate at which analog signal can be fed into the ADC. The analog signal cannot be fed faster than the rate at which it can convert it into the digital form; this rate is termed as the ADC rate.

3.5.7   Acquisition Time and Resolution

Acquisition time is defined as the time for which data is collected, in the FID. If there are N data points collected with τ being the dwell time, then the acquisition time is given by

$$t_{\text{acq}} = N\tau \tag{3.27}$$

If SW (spectral width) is the largest frequency (ω_{max}) represented by the sampling (for a single-channel detection where the carrier is placed at one end of the spectrum), then

$$\tau = \frac{1}{2\text{SW}} \tag{3.28}$$

So,

$$t_{\text{acq}} = \frac{N}{2\text{SW}} \tag{3.29}$$

After a FT of N data points, there will be $N/2$ real points and $N/2$ imaginary points in the frequency domain spectrum. Since both of these contain the same frequency information, only one of them (real) is used to display the spectrum; however (as will be described later), the imaginary points will be required for phase correcting the spectrum. Therefore, in the frequency domain, the digital resolution (Hz/point) R will be given by

$$R = \frac{\text{SW}}{\frac{N}{2}} = \frac{2\text{SW}}{N} \tag{3.30}$$

From Eqs. 3.29 and 3.30, we see that the acquisition time and digital resolution are inversely related

$$R = \frac{1}{t_{\text{acq}}} \tag{3.31}$$

In the case of a quadrature detection, the carrier is placed in the middle of the spectrum, and therefore the largest frequency is SW/2. The N data points are also divided between two channels of detection: $N/2$ for the y-component (real) and $N/2$ for the x-component (imaginary) of precessing magnetization.

In such a case, acquisition time is given by

$$t_{\text{acq}} = \frac{N}{2} \times \left(\frac{1}{2\text{SW}/2} \right) = \frac{N}{2\,\text{SW}}$$

Therefore, R will again be given by

$$R = \frac{1}{t_{\text{acq}}}$$

Thus, in order to obtain high resolution in the spectra, it is necessary to collect data for a long time; however, there is also a limit as to how long one can go, as the FID is a decaying function with time, and therefore the later points in the FID contain more noise than the earlier ones. Hence, by persisting for too long in the FID, the SNR in the spectrum decreases. The duration for which FID lasts will be determined by the transverse relaxation time (T_2), and at times greater than $3T_2$, there will be essentially only noise. Therefore, it does not help to collect data for durations longer than this time. In practice one has to strike a proper balance between SNR and resolution and optimize data collection parameters accordingly.

3.5.8   Signal Averaging and Pulse Repetition Rate

In the signal averaging process, it is importance to optimize the time interval between two successive sets of data collection (Fig. 3.20). The individual experiments are referred to as scans for historical reasons. If all of the FIDs have to be exactly identical, then the time interval T_p has to be longer enough so that the magnetization has completely relaxed back to equilibrium, before the start of the new experiment. However, this often does not yield the highest signal-to-noise ratio per unit time. This is a function of the flip angle of the pulse and relaxation times. Detailed calculations have shown that when an equilibrium or a steady state is reached, the M_x magnetization after each pulse is given by

$$M_x^+ = M_o \sin\beta \left\{ \frac{1 - E_1}{1 - E_1 \cos\beta} \right\} \tag{3.32}$$

where β is the flip angle and

$$E_1 = e^{\left(-\frac{T_p}{T_1}\right)} \tag{3.33}$$

Maximum amplitude obtained for an optimum flip angle is given by

$$\cos\beta_{\text{opt}} = E_1 \tag{3.34}$$

Figure 3.21 shows a plot of the steady-state signal amplitude (normalized intensity) as a function of β for different values of $\frac{T_p}{T_1}$.

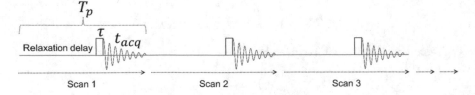

Fig. 3.20 Signal averaging in an NMR experiment

Fig. 3.21 Simulated profiles
of a steady-state signal
amplitude as a function of flip
angle (β) at different values of
$\frac{T_p}{T_1}$

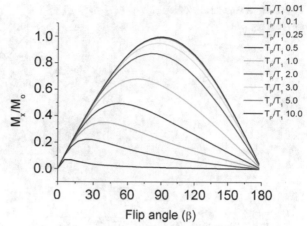

Fig. 3.22 Simulated profile
of β_{opt} vs $\frac{T_p}{T_1}$

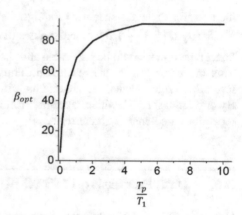

Figure 3.22 shows the optimum flip angle as a function of $\frac{T_p}{T_1}$.

The optimized sensitivity (S/N per unit time) in a repetitive pulse experiment is
given by the following equation.

$$S = M_o^{1/2} \left(\frac{T_2}{T_1}\right)^{\frac{1}{2}} (1 - E_2^2)^{\frac{1}{2}} G\left(\frac{T_p}{T_1}\right) \rho_N^{-1} \tag{3.35}$$

where $E_2 = e^{-t_{max}/T_2}$

$$G(x) = \left[\frac{2(1 - e^{-x})}{x(1 + e^{-x})}\right]^{1/2} \tag{3.36}$$

$$\rho_N = \sqrt{\text{frequency independent power spectral density}} \tag{3.37}$$

Fig. 3.23 A simulated profile
of $G\left(\frac{T_p}{T_1}\right)$ vs $\frac{T_p}{T_1}$

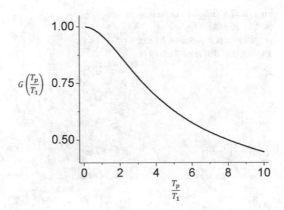

Figure 3.23 shows the function $G\left(\frac{T_p}{T_1}\right)$ vs $\frac{T_p}{T_1}$, assuming that for every value of $\frac{T_p}{T_1}$, the best flip angle has been used for signal averaging.

Clearly $G\left(\frac{T_p}{T_1}\right) \rightarrow 1.0$ for small values of $\frac{T_p}{T_1}$. It is evident that for high sensitivity faster repetitions would be preferable and the length of each FID could be restricted. This of course affects the resolution. Thus, depending upon the sensitivity and resolution requirements, the conditions will have to be optimized for every case. By and large for ^{13}C, where relaxation times are long, small flip angles and faster repetitions are generally preferred.

3.6   Data Processing in FT NMR

Unlike in CW NMR, data acquisition and processing are separate entities in FT NMR. A number of tricks have been employed to improve the quality of the spectra in terms of sensitivity or resolution. The computer has played a significant role in this regard, and manipulative approaches have been on the increase to extract the best out of the data. These include the following procedures.

3.6.1   Zero Filling

Since the length of the FID has to be restricted due to signal-to-noise ratio considerations, the number of data points in the FID will be limited, and consequently the digital resolution in the spectrum will also be limited. To circumvent this limitation, zeros are artificially added at the end of the FID to increase the total number of data points before Fourier transformation. This is schematically shown in Fig. 3.24. However, this is not to be considered as increase in acquisition time, and the inherent resolution present in the data is not affected.

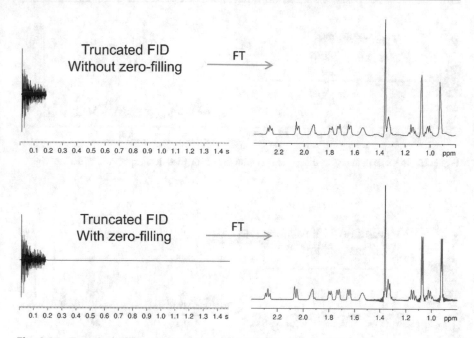

Fig. 3.24 Truncated FID without (upper) and with (bottom) zero filling of data points and the respective spectra as a result of FT

3.6.2 Digital Filtration or Window Multiplication or Apodization

This is a very important step in all data processing procedures and is invariably used to gain specific advantages. When the FID is truncated and zeros are added, the FT of the FID leads to serious distortions due to sinc $\left(\frac{\sin x}{x}\right)$ wiggles appearing on either side of every line. This is illustrated in Fig. 3.25.

To circumvent this shortcoming, digital filtering techniques are used, wherein the FID is multiplied by a suitable function, $\lambda(t)$. This operation is also called the window multiplication or apodization. The idea is to remove the abrupt discontinuity in the FID, and the window multiplication ensures that the last point in the FID is almost zero. This helps to remove the *sinc* artifacts from the spectrum. The commonly used mathematical functions for this purpose are:

(i) *Exponential Function*

$$\lambda(t) = e^{-\frac{t}{t_{\max}}} \tag{3.38}$$

Here, the FID is collected for a time equal to $t_{\max}$. This exponential multiplication results in signal-to-noise enhancement by removing the noise contribution present in

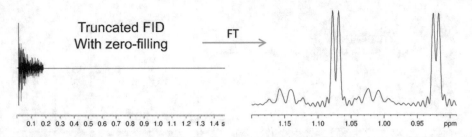

Fig. 3.25 In general, the FT of truncated FID with zero filling results in spectrum with sinc wiggles

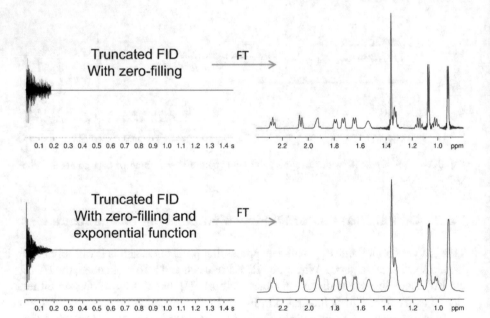

Fig. 3.26 Truncated FID with only zero filling (upper) and the combined application of zero filling and exponential functions (bottom) for the processing. This combined processing has removed the sinc wiggle patterns, which can be seen directly from the spectra

the later part of the FID (Fig. 3.26). However, this also leads to a broadening of all the lines in the spectrum by an amount $\frac{1}{\pi t_{max}}$; the resultant line widths will be given by

$$\Delta v = \frac{\left(T_2^* + t_{max}\right)}{\left(\pi T_2^* t_{max}\right)} \tag{3.39}$$

If t_{max} is equal to T_2^* (transverse relaxation time), this filter is referred to as *matched filter*.

(ii) *Cosine Function*

$$\lambda(t) = \cos\left(\frac{\pi t}{2t_{\max}}\right) \tag{3.40}$$

The data is multiplied by a cosine function of a period, which is adjusted such that the function falls to zero at the last point of the FID that is at $t_{\max}$. This function does not cause significant changes in the line widths, but contributes to a significant enhancement in the signal-to-noise ratio. The function can be appropriately optimized after choosing the appropriate number of data points in the FID for the desirable signal-to-noise enhancement: Note that the initial points in the FID have higher signal-to-noise ratios compared to the later ones (Fig. 3.27).

(iii) *Sine-Bell Function*

$$\lambda(t) = \sin\left(\phi + \frac{\pi t}{2t_{\max}}\right) \tag{3.41}$$

This is essentially the cosine function shifted by phase $\phi = 90°$. In this function, if $\phi = 0°$, it results in making the first point in the FID zero, as a result of which the signal-to-noise ratio will be severely reduced, but the resolution will be significantly

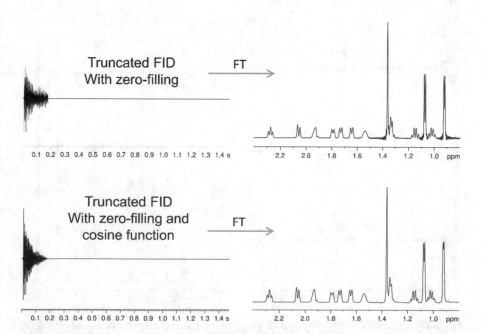

Fig. 3.27 Truncated FID with only zero filling (upper) and the combined application of zero filling and cosine functions (bottom) for the processing. This combined processing has removed the sinc wiggle patterns, which can be seen directly from the spectra

enhanced. So, for optimizing the value of ϕ, a balance has to be struck between the signal-to-noise ratio loss and resolution enhancement. This function will also result in line shape distortion, so one often prefers to use ϕ values which are closer to 90°. The appearance of the spectra for varying values of ϕ is shown in Fig. 3.28.

(iv) *Lorentz-Gauss*

$$\lambda(t) = e^{\left(\frac{t}{T_2^*} - \frac{\sigma^2 t^2}{2}\right)} \tag{3.42}$$

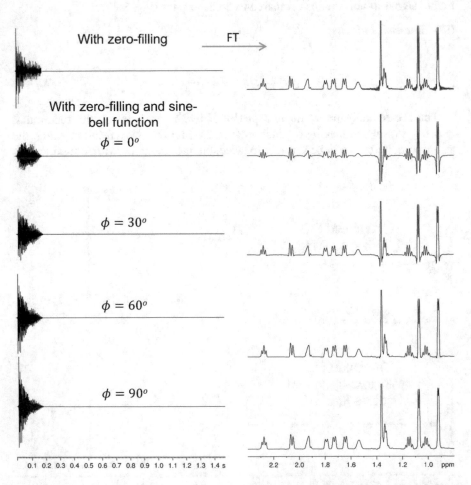

Fig. 3.28 Truncated FID with only zero filling (upper) and the combined application of zero filling and sine-bell functions (at different ϕ angles) for the processing. This combined processing has removed the sinc wiggle patterns, which can be seen directly from the spectra

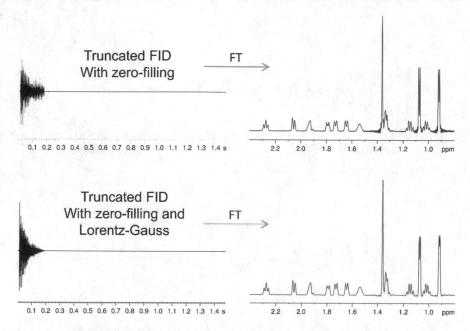

Fig. 3.29 Truncated FID with only zero filling (upper) and the combined application of zero filling and Lorentz-Gauss functions (bottom) for the processing. This combined processing has removed the sinc wiggle patterns, which can be seen directly from the spectra

Here σ is a parameter which has to be adjusted for a given $t_{\max}$ and an estimated T_2^*, for optimum performance. This actually has an effect similar to that of the optimized sine-bell. This is illustrated in Fig. 3.29.

3.7 Phase Correction

A real Fourier transformation (or cosine transformation) of decaying cosine or sine form of FID generates absorptive and dispersive signals, respectively, as shown in Fig. 3.30. Mathematically, this is given by Eqs. 3.43 and 3.44.

$$2 \int e^{-at} \cos{(\omega_o t)} \cos{(\omega t)} dt = \frac{a}{\left[a^2 + (\omega - \omega_o)^2\right]} + \frac{a}{\left[a^2 + (\omega + \omega_o)^2\right]} \quad (3.43)$$

$$2 \int e^{-at} \sin{(\omega_o t)} \cos{(\omega t)} dt = \frac{(\omega - \omega_o)}{\left[a^2 + (\omega - \omega_o)^2\right]} + \frac{(\omega + \omega_o)}{\left[a^2 + (\omega + \omega_o)^2\right]} \quad (3.44)$$

However, for certain reasons, due to spectrometer hardware limitations, one does not obtain pure absorptive or dispersive line shapes but some mixtures of the two. These are briefly described in the following.

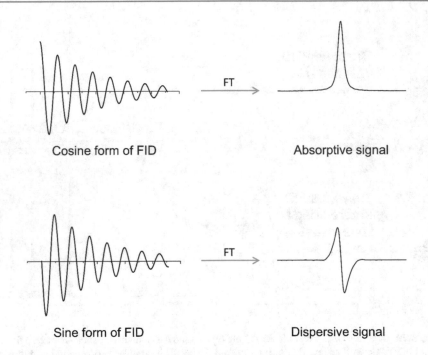

Fig. 3.30   Cosine and sine form of FIDs and the respective FT NMR spectra

(i) *Improper Phase of the RF Pulse*

Instead of the RF pulse being applied exactly along the x-axis or the y-axis, if its axis is deviated slightly (say by angle θ), then, the magnetization is also rotated away from the y-axis or the x-axis, as the case may be. As a result, the total magnetization has a phase θ with respect to the receiver at time $t = 0$ (Fig. 3.31a). The FID for a single resonance has therefore the form

$$S(t) = e^{-at} \cos(\omega_o t + \theta) \tag{3.45}$$

or

$$S(t) = e^{-at} \sin(\omega_o t + \theta) \tag{3.46}$$

Even when there are many resonances in the spectrum, the initial phase will be θ for all the individual resonances. This is called the *zero-order* phase of the spectrum and will be the same throughout the spectrum (Fig. 3.31b). The complex Fourier transformation of the cosine-dependent FID leads to real and imaginary parts, both of which have absorptive and dispersive contributions:

$$\text{Real part } (R) = A\cos\theta - D\sin\theta \tag{3.47}$$

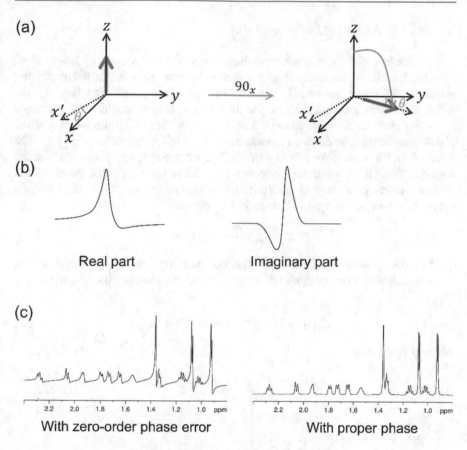

Fig. 3.31 (a) The deviation in the RF application direction results in a deviation of magnetization alignment in the transverse plane. (b) This leads to phase errors in both the real and imaginary components. (c) 1D-NMR spectra with zero-order phase error and with proper phase correction

$$\text{Imaginary part } (I) = D\cos\theta + A\sin\theta \tag{3.48}$$

where A and D represent the absorptive and dispersive line shapes. From these equations, it follows that

$$A = R\cos\theta + I\sin\theta \tag{3.49}$$

Thus, the zero-order phase correction involves multiplication of the real and imaginary parts of the spectrum by constants dependent on the phase error θ and adding them together. The value of θ is determined by continuously altering it while monitoring the resultant shapes of the lines.

(ii) *Delay in the Start of Data Acquisition*

In order to avoid direct interference between transmitter and the phase-sensitive detector, it becomes necessary to give some delay after the pulse for the transmitter effects to die down before FID can be acquired. During this time (say Δ), the different frequency components of the total magnetization would have precessed in the *x-y* plane to different extents, and thus, in the net FID collected, the initial phases are different for different resonance lines. This is illustrated in Fig. 3.32. The effect of such a phase change is to produce frequency-dependent phase errors in the final spectrum after Fourier transformation (Fig. 3.33). Quantitatively, the errors can be understood by referring to the theorems on Fourier transforms; a shift of origin results in a ω-dependent phase change in the spectrum.

$$S(t + t_o) \xrightarrow{FT} e^{i\omega t_o} F(\omega) \tag{3.50}$$

If $F(\omega)$ is represented as a complex function with $R(\omega)$, the real part representing absorptive line shapes and $I(\omega)$, the imaginary part representing the dispersive line shapes, then

$$F(\omega) = R(\omega) + i\,I(\omega) = |F(\omega)|\,e^{i\phi} \tag{3.51}$$

where ϕ is the phase

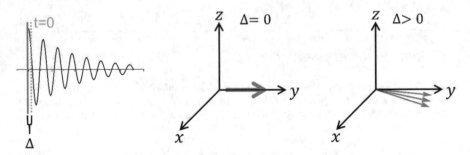

Fig. 3.32 The delay in data acquisition (left) leads to different initial phases for different spins (right)

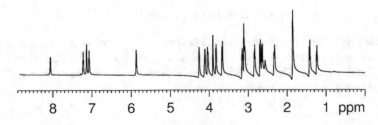

Fig. 3.33 First-order phase error (also called frequency-dependent phase error) in the 1D-NMR spectrum

$$|F(\omega)| = \sqrt{R(\omega)^2 + I(\omega)^2} \qquad (3.52)$$

and

$$\tan \phi = \frac{I(\omega)}{R(\omega)} \qquad (3.53)$$

Now as a result of the shift of origin in the FID, the modified spectrum $F'(\omega)$ will be

$$F'(\omega) = e^{i\omega t_o} F(\omega) \qquad (3.54)$$

$$= e^{(\phi + i\omega t_o)} \mid F(\omega) \mid \qquad (3.55)$$

If $R'(\omega)$ and $I'(\omega)$ are the new real and imaginary parts, then

$$\tan (\phi + \omega t_o) = \frac{I'(\omega)}{R'(\omega)} \qquad (3.56)$$

$$R'(\omega) = |F(\omega)| [\cos (\phi + \omega t_o)] \qquad (3.57)$$

$$= |F(\omega)| \cos \phi \cos (\omega t_o) - \sin \phi \sin (\omega t_o) \qquad (3.58)$$

$$= \cos (\omega t_o) R(\omega) - \sin (\omega t_o) I(\omega) \qquad (3.59)$$

This shows the absorptive and dispersive components getting mixed. By following the same procedure as per the zero-order phase, proper phases can be obtained, except that the phase errors are different for different frequencies. It is possible to calculate these phase constants, since t_o and frequencies are known quantities in an experiment. However, a simplification occurs if $\omega t_o \ll 1$ for the whole range of frequencies. Then $e^{i\omega t_o}$ can be expanded up to first order in ω .

$$F'(\omega) = (1 + i\omega t_o) F(\omega) \qquad (3.60)$$

and

$$R'(\omega) = R(\omega) - \omega t_o I(\omega) \qquad (3.61)$$

Similarly,

$$I'(\omega) = I(\omega) + \omega t_o R(\omega) \qquad (3.62)$$

Therefore,

$$R(\omega) = \frac{\{R'(\omega) + \omega t_o\, I'(\omega)\}}{\left\{1 + (\omega t_o)^2\right\}} \tag{3.63}$$

Under the condition $\omega t_o \ll 1$, this simplifies to

$$R(\omega) = R'(\omega) + \omega t_o\, I'(\omega) \tag{3.64}$$

Since this admixture is linearly dependent on ωt_o, the phase constants for all the frequencies can be relatively easily calculated, and pure phase spectra can be obtained.

3.8   Dynamic Range in FTNMR

A dynamic range is a special feature in FTNMR that limits the range of intensities of lines that can be properly recorded. This is the consequence of limited digitizer (ADC: analog-to-digital converter) resolution.

We know that the FID signal as it comes out as a function of time is digitized in real-time and the strength of each data point is outputted as a binary number which is then fitted into a computer word. We also know that each data point in the FID has contributions from all lines present in the spectrum. Thus, the largest number that the ADC can output determines the maximum intensity ratio between the largest and the smallest line present in the spectrum. The smallest number is obviously 1.0. Explicitly, if an ADC has 12 bits (11 bits excluding the sign bit), the largest number that can be stored in it is 2047. This limits the dynamic range of the digitizer, that is, if there are only two lines in a spectrum, the largest intensity for the strong signal is 2046, and the intensity of the small signal is 1. Note that these are relative numbers, if the actual intensities are more they can be scaled down by adjusting the receiver gains, the receiver gain is adjusted such that the maximum in the FID points fills the ADC. But, if the intensity ratio of the two signals is greater than 2046, then the big signal fills the whole ADC, and the small signal will not be represented at all. If the spectrum has more than two signals which also contribute to every FID point, the intensity ratio of the largest to the smallest signal will reduce accordingly. In the event of no representation of the weakest signal in the ADC, signal averaging will not help in recovering the weak signals (Fig. 3.34).

3.9   Solvent Suppression

As a consequence of the dynamic range problems discussed, it often becomes necessary to suppress signals coming from the solvents which are generally very strong and are of no particular value from the analysis point of view. For example, while recording the spectrum in water (H_2O), the proton concentration is $\sim$110 M. If the sample concentration is $\sim$1 mM, then the intensity ratio of water to sample will be

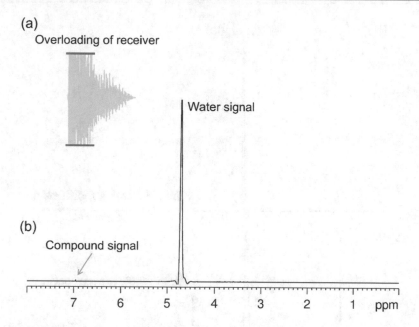

Fig. 3.34 (a) The overloading of FID is due to the strongest intensity of signal. (b) For example, if a compound is present at a very low concentration in the aqueous solvent medium, the strong solvent signal fills the ADC, and the small signal will have no representation in the receiver

the order of 1.1×10^5; this is much more than a 12-bit or even a 16-bit digitizer can accommodate: Note that it is not practical to increase the digitizer resolution arbitrarily because it contributes to noise and also slows down the digitization process. Because of this, it becomes essential to selectively suppress such strong signals (Fig. 3.35). Several strategies have been used for this purpose, and some of the common ones are listed in the following.

(i) *Presaturation*

The strong solvent signal is suppressed by continuous irradiation prior to the application of the observe pulse (Fig. 3.36). This has proved useful, but it also has some shortcomings. First, the sample signals buried under the solvent will also get suppressed. Second, saturation transfer can occur to protons which exchange with the solvent resulting in the reduction of their intensities. The success of this suppression will also depend upon the T_1 relaxation of the solvent.

(ii) *Inversion Recovery Sequence*

Here, a pulse sequence $180° - \tau - 90°$ of the type *acquisition* (Fig. 3.37) is used to suppress the solvent signals. The first $180°$ pulse inverts the magnetization and puts along the negative z-axis, then the signals will relax by the spin-lattice relaxation process, and these are different for different spins in the system. So, the period τ is adjusted such that the solvent magnetization is zero at the end of τ so that this will

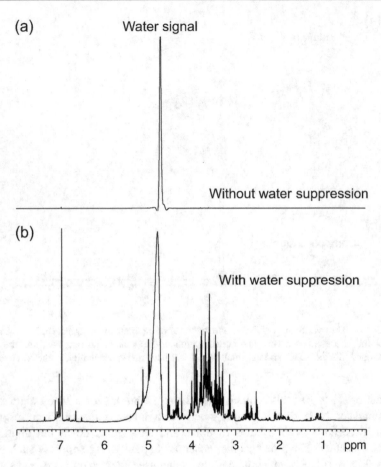

Fig. 3.35 Triphala 1D-NMR spectra (**a**) without water suppression and (**b**) with water suppression

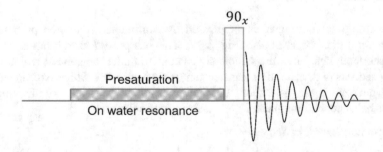

Fig. 3.36 A schematic representation of a presaturation NMR pulse sequence

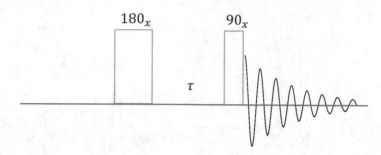

Fig. 3.37 A schematic of the inversion recovery pulse sequence

Fig. 3.38 In the jump-return sequence of water suppression, the offset is placed on water signal. The delay τ is calculated depending on the region of interest in the spectrum for maximal excitation

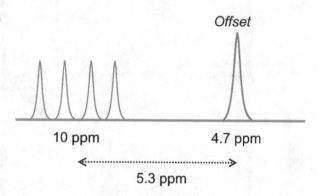

not appear after the following 90° pulse. The other signals which may have widely different T_1 relaxation times would be either on the negative z-axis, if they are slowly relaxing compared to solvent or on the positive z-axis, if they are relaxing faster than the solvent. Of course, if by coincidence any of the sample spin relaxes at the same rate as the solvent, then that also will be suppressed.

(iii) *Jump and Return*

This uses the pulse sequence $90^{\circ}_x - \tau - 90^{\circ}_{-x} - acquisition$. τ is the period to be optimized for getting the best excitations in desired region of the spectrum. The offset is placed on the water resonance, and the illustration is shown in Fig. 3.38. If the excitation is to be maximized at a frequency ν_{max} away from the offset, then τ is set equal to $\frac{1}{4\nu_{max}}$. The evolution of the magnetization in the transverse plane is indicated in Fig. 3.39. It follows that the signals on the two sides of the water resonance appear with opposite signs. However, one particular disadvantage is that the baseline appears highly distorted, especially close to the water resonances. This sequence is very commonly used for exciting imino protons in the DNA, whose resonances usually lie far away from the water resonance.

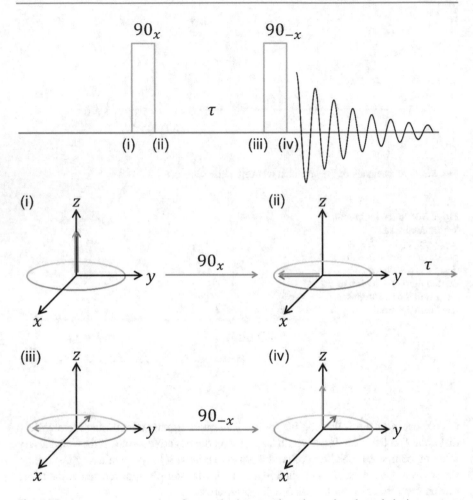

Fig. 3.39 A vector representation of magnetization at various time points through the jump-return pulse sequence. Orange and blue arrows represent the sample and water magnetizations, respectively

3.10 Spin Echo

Discovered by Erwin Hahn in the early 1950s, spin echo is a simple multipulse sequence which has become the most crucial and integral part of many sophisticated NMR pulse sequences. It is the simple extension from the one pulse FT NMR experiment. The spin echo employs the pulse sequence shown in Fig. 3.40, and the evolution of the magnetization components is shown in Fig. 3.41.

Considering two spins A and B, the first 90_x° degree pulse rotates the z-magnetization on to the negative y-axis; during the time τ, the two spins dephase rotating with their characteristic frequencies ν_A and ν_B . The 180_x° pulse rotates the

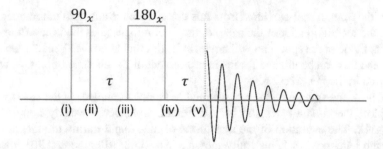

Fig. 3.40 A schematic representation of the spin echo pulse sequence

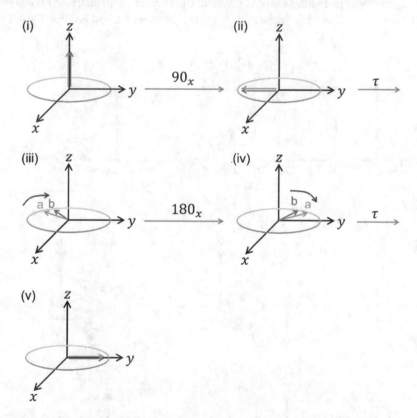

Fig. 3.41 A vector representation of the magnetization at various time points through the spin echo pulse sequence. Two spins "*A*" and "*B*" with different precessional frequencies are considered for illustration

two spins by 180° in the transverse plane and as they continue to move and rephase at the end of the τ period, along the positive y-axis. This is called the spin echo. It is clear that the actual frequencies v_A and v_B do not matter for the refocusing at the end of the 2τ period. In other words, on the whole the chemical shifts get refocused at the time of the echo. It is also evident from this that if there are any field inhomogeneities across the length/breadth of the sample, they get refocused at the time of the echo. The amplitude of the spin echo will however be dependent on the T_2 relaxation of the spins, and this can be utilized for the measurement of T_2 relaxation times, as will be discussed in the next section.

Till now, the spins A and B are assumed to be not J-coupled. In the case they are J-coupled, the situation will be very different. Consider a weakly coupled spin system AX. The evolution of the transitions of the A spin during the length of the spin echo sequence is schematically shown in Fig. 3.42. After the first 90° pulse, the magnetization vectors corresponding to the transitions A_1 and A_2 of spin A are along the negative y-axis. During the next τ period, they precess with different frequencies in the transverse plane and dephase. The angle θ between the two vectors is given by

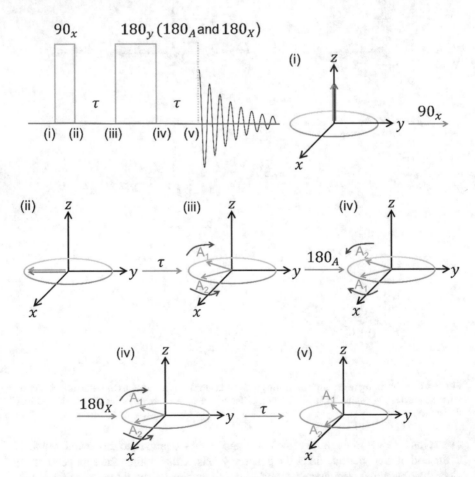

Fig. 3.42 A vectoral representation of the magnetization evolution of A spin in the AX weakly coupled spin system. Here, one is looking at the two magnetizations (A_1 and A_2) while sitting at the two transitions (in the rotating frame of A spin). Therefore, they seem to move in opposite directions

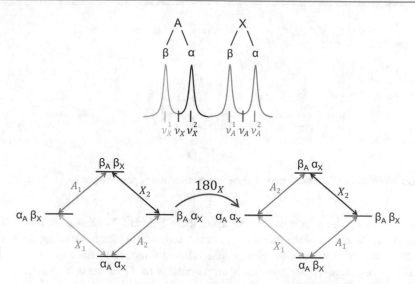

Fig. 3.43 An energy level diagram of a weakly coupled AX spin system (left). The 180° pulse on the X spin interchanges transitions A_1 and A_2

$2\pi J\tau$. The center of these two vectors represents the chemical shift of the A spin. A 180° pulse on the A spin rotates the two components to new positions in the transverse plane. At the same time, a 180° pulse on the X spin which interchanges the spin states of X also interchanges the labels A_1 and A_2 (see the energy level diagram in Fig. 3.43). As a consequence, during the next τ period, the two transitions continue to dephase further, and at the end of 2τ, the angle between them will be $4\pi J\tau$. The chemical shift vector of A will be along the positive y-axis. This indicates that in the spin echo, the chemical shifts are refocused, but coupling constants are not refocused.

3.11   Measurement of Relaxation Times

The pulsed methods provide convenient ways of measuring the relaxation times T_1 and T_2 of any given system. These are described in the following paragraphs.

3.11.1 Measurement of T_1 Relaxation Time

(i) *Inversion Recovery*

The most common method of T_1 measurement is called the "inversion recovery" technique. This uses the pulse sequence shown in Fig. 3.44.

The first 180° pulse inverts the magnetization on to the negative z-axis; as the spins relax back during the next τ period, they would have reached different

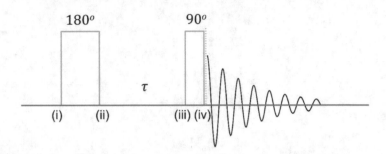

Fig. 3.44 A schematic representation of the inversion recovery pulse sequence

positions along the z-axis depending upon their different T_1 relaxation times. The next 90° read pulse and the following data collection allow for monitoring the status of the recovery at the end of the τ period. Thus, by repeating the experiments, for different values of τ, one can get a measurement of T_1 relaxation times, for the various spins (Fig. 3.45).

Mathematically this can be analyzed using the rate equation:

$$\frac{dM_z(t)}{dt} = -\frac{(M_z(t) - M_o)}{T_1} \tag{3.65}$$

where M_z is the z-magnetization at time t and M_o is the equilibrium magnetization. Integrating with the condition $M_z = -M_o$, at $t = 0$, we get

$$M_z(t) = M_o\left(1 - 2e^{-\frac{t}{T_1}}\right) \tag{3.66}$$

This recovery is shown schematically in Fig. 3.46, and an experimental demonstration for different transitions in a spectrum is shown in Fig. 3.47. Clearly the different spins relax differently.

At a particular value of $t = \tau_{null}$, $M_z = 0$,
this means

$$e^{-\frac{\tau_{null}}{T_1}} = \frac{1}{2} \text{ or } T_1 = \frac{\tau_{null}}{\ln 2} \tag{3.67}$$

This equation allows a quick estimation of the T_1 relaxation times. Alternatively, Eq. 3.66 can be recast as

$$M_o - M_z = 2M_o e^{-\frac{t}{T_1}} \tag{3.68}$$

$$\ln(M_o - M_z) = \ln 2M_o - \frac{t}{T_1} \tag{3.69}$$

A plot of $\ln(M_o - M_z)$ vs t is a straight line whose slope yields the value of T_1.

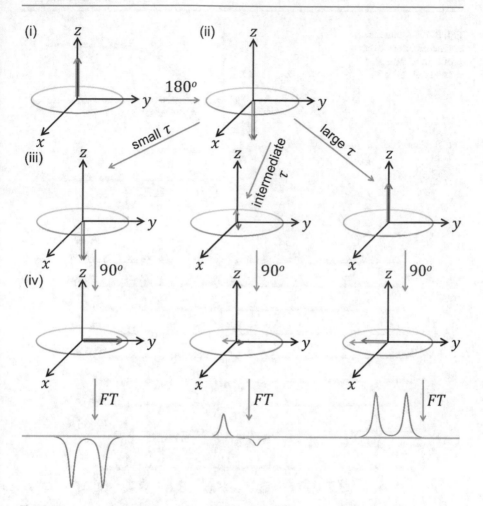

Fig. 3.45 A vector representation of magnetization at different time points of inversion recovery pulse sequence. The resultant spectra are shown at the bottom considering two spins having different relaxation rates. The spin that relaxes faster (cyan color) recovers to the z-axis faster than the slow-relaxing spin (red color) (middle picture)

(ii) Progressive Saturation

This sequence uses a train of 90° pulses separated by a constant time period τ, and the steady-state magnetization is then measured, as shown in Fig. 3.48.

τ is selected in such a way that transverse magnetization has died down at the end of each τ period, while the longitudinal magnetization recovers towards the z-axis. The τ period is varied such that the steady-state magnetization increases progressively from zero to M_o. So the rate equation here is

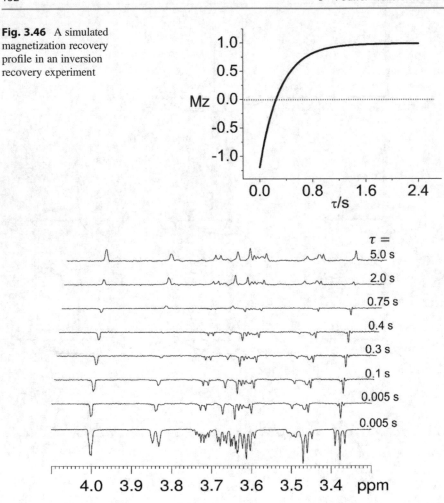

Fig. 3.46 A simulated magnetization recovery profile in an inversion recovery experiment

Fig. 3.47 Experimental magnetization recovery profiles in an inversion recovery experiment

$$\frac{\mathrm{d}(M_o - M_z)}{\mathrm{d}t} = -\frac{M_o - M_z}{T_1} \qquad (3.70)$$

with the initial condition $M_z = 0$ at $t = 0$. On integration this yields the relation

$$\ln \frac{(M_o - M_z)}{M_o} = -\frac{t}{T_1} \qquad (3.71)$$

$$\ln (M_o - M_z) = \ln M_o - \frac{t}{T_1} \qquad (3.72)$$

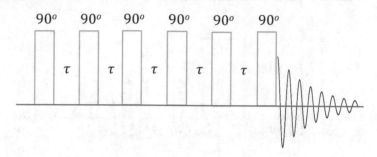

Fig. 3.48 A schematic representation of progressive saturation pulse sequence

Experimental data measured as a function of increasing τ can be fitted to Eqs. 3.71 and 3.72 to extract the value of T_1.

3.11.2 Measurement of T_2 Relaxation Time

The echo amplitude in the spin echo experiment is proportional to $e^{-\frac{2\tau}{T_2}}$. Thus, fitting the echo amplitude to the data obtained for different values of τ allows the measurement of T_2 relaxation times.

It must be mentioned here that precise refocusing of the signal in the spin echo experiment is dependent on each nucleus remaining in constant magnetic field during the time 2τ. If, however, there is molecular diffusion during this time between regions of different field strengths because of field inhomogeneities, then the echo amplitude will be modified, and we do not get the true value of T_2. To overcome this problem, Carr-Purcell modified the Hahn sequence to consist of a train of spin-echoes as

$$90 - \tau - 180 - \tau - \text{echo} - \tau - 180 - \tau - \text{echo} - \text{data collection}$$

Here the τ value is kept small so that there is no significant diffusion during each of the spin echoes. The number of echoes is varied so as to get different time points for the exponential fitting procedures for T_2 estimation. This yields a more reliable value of T_2.

Meiboom-Gill modified the sequence further by changing the phase of the 180° pulses after the first one by 90°, and this helps to average out the imperfections in the pulses.

3.12   Water Suppression Through the Spin Echo: Watergate

Spin echo provides an elegant method for effective water suppression for running spectra in aqueous solutions. The corresponding pulse sequence is termed as Watergate and is indicated in the Fig. 3.49. The offset is placed on water, and it combines

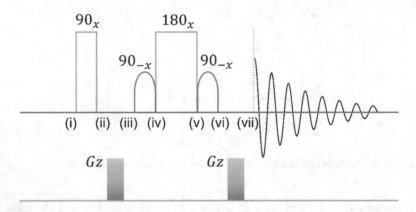

Fig. 3.49 A schematic representation of Watergate pulse sequence. G_Z represents z-field gradient pulses. A z-field gradient changes the field along the z-axis through the height of the sample and thus spins in different positions of the sample precess with different frequencies which leads to dephasing of the spins

linear gradient pulses with the spin echo sequence; the principles and uses of linear field gradients in high-resolution NMR have been described in some detail in Appendix A4. The gradients placed on either side of the central pulse train $[90_{-x}(\text{sel}) - 180_x - 90_{-x}(\text{sel})]$, where $90_{-x}(\text{sel})$ is a selective pulse on water, help to dephase and rephase the signals of water and from the sample differently. The sample resonances see a $180°$ rotation by the central hard pulse, whereas the net rotation for water is zero. As a result, the dephasing of the sample signals caused by the first linear field gradient is refocused by the second identical field gradient, whereas the two gradients add on to completely dephase the water resonance. Thus, the water signals are not refocused at all at the end of the spin echo, whereas the signals from the sample are refocused. The vector diagram to explain this phenomenon is shown in Fig. 3.50. In this case, one obtains clean in-phase spectra with a much better baseline.

3.13 Spin Decoupling

The J-decoupling interaction between two spins A and M is given by

$$J\, I_A \cdot I_M$$

If the two spins are locked along orthogonal axes, then the dot product vanishes, resulting in decoupling of A and M, which is shown in Fig. 3.51. Generally, spin locking involves a complex sequence of pulses with proper adjustment of power and phases. However, in the case of a simple two-spin system, a continuous saturation of the M-magnetization also achieves the same result of selective decoupling.

Spin decoupling arising from continuous saturation can be understood qualitatively in the following manner. Referring to Fig. 2.7, we see that the two energy

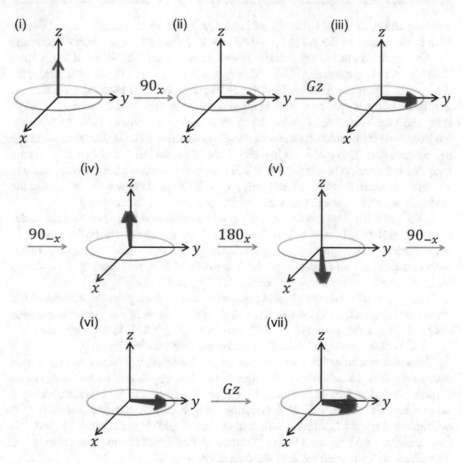

Fig. 3.50 A vectoral representation of water magnetization at different time points of Watergate experiment. The water signal gets progressively dephased, leading to its elimination in the transverse plane, whereas the sample signals will get refocused (see text for details)

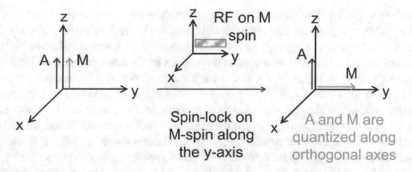

Fig. 3.51 A schematic representation of spin decoupling concept in NMR

levels, α and β, of the A spin split into two levels due to coupling with the X-spin. That is, the α-state of the A-spin has different energies depending upon whether the X-spin is in the α-state or the β-state. The same is true for the β-state of the A spin. This results in two transitions (labeled as $\delta_A{}^1$ and $\delta_A{}^2$), as per the *selection rule*, for the A-spin which are separated by the J-coupling constant. Now, the continuous irradiation of the X spin causes rapid flipping of the X spin from the α-state to the β-state and vice versa. As a result the A spin undergoes rapid exchange between the two transitions, which causes an averaging, resulting in a single transition at the average position. This is the original single transition for the A spin. That means the coupling information is removed or the X spin is decoupled. Alternatively, one can also visualize that due to the rapid exchange of the X spin between the two states, the A spin is not able to see the X spin, and therefore there is no coupling.

In heteronuclear cases, spin decoupling can be achieved by refocusing the scalar coupling evolution by the application of 180° pulse selectively on M spin. The heteronuclear spin decoupling pulse sequence and the respective vector representations are shown in Figs. 3.52a and Fig. 3.52b, respectively. The A and M spins are effectively decoupled during the 2τ period.

Note that in this case the coupling is removed only during the 2τ period and not during data acquisition. Therefore, the coupling will appear in the spectrum obtained after Fourier transformation of the FID. However, this kind of decoupling sequences are used in many multidimensional experiments (see later in Chap. 6).

The important to point note here is that in standard FTNMR spectra, if spin decoupling is to be achieved in the spectrum, the steps discussed will have to be carried out during data acquisition. This imposes constraints for homonuclear spin decoupling. But heteronuclear decoupling can be carried out (e.g., selective ^{1}H decoupling during ^{13}C acquisition), without much problem since the ^{1}H and ^{13}C frequencies are very far apart and continuous application of RF on proton during ^{13}C acquisition will not interfere with the data collection.

3.14   Broadband Decoupling

Carbon-13 NMR plays a key role in the determination of molecular structures in organic chemistry. All the carbons which have protons attached to them display fine structures due to C-H coupling; e.g., a methyl group (CH_3) which has three protons attached to the carbon will be a quartet with peaks in the intensity ratio of 1:3:3:1; a methylene group (CH_2) carbon will show a triplet with peaks in the intensity ratio of 1:2:1; and a C-H group will show a doublet with intensity ratio of 1:1. Quaternary carbons will be singlets. If the spectral dispersion is not adequate, the peaks may overlap, and then the identification of the multiplets will become difficult. Therefore, it is desirable to remove all proton couplings to all the carbons, and this is termed as broadband decoupling. Then, there will be one signal for each carbon, and the total number of carbon atoms in the molecule can be counted. Using proton-coupled spectra, the molecular structures can be derived.

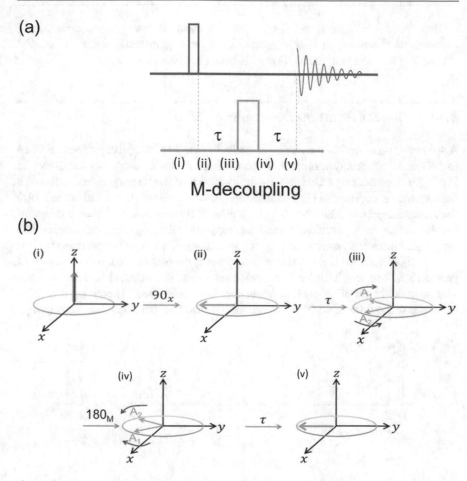

Fig. 3.52 (a) A schematic representation of heteronuclear spin decoupling pulse sequence. (b) The magnetization at different time points of heteronuclear spin decoupling experiment is represented vectorially. The magnetization vectors of the A spin are shown which get refocused before detection. Note one is looking at the two transitions (A_1 and A_2) of A spin, sitting at the center of the two transitions

Referring to Fig. 3.51, it is clear that during data acquisition on the carbon channel, it becomes necessary to saturate (or invert as per Fig. 3.52) all the protons at the same time. This cannot be achieved by a single radiofrequency. Elaborate pulse sequences have been developed which involve repetitive application of the so-called composite pulses which consist of dozens of pulses with properly chosen amplitudes and phases applied in quick succession, and these achieve saturation (or inversion) of all the protons at the same time, while carbon data is being acquired. This discussion is beyond the scope of this book.

Broadband decoupling is also a key element in many multidimensional experiments discussed in later chapters. In these experiments, often ^{13}C or ^{15}N decoupling is employed, while ^1H data is being acquired.

3.15 Bilinear Rotation Decoupling (BIRD)

A spin echo based technique can be effectively used to distinguish protons attached to ^{13}C and ^{12}C in any molecule. It consists of a cluster of pulses as shown in Fig. 3.53a. Considering a C-H system, the evolution of the ^1H magnetization through the sequence is depicted in Fig. 3.53b using vector diagrams. At the end of the BIRD sequence, the protons attached to ^{12}C are selectively inverted, and such a discrimination can be very effectively used to suppress ^1H magnetization components originating from protons attached to ^{12}C, in many heteronuclear experiments.

From Fig. 3.53, we also see that at time point 5, the magnetization components of protons attached to ^{13}C have also refocused along the negative y-axis. In other words, the C-H coupling evolution has also been refocused. Therefore, this pulse sequence without the last 90° pulse can also be used for heteronuclear decoupling.

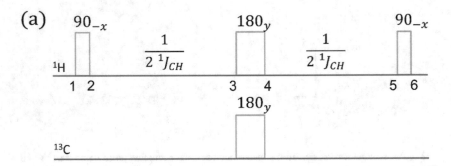

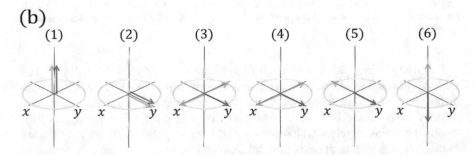

Fig. 3.53 (**a**) A schematic of the BIRD NMR pulse sequence, where the narrow and wide rectangles, respectively, represent the 90° and 180° pulses. (**b**) The vector depiction of ^{13}C-attached and ^{12}C-attached ^1H magnetizations at different time points of the BIRD pulse sequence are shown in cyan and red colors, respectively, under the assumption of chemical shift refocusing in the spin echo period

3.16 Summary

- The principle of Fourier transform (FT) NMR is described. Advantages over the previously used continuous wave (CW) are described.
- Some mathematical theorems regarding FT are presented.
- The concepts and relation between RF phase and spectral phase are presented.
- The concepts of single-channel detection, quadrature detection, phase correction, dynamic range, and various aspects of data processing are described.
- The spin echo has been described.
- Methods of relaxation time measurements are described.
- Some common methods of solvent peak suppression to get over dynamic range problems are described.

3.17 Further Reading

- Principles of NMR in one and two dimensions, R. R. Ernst, G. Bodenhausen, A. Wokaun, Oxford, 1987
- Spin Dynamics, M. H. Levitt, 2nd ed., Wiley 2008
- High Resolution NMR Techniques in Organic Chemistry, T. D. W. Claridge, 3rd ed., Elsevier, 2016
- NMR Spectroscopy: Basic Principles, Concepts and Applications in Chemistry, H. Günther, 3rd ed., Wiley, 2013
- Understanding NMR Spectroscopy, J. Keeler, Wiley, 2005
- Protein NMR Spectroscopy, J. Cavanagh, N. Skelton, W. Fairbrother, M. Rance, A, Palmer III, 2nd ed., Elsevier, 2006

3.18 Exercises

3.1. For a RF strength of 25 KHz in frequency units (γB_1), the flip angle for a pulse of duration 10 μs is
   (a) $\pi/4$
   (b) π
   (c) $\pi/2$
   (d) $\pi/3$

3.2. In a quadrature detection, the carrier is placed at
   (a) the high-frequency end of the spectrum
   (b) the low-frequency end of the spectrum
   (c) the middle of the spectrum
   (d) either high-frequency or low-frequency end of the spectrum

3.3. The dwell time in the quadrature detection schemes is
    (a) the same as the single-channel detection
    (b) twice that of the single-channel detection
    (c) $\frac{1}{4}$ th of the single-channel detection
    (d) half the single-channel detection
3.4. In a free induction decay, the first data point has zero intensity; the Fourier transformation of this FID leads to
    (a) positive absorptive signals
    (b) negative absorption signals
    (c) dispersive signals
    (d) a mixture of absorptive and dispersive signals
3.5. Folding of signals in the NMR spectrum occurs because of
    (a) wrong pulse width
    (b) smaller spectral width than required
    (c) wrong choice of pulse phase
    (d) shorter dwell time than the Nyquist equation
3.6. In the slow passage experiment, a signal sweep through the spectrum of 5000 Hz takes 23 min and 20 s. In a Fourier transformation experiment of the same sample, a single scan takes 1.5 s and the time for Fourier transformation is 50 s; then the signal-to-noise gain in the Fourier transformation NMR experiment is
    (a) 10
    (b) 5
    (c) 30
    (d) 25
3.7. A 90° x pulse rotates the z-magnetization to $-y$ axis, a 90° y pulse rotates the z-magnetization to
    (a) $-x$ axis
    (b) $-z$ axis
    (c) $+x$ axis
    (d) does not affect the z-magnetization
3.8. In NMR spectrum, for a highest frequency of 500 Hz with respect to offset the maximum dwell time in the FID should be
    (a) 0.2 ms
    (b) 1 ms
    (c) 0.1 ms
    (d) 0.05 ms
3.9. In a standard ^{13}C NMR data acquisition, which of the following flip angles yields the best signal-to-noise ratio per unit time by signal averaging?
    (a) 90°
    (b) 30°
    (c) 60°
    (d) 120°

3.10. Apodization by exponential multiplication causes
   A. increase in line width
   B. improvement in resolution
   C. improves line shape
   D. no effect on the spectra
   (a) A and B
   (b) A and C
   (c) B and D
   (d) C and B
3.11. Frequency-independent phase error in the spectrum is because of the
   (a) wrong pulse width
   (b) improper phase of the RF pulse
   (c) wrong dwell time
   (d) delay in the data acquisition
3.12. First-order phase error in the NMR spectrum
   (a) is the same throughout the spectra
   (b) decreases as we move away from the offset
   (c) increases as we move away from the offset
   (d) is proportional to zero-order phase error
3.13. For a sample consisting of two lines with intensities 10,000 and 1 in arbitrary units, what is the minimum ADC resolution required to represent both the signals correctly, assuming 1 bit for sign representation for the signal?
   (a) 8 bits
   (b) 12 bits
   (c) 14 bits
   (d) 16 bits
3.14. In a spin echo experiment, the echo amplitude for a given signal depends on
   (a) inhomogeneity in the magnetic field
   (b) chemical shift of the signal
   (c) T_1 relaxation time of the spin
   (d) T_2 relaxation time of the spin
3.15. In a spin echo (90-τ-180-τ-aq) experiment on a coupled two spin system AX with coupling constant (J), the phase difference between the two components of the doublet at the time of echo will be
   (a) 0
   (b) $2\pi J\tau$
   (c) $4\pi J\tau$
   (d) $6\pi J\tau$
3.16. The echo amplitude in a spin echo experiment decreases due to
   (a) inhomogeneity in the field
   (b) translational diffusion in a homogenous field
   (c) translational diffusion in an inhomogeneous field
   (d) chemical shift anisotropy

3.17. In an FTNMR experiment with the carrier placed in the middle of the spectrum and with quadrature detection, what will be the phase shift of the signal at half the maximum frequency, if the acquisition is delayed by 1 dwell time? What will be the phase shift at the maximum frequency? Assume that in the absence of any delay, all the lines will have positive absorptive shapes.

3.18. Derive an expression for the flip angle due to a RF of amplitude B_1 and pulse width of T_p at an offset of Ω from the RF frequency. For a pulse width T_P giving an on-resonance rotation of $90°$, calculate the offset at which the rotation angle will be $360°$.

3.19. For the spin echo sequence 90_x-τ-180_y-τ acquisition, plot the echo amplitude as a function of time for (i) single spin and (ii) observe spin A of the coupled heteronuclear AX system with a coupling constant of 100 Hz.

3.20. In a jump-return experiment, $(90_x$-t-$90_x)$, what should be the value of t for the maximum water suppression and maximum excitation at 10 ppm away from the water on a 500 MHz spectrometer.

3.21. In an inversion recovery T_1 measurement experiment, if the τ_{null} is 2 s, calculate the T_1 value.

Polarization Transfer

4

Contents

4.1	Introduction	143
4.2	The Nuclear Overhauser Effect (NOE)	144
	4.2.1 Experimental Schemes	145
4.3	Origin of NOE	147
	4.3.1 A Simplified Treatment	147
	4.3.2 A More Rigorous Treatment	150
4.4	Steady-State NOE	152
4.5	Transient NOE	155
4.6	Selective Population Inversion	157
4.7	INEPT	159
	4.7.1 INEPT Has the Following Disadvantages	161
4.8	INEPT$^+$	162
4.9	Distortionless Enhanced Polarization Transfer (DEPT)	163
4.10	Summary	165
4.11	Further Reading	166
4.12	Exercises	166

Learning Objectives
- Introducing magnetization transfer between spins
- Nuclear Overhauser effect
- Sensitivity enhancement by polarization transfer

© The Author(s), under exclusive license to Springer Nature Switzerland AG 2022
R. V. Hosur, V. M. R. Kakita, *A Graduate Course in NMR Spectroscopy*,
https://doi.org/10.1007/978-3-030-88769-8_4

4.1   Introduction

We have seen in Chap. 1 that the intensities of the resonance lines in the NMR spectra are proportional to the population differences between the energy levels they connect. These differences are dictated by Boltzmann's statistics and are dependent on the energy difference between the levels. Given that these energy differences are in the radio frequency regime, the population differences are minimal, and this makes NMR an intrinsically insensitive technique when compared to other spectroscopies such as optical spectroscopy. Different NMR active nuclei have different intensities since they possess different magnetic moments, and among the various nuclei proton is the most sensitive.

Polarization transfer is a process that affects the transfer of magnetization between "like" or "unlike" nuclei. The term "like" implies nuclei have the same gyromagnetic ratio but different chemical shifts, and the term "unlike" refers to nuclei having different gyromagnetic ratios. These two are referred to as homonuclear and heteronuclear transfers, respectively. For example, the transfer between two protons having different resonance frequencies (^1H-^1H) is a homonuclear transfer, and the transfer from ^1H to ^{13}C or ^{15}N is a heteronuclear transfer. One way of achieving this is by manipulating the population differences between different energy levels in a given system of spins. In this case there is an exchange via the z-magnetizations of the individual nuclei. The magnetization transfer can also be effected via transverse (x or y) magnetizations of the individual nuclei. While the former is largely used for structure determination as will be shown later, the latter is mostly used to enhance sensitivities of less sensitive nuclei, for example, ^1H to ^{13}C, ^1H-^{15}N, etc. In fact, both these transfers have useful applications to derive information about the systems under study. Several experimental strategies have been used to achieve such polarization transfer and that will be the subject matter of this chapter.

4.2   The Nuclear Overhauser Effect (NOE)

The nuclear Overhauser effect (NOE) occupies a special place in the collection of techniques in NMR spectroscopy. It is an invaluable tool for chemists and biologists for the assignment of resonances and elucidation of three-dimensional structures of molecules in solution media. This has enabled in recent years the so-called impossible task of determination of structures of large molecules such as proteins, nucleic acids, and carbohydrates to atomic resolution and thus has rendered NMR to be a complementary technique to X-ray crystallography in structural biology. For organic chemists, NOE has been a valuable tool for deciphering the stereochemistry at specific places in the molecules. It allows filtering out specific regions of spectra belonging to nuclei in close proximity and thus has been a useful tool for resonance assignments.

The NOE can be defined as a change in the intensity of one NMR resonance line when another resonance line in the NMR spectrum is in some way perturbed, say by saturation, inversion by a pulse, etc.

$$\eta_i(s) = \frac{I - I_o}{I_o} \tag{4.1}$$

where $\eta_i(s)$ is the NOE at resonance i due to a perturbation at resonance s, I_o is the intensity of the resonance i in the absence of the perturbation, and I is the intensity after the perturbation at s. The perturbation may be by a continuous radio frequency irradiation or by a radio frequency pulse. In the former case, the NOE is termed as steady-state NOE, and the latter case leads to the so-called transient NOE. It is often a practice to talk of NOE between two nuclei, rather than between two resonances. The NMR spectra of the nuclei may have multiplet structures, and then the NOE experiment involves the perturbation of the whole group of resonances belonging to one nucleus and observation of the total intensity change of the resonances in the multiplet of the other nucleus. The intensities I and I_o in Eq. 4.1 would then refer to the integrals of the individual multiplets rather than single resonances.

4.2.1   Experimental Schemes

Figure 4.1 shows the standard experimental schemes for obtaining NOE spectra.

Scheme (a) in Fig. 4.1 is for the steady-state experiment in which the RF perturbation is continuously applied for a period of time at resonance or at the midpoint of a group of resonances belonging to a multiplet of a spin. The time period of the irradiation is determined by the spin-lattice relaxation time of the spin and is long enough to ensure that the spin system has reached a new equilibrium distribution of the populations of the different energy levels. Then the state of the

Fig. 4.1 Schematic representations of steady-state (**a**) and transient (**b**) NOE pulse schemes. In (**a**), low power RF is continuously applied to saturate a signal. In (**b**), the Gaussian-shaped 180° pulse indicates a selective pulse applied to a particular transition. τ_m represents the mixing time during which transfer of magnetization occurs

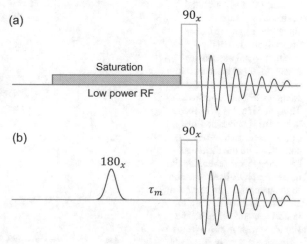

system is monitored by applying an observe pulse and recording the spectrum. This spectrum is called the on-resonance spectrum. A control experiment is performed by keeping the irradiation far away from any resonance, and this spectrum is referred to as the off-resonance spectrum. The difference between the on-resonance and off-resonance spectra eliminates other effects of irradiation and shows only those resonances which are affected by the RF via NOE. Such a spectrum is known as the NOE difference spectrum.

Scheme (b) in Fig. 4.1 is for the transient NOE experiment. The steps here are similar to those of the steady-state experiment except that the perturbation is by a selective 180° pulse to invert a particular resonance or a group of resonances as before. A certain time period t_m called the "mixing time" is allowed to elapse during which the spins exchange magnetization leading to NOE. The spectrum is then recorded by applying an observe pulse to obtain the on-resonance spectrum. Again an off-resonance spectrum is recorded by applying the inversion pulse at a frequency where there is no resonance in the spectrum. The difference between the two yields is the difference NOE spectrum. The mixing time in the experiment can be varied to monitor the time course of magnetization transfer between spins in a molecule, and this permits the identification of spins that are closest to the perturbed spin, spins which are slightly farther away from the perturbed spin and so on. This time is again dictated by the spin-lattice relaxation time of the perturbed spin and can vary from a few tens of milliseconds to seconds in different systems. The knowledge of the relaxation times of the spins in the molecule will be a great help in deciding on the ranges of mixing times useful for the purpose.

Figure 4.2 shows a typical transient ^1H NOE difference spectrum of a large molecule. The signals present in the spectrum belong to protons which are less

Fig. 4.2 A schematic transient NOE spectrum of a large molecule case (**c**), wherein off-resonance and on-resonance spectra are shown in (**a**) and (**b**), respectively. In (**b**) the inverted signal appears negative. In (**a**) the selective pulse is at a location far away from any of the resonances. Thus, no signal is perturbed. When (**b**) is subtracted from (**a**), the perturbed resonance coadds, while the differences appearing in the other signals reflect the NOE-based transfer of magnetization. In this case NOE is taken to decrease the intensities, and hence all the signals appear positive in the difference spectrum (**c**)

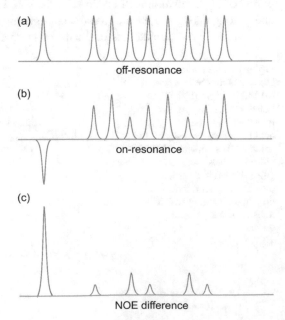

than 5 Å apart in space in the molecule. The strong signal identifies the perturbed resonance, and the weak signals seen are the NOEs. All the other signals in the unperturbed spectrum have been subtracted out. A quantitative estimation of the intensities of the NOE signals allows the estimation of distances between the interacting protons. This is the fundamental information provided by the nuclear Overhauser effect and allows specific identification of resonances and estimation of a large number of such distances in a given molecule that enables the determination of its three-dimensional structure. In the following sections, we shall discuss the theoretical and experimental aspects of the NOE and also its use for estimating inter-proton distances.

4.3   Origin of NOE

The nuclear Overhauser effect arises due to dipolar interactions between the two nuclei. Whenever a nucleus is perturbed from its equilibrium state, the perturbation is transferred to its neighboring nucleus via dipolar interactions, and this provides a mechanism for the relaxation of the perturbed nucleus. The extent of perturbation measured by the disturbance in the population distributions in various energy levels in the system is a function of several parameters of the system, such as relaxation times, the strength of the irradiation, and the duration of the irradiation, and consequently the extent of NOE is also dependent on the relaxation properties of the spin system. In what follows, we first give a qualitative and simplified explanation for the phenomenon, and it will be followed by a more rigorous treatment in the next subsection.

4.3.1   A Simplified Treatment

Let us consider a two spin AX system with the energy level diagram, as shown in Fig. 4.3.

P_1, P_2, P_3, and P_4 are the populations of the four states 1–4, respectively; $\alpha\alpha$, $\alpha\beta$, $\beta\alpha$, and $\beta\beta$ are the spin polarizations of the four states; $A1$ and $A2$ are the transitions

Fig. 4.3 An energy level diagram of a two-spin system. The different transition probabilities, namely, zero-quantum (W_0), single-quantum (W_1^X, W_1^A), and double-quantum (W_2), between the different energy levels are indicated

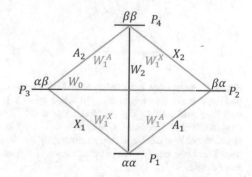

of the A spin; $X1$ and $X2$ are the transitions of the X spin. $W_1's$ are the single-quantum transition probabilities for the A and X spins, and W_0 and W_2 are the zero-quantum and double-quantum transition probabilities, respectively. The four transitions are represented as follows:

$$A1 : P_1 \rightarrow P_2 \quad A2 : P_3 \rightarrow P_4$$
$$X1 : P_1 \rightarrow P_3 \quad X2 : P_2 \rightarrow P_4 \tag{4.2}$$

At equilibrium, all these have equal population differences resulting in identical intensities for all the four transitions. Let these population differences be equal to d (Fig. 4.4a).

The total intensities of A transitions, as well as of X transitions, are each equal to $2d$. Now, let us consider that the A transitions are saturated by continuous radio

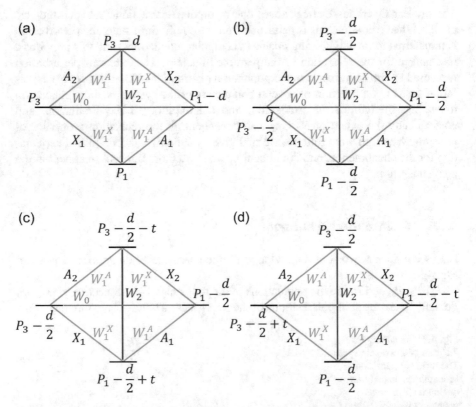

Fig. 4.4 An energy level diagram and the populations of a two-spin AX system: (**a**) when the populations are at equilibrium, (**b**) when spin-A is saturated, (**c**) if only the double-quantum transition probability (W_2) is effective, and (**d**) if only the zero-quantum transition probability (W_0) is effective. These are simply indicative, and in reality, all the transition probabilities will be simultaneously effective, although some may be more dominant than the others. P_1, P_3, and d are used to indicate equilibrium populations, and t is used to indicate the extent of transfer. See text for more details

frequency irradiation. This results in an equalization of the populations P_1 and P_2 and of P_3 and P_4. The populations of the four states will now be (Fig. 4.4b)

$$P'_1 = P'_2 = P_1 - \frac{d}{2}$$
$$P'_3 = P'_4 = P_3 - \frac{d}{2}$$
(4.3)

At this stage, energy levels 2 and 4 gain populations, and hence P'_2 and P'_4 would be higher than the corresponding starting equilibrium populations P_2 and P_4, respectively.

Now, the A transitions will have zero intensities and the X transitions will have the intensities

$$X1 = X2 = P_1 - P_3$$
(4.4)

It appears that the X transitions still have the same intensities as in the unperturbed case. Nevertheless, we have to remember that while the irradiation tends to saturate the A transitions, the relaxation processes tend to drive the system towards equilibrium. This is brought about by the three transition probabilities indicated in Fig. 4.3. In this process, levels whose populations are higher than their equilibrium populations would lose, and others would gain. Now, let us see which of these are most effective in altering the populations. The single-quantum transition probability of the A spin cannot cause any changes since this will be overridden by the RF. Similarly, W_1^x also cannot cause any population transfer since the population differences between states 1 and 3 and 2 and 4 are already equal to the equilibrium differences. Now, let us see the effect of W_2 alone, and let the steady-state populations be as shown in Fig. 4.4c. It is, of course, necessary to assume that W_2 changes the population faster than the saturating field, and consequently, there is a net gain by state 1 and a net loss by state 4. The population differences corresponding to the X transitions are now

$$X1 = P_1 - P_3 + t; \quad X2 = P_1 - P_3 + t$$
(4.5)

Thus, there is a net increase in the intensities of the X transitions. The effect of W_0 alone can be considered in a similar manner. The zero-quantum transition causes population transfer between states 2 and 3, and this can be represented as shown in Fig. 4.4d. It is important to note here that the population transfer occurs from state 2 to state 3 and not vice versa since state 2 has a higher population than the equilibrium value, and this is higher than the population of state 3; state 3 has a lower population than its equilibrium value actually. Thus, the intensities of the X transitions are proportional to the populations as indicated in Eq. 4.6:

$$X1 = P_1 - P_3 - t; \quad X2 = P_1 - P_3 - t$$
(4.6)

In this case, the X transitions have lost some intensity compared to the unperturbed state. In reality, however, all the transition probabilities will be operating simultaneously, and depending on the relative efficiencies, there will be net gain or loss in the intensities of the X transitions. The two situations are referred to as positive NOE and negative NOE, respectively. The magnitude of the NOE is proportional to the difference in the two relaxation rates, which is referred to as the "cross-relaxation rate" σ.

$$\sigma = W_2 - W_0 \tag{4.7}$$

Positives NOEs generally occur in small molecules, while the negative NOE is a common feature of macromolecules. The reasons are fairly simple. As discussed in Chap. 1, in large molecules, there is a very efficient redistribution of magnetizations by zero-quantum transition processes among various nuclei within the molecule, whereas in small molecules, the population changes and hence the magnetization changes will have to occur by dissipative processes to the lattice by double-quantum transitions. Thus, in large molecules W_0 is a very dominating process leading to negative NOEs, and in small molecules W_2 is a dominating process leading to positive NOEs.

4.3.2   A More Rigorous Treatment

We have seen in the previous section that the NOE arises as a consequence of the interplay of several relaxation processes following a perturbation of the spin system tending to bring back the populations of various energy levels to their equilibrium values. Therefore, the time evolution of these populations is the crucial process to be looked at in getting a deeper understanding of the NOE phenomenon. Let us consider an N level system as shown schematically in Fig. 4.5.

Following a perturbation of the spin system, the recovery of the population of a state i will be governed by the equation

$$\frac{\mathrm{d}P_i}{\mathrm{d}t} = \sum_j W_{ij}\left(P_j - P_j^0\right) - \left(P_i - P_i^0\right)\sum_j W_{ij} \tag{4.8}$$

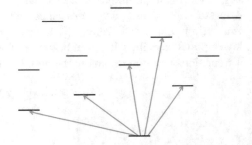

Fig. 4.5 A schematic of an N energy level system, wherein upward transitions are indicated from the lowest energy level to the other higher levels

where W_{ij} is the transition probability from state i to state j, Ps are the populations at any time t, and P^0s are the equilibrium populations of the different states. This equation is referred to as the *master equation*. For uncoupled or weakly coupled spin systems, the populations can also be related to the magnetizations, and then Eq. 4.8 can be recast in terms of magnetizations of individual spins. We will illustrate the procedure for solving the master equation by considering a two-spin AX system and referring to the energy level diagram in Fig. 4.3.

If M_A and M_X are the z-magnetizations of the spins A and X, respectively, they are related to populations P_1 to P_4 as follows:

$$M_A \alpha \frac{1}{2} [P_1 + P_2 - P_3 - P_4]$$

$$M_X \alpha \frac{1}{2} [P_1 + P_3 - P_2 - P_4] \tag{4.9}$$

The rate equations for various population changes are

$$\frac{dP_1}{dt} = -\left(W_1^A + W_1^X + W_2\right)\left(P_1 - P_1^0\right) + W_2\left(P_4 - P_4^0\right) + W_1^X\left(P_2 - P_2^0\right)$$
$$+ W_1^A\left(P_3 - P_3^0\right) \tag{4.10}$$

$$\frac{dP_2}{dt} = -\left(W_1^A + W_1^X + W_0\right)\left(P_2 - P_2^0\right) + W_1^A\left(P_4 - P_4^0\right) + W_1^X\left(P_1 - P_1^0\right)$$
$$+ W_0\left(P_3 - P_3^0\right) \tag{4.11}$$

$$\frac{dP_3}{dt} = -\left(W_1^A + W_1^X + W_0\right)\left(P_3 - P_3^0\right) + W_1^A\left(P_1 - P_1^0\right) + W_1^X\left(P_4 - P_4^0\right)$$
$$+ W_0\left(P_2 - P_2^0\right) \tag{4.12}$$

$$\frac{dP_4}{dt} = -\left(W_1^A + W_1^X + W_2\right)\left(P_4 - P_4^0\right) + W_1^A\left(P_2 - P_2^0\right) + W_1^X\left(P_3 - P_3^0\right)$$
$$+ W_2\left(P_1 - P_1^0\right) \tag{4.13}$$

If we define equilibrium magnetizations M_A^0 and M_X^0 as

$$M_A^0 \alpha \frac{1}{2} \left[P_1^0 + P_2^0 - P_3^0 - P_4^0\right] \tag{4.14}$$

$$M_X^0 \alpha \frac{1}{2} \left[P_1^0 - P_2^0 + P_3^0 - P_4^0\right] \tag{4.15}$$

then the rate of change of magnetizations M_A and M_X of A and X spins can be written after some tedious algebra as

$$\frac{dM_A}{dt} = -\left(2W_1^A + W_0 + W_2\right)\left(M_A - M_A^0\right) - \left(W_2 - W_0\right)\left(M_X - M_X^0\right) \tag{4.16}$$

$$\frac{dM_X}{dt} = -\left(2W_1^X + W_0 + W_2\right)\left(M_X - M_X^0\right) - \left(W_2 - W_0\right)\left(M_A - M_A^0\right) \quad (4.17)$$

Defining further,

$$\rho_A = \left(2W_1^A + W_0 + W_2\right)$$

$$\rho_X = \left(2W_1^X + W_0 + W_2\right)$$

$$\sigma_{AX} = W_2 - W_0 \quad (4.18)$$

we obtain

$$\frac{dM_A}{dt} = -\rho_A\left(M_A - M_A^0\right) - \sigma_{AX}\left(M_X - M_X^0\right) \quad (4.19)$$

$$\frac{dM_X}{dt} = -\rho_X\left(M_X - M_X^0\right) - \sigma_{AX}\left(M_A - M_A^0\right) \quad (4.20)$$

The entities ρ_A and ρ_X are called auto-relaxation rates of the spins A and X, respectively, and the entity σ_{AX} is called the cross-relaxation rate between the two spins. The two Eqs. 4.19 and 4.20 can be clubbed and written in the form of a matrix equation as

$$\frac{dm}{dt} = -Rm \quad (4.21)$$

where m is a column vector representing the deviations in the magnetizations from equilibrium values and R is a matrix of relaxation rates called the relaxation matrix. This equation is actually a very general one and is not limited to two spins alone. Accordingly, the dimensions of the vector and the matrix will be different. For the two-spin case explicitly considered,

$$m = \begin{bmatrix} \left(M_A - M_A^0\right) \\ \left(M_X - M_X^0\right) \end{bmatrix}; \quad R = \begin{bmatrix} \rho_A & \sigma_{AX} \\ \sigma_{AX} & \rho_X \end{bmatrix} \quad (4.22)$$

Equation 4.21 will have to be solved to calculate the NOE. Since the treatments are different for the steady-state and transient NOE, we will consider the two separately in the following paragraphs.

4.4   Steady-State NOE

For the steady-state NOE on spin X due to irradiation of spin A, we can set

$$\frac{dM_x}{dt} = 0; M_A = 0 \qquad (4.23)$$

From Eq. 4.20, this leads to the expression for NOE as

$$\eta_X^A = \frac{M_X - M_X^0}{M_X^0} = \left(\frac{M_A^0}{M_X^0}\right)\left(\frac{\sigma_{AX}}{\rho_X}\right) \qquad (4.24)$$

If dipole-dipole interaction between spins A and X is the sole mechanism of relaxation, then the auto- and cross-relaxation rates can be calculated by using explicit expressions for the transition probabilities contributing to the respective rates. These are related to spectral densities discussed in Chap. 1, and without going into the details of these calculations, we shall simply accept the relations for the transition probabilities as given in Eqs. 4.25, 4.26 and 4.27:

$$W_o = \left(\frac{K}{20}\right)\left(\frac{2\tau_c}{1 + ((\omega_A - \omega_X)\tau_c)^2}\right) \qquad (4.25)$$

$$W_1^A = \left(\frac{K}{20}\right)\left(\frac{3\tau_c}{1 + (\omega_A\tau_c)^2}\right)$$

$$W_1^X = \left(\frac{K}{20}\right)\left(\frac{3\tau_c}{1 + (\omega_X\tau_c)^2}\right) \qquad (4.26)$$

$$W_2 = \left(\frac{K}{20}\right)\left(\frac{12\tau_c}{1 + ((\omega_A + \omega_X)\tau_c)^2}\right) \qquad (4.27)$$

where K is a constant given by

$$K = \left(\frac{\mu_0}{4\pi}\right)^2 (\gamma_A \gamma_X)^2 \left(\frac{h}{2\pi}\right)^2 r_{AX}^{-6} \qquad (4.28)$$

r_{AX} is the distance between the spins A and X, and the other constants have the usual meanings (h refers to Planck's constant and μ_o refers to permeability of the vacuum). It is clear that the NOE will depend upon the rates of molecular motions since the spectral densities, J, are dependent on the reorientational correlation times in the system. Figure 4.6 shows plots of maximum NOE vs correlation time for homonuclear spin systems.

Two limiting cases can be specifically discussed.

Extreme Narrowing Limit
Under the conditions of extreme narrowing which are prevalent in small molecules as discussed in Chap. 1, the spectral densities J in Eqs. 4.25, 4.26 and 4.27 become equal, and the transition probabilities will be in the proportion

Fig. 4.6 The dependence of the homonuclear NOE on molecular sizes (i.e., correlation times). Large molecules have large correlation times and small molecules have short correlation times

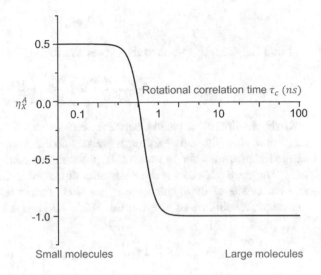

$$W_o : \left(W_1^A \, or \, W_1^X\right) : W_2 : 2 : 3 : 12 \tag{4.29}$$

Then,

$$\left(\frac{\sigma_{AX}}{\rho_X}\right) = \frac{(W_2 - W_0)}{(2W_1^X + W_0 + W_2)} \tag{4.30}$$

$$= \frac{1}{2} \tag{4.31}$$

Now, for spin ½ systems being considered here, the equilibrium magnetizations M_A^0 and M_X^0 defined by the population differences $(p_\alpha - p_\beta)$ are proportional to μ (see Chap. 1, Eq. 1.18) and hence to the individual magnetogyric ratios, and hence,

$$\frac{M_A^0}{M_X^0} \propto \gamma_A / \gamma_X \tag{4.32}$$

Thus, the steady-state NOE is given by

$$\eta_X{}^A = \frac{1}{2} \left(\frac{\gamma_A}{\gamma_X}\right) \tag{4.33}$$

If, for example, A is a proton and X is a carbon-13 nucleus, then the NOE will have a value of 2. This represents the maximum NOE obtainable in the system. If X is ^{15}N, then the maximum NOE will close to -5.0; the negative sign arises because ^{15}N has negative magnetogyric ratio. In practice, however, the observed NOE will be smaller than this value because mechanisms other than dipole-dipole interaction also contribute to the relaxation of the spin system. Similarly, for the homonuclear system, the maximum NOE is 0.5.

Slow Motion Limit
This situation occurs in large molecules, and the spectral densities will now have an explicit dependence on the correlation times, and the transition probabilities will be given by

$$W_0 = \left(\frac{K}{10}\right)\tau_c \tag{4.34}$$

$$W_1^A = 0 \tag{4.35}$$

$$W_2 = 0 \tag{4.36}$$

The NOE for spin ½ systems is given by

$$\eta_X{}^A = \left(\frac{M_A^0}{M_X^0}\right)\left(\frac{\sigma_{AX}}{\rho_X}\right) = -\left(\frac{M_A^0}{M_X^0}\right) = -\left(\frac{\gamma_A}{\gamma_X}\right) \tag{4.37}$$

Thus, for homonuclear systems, the maximum NOE is −1 and is higher for heteronuclear systems when the protons are saturated, and the X nucleus is observed.

4.5 Transient NOE

Figure 4.1b shows the pulse scheme for the one-dimensional transient NOE experiment. The experiment begins with selective inversion ($180°$ pulse) of one particular spin, say i. This is followed by a constant delay, τ_m, and then an observed hard $90°$ pulse is applied to generate transverse magnetization, which is then detected as FID. During the time τ_m, the transfer of magnetization occurs from the inverted spin to the other spins via dipole-dipole interaction as the inverted spin relaxes back to equilibrium, which generates the NOE on the other spins. This time period is referred to as the "mixing time."

The analytical treatment of transient NOE begins with Eq. 4.21 extended to include multiple spins. Rewriting in a generalized sense,

$$\frac{dM}{dt} = -RM \tag{4.38}$$

with M now representing multiple spins in a column vector

$$M = \{m_1, m_2, m_3, \ldots, m_n\} \tag{4.39}$$

$$m_i = M_{iz} - M_i^0 \tag{4.40}$$

represents the deviation of z-magnetization of ith spin from equilibrium value, and

$$R = \begin{bmatrix} \rho_{11} & \sigma_{12} & \sigma_{13} & \sigma_{14} & \cdots & \sigma_{1n} \\ \sigma_{21} & \rho_{22} & \sigma_{23} & \sigma_{24} & \cdots & \sigma_{2n} \\ \sigma_{31} & \rho_{32} & \sigma_{33} & \sigma_{34} & \cdots & \sigma_{3n} \\ \sigma_{41} & \rho_{42} & \sigma_{43} & \sigma_{44} & \cdots & \sigma_{4n} \\ \vdots & \vdots & \vdots & \vdots & \ddots & \vdots \\ \sigma_{n1} & \rho_{n2} & \sigma_{n3} & \sigma_{n4} & \cdots & \sigma_{nn} \end{bmatrix}$$ (4.41)

The solution of Eq. 4.38 will be

$$M(t) = e^{-(Rt)} M(0)$$ (4.42)

Explicitly, for $t = \tau_m$,

$$M(\tau_m) = \left\{ 1 - R\tau_m + \frac{1}{2!} R^2 \tau_m^2 - \frac{1}{3!} R^3 \tau_m^3 + \ldots\ldots\ldots \right\} M(0)$$ (4.43a)

For short τ_m the second- and higher-order terms in Eq. 4.43a can be neglected. Under these conditions,

$$M(\tau_m) = \{ 1 - R\tau_m \} M(0)$$ (4.43b)

This solution in the matrix form will look like

$$\begin{bmatrix} m_1(\tau_m) \\ m_2(\tau_m) \\ m_3(\tau_m) \\ m_4(\tau_m) \\ \vdots \\ m_n(\tau_m) \end{bmatrix} = \left\{ 1 - \tau_m \begin{bmatrix} \rho_{11} & \sigma_{12} & \sigma_{13} & \sigma_{14} & \cdots & \sigma_{1n} \\ \sigma_{21} & \rho_{22} & \sigma_{23} & \sigma_{24} & \cdots & \sigma_{2n} \\ \sigma_{31} & \rho_{32} & \sigma_{33} & \sigma_{34} & \cdots & \sigma_{3n} \\ \sigma_{41} & \rho_{42} & \sigma_{43} & \sigma_{44} & \cdots & \sigma_{4n} \\ \vdots & \vdots & \vdots & \vdots & \ddots & \vdots \\ \sigma_{n1} & \rho_{n2} & \sigma_{n3} & \sigma_{n4} & \cdots & \sigma_{nn} \end{bmatrix} \right\}$$

$$\times \begin{bmatrix} m_1(0) \\ m_2(0) \\ m_3(0) \\ m_4(0) \\ \vdots \\ m_n(0) \end{bmatrix}$$ (4.44)

In other words, for $j \neq i$

$$m_i(\tau_m) = (1 - \rho_{ii}\tau_m)m_i(0) - \sum \sigma_{ij}\tau_m m_j(0) \tag{4.45}$$

The expression for NOE will then become

$$\eta_i(\tau_m) = \frac{m_i(\tau_m)}{M_i^0} = \left(\frac{1}{M^0}\right)\left\{(1 - \rho_{ii}\tau_m)m_i(0) - \sum \sigma_{ij}\tau_m m_j(0)\right\} \tag{4.46}$$

For a selective inversion of jth spin,

$$m_j(0) = -2M_j^0 \text{ and } m_i(0) = 0, i \neq j \tag{4.47}$$

Thus,

$$\eta_i(\tau_m) = -\left(\frac{1}{M_i^0}\right)(-2M_j^0)\sigma_{ij}\tau_m \tag{4.48}$$

$$= 2\sigma_{ij}\tau_m, \text{ for homonuclear case}; M_i^0 = M_j^0, \text{ for homonuclear case.} \tag{4.49}$$

Thus, it is seen that for short mixing times, the transient NOE at spin i due to the inversion of spin j is directly proportional to the cross-relaxation rate between the two spins and is also linearly dependent on the mixing time. Hence this approximation of short mixing time is referred to as the "isolated spin pair approximation" (ISPA). Since the cross-relaxation rate is inversely proportional to the inverse sixth power of the internuclear distance as discussed earlier, the transient NOE experiment provides a powerful tool for estimating internuclear distances in molecules and hence for structure determination. As we will see in later chapters, this strategy has been elegantly used for the structure determination of large molecules in solution.

One might ask at this stage how do we know "what is short mixing time." Indeed, this is a complex question, and a simple answer to this would be it should be much shorter compared to the spin-lattice relaxation time of spin j.

In a more general case, where no approximations are made, all the cross-relaxation rates and leakage to the lattice will contribute to the transient NOE at any particular spin, and the dependence of $\eta_i(\tau_m)$ on mixing time will appear, as shown in Fig. 4.7.

The nonlinearity and decay of the NOE at higher mixing times are a consequence of leakage to the lattice and contributions from various cross-relaxation rates, which is termed as "spin diffusion" in the NMR jargon.

4.6 Selective Population Inversion

Consider a two spin-system, AX (X represents the insensitive hetero nuclei and A represents the ^1H), with the energy level diagram shown in Fig. 4.8a. The populations of individual levels are as indicated in Fig. 4.8b. This population distribution leads to two transitions for each spin, $A1 : (1 \rightarrow 2$ transition),

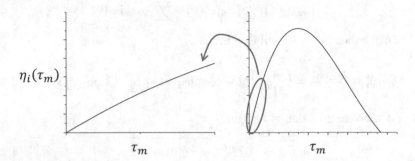

Fig. 4.7 The dependence of NOE ($\eta_i(\tau_m)$) on mixing time, τ_m, is shown schematically

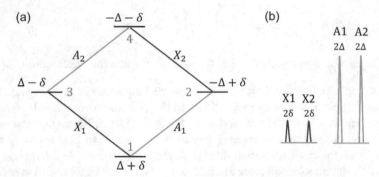

Fig. 4.8 (a) Populations of the four energy levels (1–4) of a two-spin, AX spin system (Δ and δ are used to indicate the populations), (b) the resultant population differences for the transitions A_1 and A_2 and X_1 and X_2 and the corresponding spectral intensities

$A2$: ($3 \rightarrow 4$ transition), and two transitions for the X : $X1$ ($1 \rightarrow 3$) and $X2$ ($2 \rightarrow 4$). Intensities of the $A1$ and $A2$ transitions are 2Δ each, and similarly intensities of X transitions are 2δ each. Let us assume a 180° pulse is applied to the $A1$ transition (Fig. 4.9a); as a consequence, the populations of the levels 1 and 2 get interchanged. In this scenario, the intensities of the $X1$ and $X2$ transitions will become $-(2\Delta - 2\delta)$ and $2\Delta + 2\delta$, respectively (Fig. 4.9b). Clearly, there is a substantial change in the intensities of the X transitions; if one calculates the changes in the intensities of two transitions from the equilibrium situation, this will become $+2\Delta$ and -2Δ, respectively, and the magnitude of change is proportional to $\frac{\Delta}{\delta}$, and this is equal to $\frac{\gamma_A}{\gamma_X}$. Thus, for a ^{13}C-^{1}H system, for example, there will be an intensity gain of factor of 4, although the two transitions will have opposite signs. Similarly, for a ^{15}N-^{1}H system, a sensitivity gain of factor 10 will be observed when the populations of the energy levels corresponding to ^{1}H transition are inverted. So, this provides a significant advantage for sensitivity enhancement. However, the difficulty lies in obtaining a clean selective 180° pulse, especially when the resolutions are poor or the spectra are crowded.

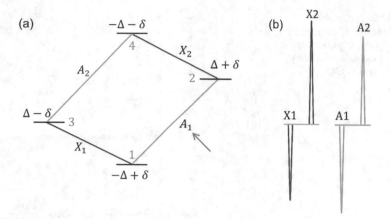

Fig. 4.9 (a) Population distribution among the four (1–4) levels of an AX spin system after selective inversion of the A_1 transition by the application of selective 180° pulse on the A_1 transition, as indicated by the arrow (the populations of levels 1 and 2 are interchanged), (b) the resultant spectral lines (see text for intensities)

4.7   INEPT

To circumvent the issues related to selective inversions, this method called INEPT (insensitive nuclei enhance polarization transfer), which uses nonselective pulses for the transfer, was developed. The basic pulse sequence for the INEPT experiment is as shown in Fig. 4.10. The vector diagram showing the evolution of the magnetization components is shown in Fig. 4.11. The first 90_x pulse on the A spin rotates the z-magnetization of the A spin on to the $-y$ axis. During the next $\tau = \frac{1}{4J}$ period, the two components dephase to the extent of 90° phase difference. A 180^o_x pulse on the A transitions rotates the two transitions in the opposite quadrants; a simultaneous 180° pulse on X nucleus interchanges the labels of A transitions, as a result of which they continue to dephase further and at the end of next $\frac{1}{4J}$ period they are aligned in opposite directions along the x-axis. The following 90^o_y pulse on the A spin puts these two transitions along the $+z-$ and $-z$-axes. This scenario is equivalent to a selective inversion of one of the A transitions described in the previous section; consequently, this contributes to an enhancement of the X-transition intensity by the ratio $\frac{\gamma_A}{\gamma_X}$. Following a final 90^o_x pulse on X-channel, the signal is detected and Fourier transformed.

The success of the polarization transfer is determined largely by the relaxation time of the sensitive nuclei, which should be short enough to re-establish the equilibrium during the pulse sequence. Intensity enhancements by this method can be much higher than those obtained by NOE methods.

Fig. 4.10 A schematic
representation of the primary
INEPT pulse sequence for the
spin system AX. τ is a delay
period equal to $1/4J_{AX}$

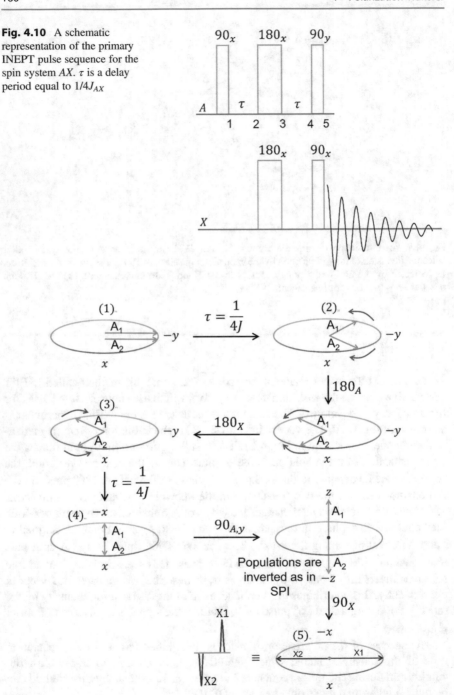

Fig. 4.11 The vector evolution of A spin magnetization components at different time points 1–5 in
the INEPT pulse scheme. Note one is looking at the two transitions A_1 and A_2, sitting at the center of
the two

4.7.1   INEPT Has the Following Disadvantages

(i) *Intensity and Multiplet Anomalies*

Incorrect relative intensities occur in different spin multiplets, and this depends on the following:

(a) Strength of heteronuclear J-coupling
(b) Multiplicity of spin multiplets

Multiplet pattern distortions occur in systems such as A_2X and A_3X (Fig. 4.12). These distortions are a consequence of cancellation of intensities in the central lines after evolution of the X-magnetization under more than one J-couplings to heteronuclei. For a simple two-spin system, AX, the X-lines appear as a simple antiphase doublet after X-evolution during detection. When there are multiple J-couplings with the X spin, such as in A_2X, A_3X, etc., the X spin evolves under all those couplings resulting in in-phase splitting of the antiphase doublet, and this results in the cancellation of the central components since all the couplings are equal in magnitude. Thus, in an A_2X system, the X-multiplet appears as $(-1, 0, 1)$ intensity pattern, as against the $(1, 2, 1)$ pattern in the conventional spectrum. Similarly, in an A_3X system, the X-multiplet appears as $(-1, -1, 1, 1)$ intensity pattern as against the $(1, 3, 3, 1)$ pattern in the conventional spectrum.

Further, as can be seen from the previous discussion, the pulse sequence contains a delay period of 1/2J during which time the magnetization is on the A nucleus. This transverse magnetization undergoes transverse relaxation, which leads to some loss of intensity. The extent of loss depends on the transverse relaxation rate of the A spin and the magnitude of the coupling constant since the delay for effective transfer is 1/2J for a two-spin system.

Fig. 4.12 Schematic INEPT-resultant multiplet patterns of X spin in the A_2X (**a**) and A_3X (**b**) spin systems

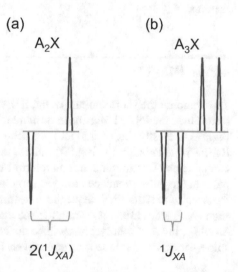

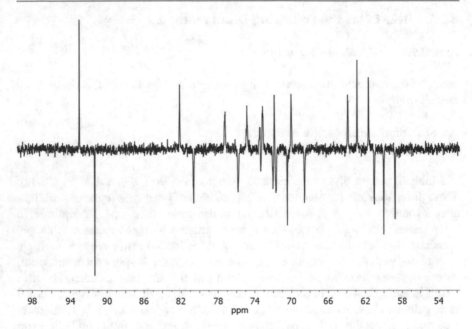

Fig. 4.13 An illustrative INEPT spectrum of sucrose. The positive and negative signals are clearly visible

(ii) *Positive and Negative Signals in the Multiplets*

As can be seen in Fig. 4.12, the multiplets have positive and negative signals. These not only result in the cancellation of intensities but also are inconvenient to analyze when many multiplets are close to each other in a spectrum.

Figure 4.13 shows an illustrative INEPT spectrum displaying the different features.

4.8   INEPT$^+$

To overcome the shortcomings of the INEPT sequence with regard to the multiplet distortions, the INEPT$^+$ has been designed. The pulse sequence for this is shown in Fig. 4.14. Firstly, an additional spin echo segment similar to the first one in the INEPT is added after the last 90° pulses; this helps to refocus the positive-negative components. This modification is referred to as the refocused INEPT. The last 90° pulse in the pulse sequence removes some artifacts and results in clean multiplets in the spectrum. This final sequence is referred to as INEPT$^+$. The delay τ' in this segment is adjustable for editing the spectra since refocusing of the A multiplets in AX, A_2X, and A_3X spin systems will be different. The peak intensities (I) for the three spin systems are dependent on τ' as given Eqs. 4.50, 4.51 and 4.52:

Fig. 4.14 A schematic of the INEPT$^+$ pulse sequence for the spin system AX

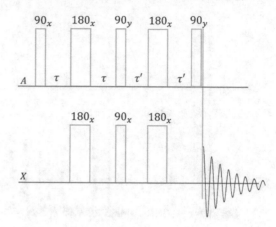

$$I(AX) \propto \sin \theta \qquad (4.50)$$

$$I(A_2X) \propto 2\sin \theta \cos \theta \qquad (4.51)$$

$$I(A_3X) \propto 3\sin \theta \cos^2 \theta \qquad (4.52)$$

where $\theta = \pi J_{AX}\tau'$. The intensities for the different multiplets can be positive or negative depending upon the choice of the θ. For example, for $\theta = 45°$, all multiplets will have a positive sign. For $= 135°$ AX and A_3X will have a positive sign, while A_2X will have a negative sign. Thus, spectra acquired with different θ values can be coadded to filter out desired multiplets. The 90° proton pulse before the detection period in the INEPT$^+$ is called a "purge" pulse which helps to eliminate unwanted antiphase magnetization components and helps to restore the binomial distribution of intensities in the multiplets.

An experimental ^{13}C INEPT$^+$ spectrum of a small molecule is shown in Fig. 4.15 as an illustration.

The signal-to-noise ratio can be further enhanced by employing heteronuclear decoupling in the detection period of the experiment. An example of comparison of coupled and decoupled spectra is shown in Fig. 4.16.

4.9   Distortionless Enhanced Polarization Transfer (DEPT)

DEPT is the most widely used polarization transfer experiment, as it is free from many of the shortcomings of the INEPT sequences. It produces ^{13}C spectra identical to those obtained by direct ^{13}C observation, which means the multiplet structures are well preserved, along with the advantage of polarization transfer from ^{1}H to ^{13}C. The pulse sequence of the DEPT experiment is given in Fig. 4.17. In this pulse scheme, changing the flip angle (θ) of the final ^{1}H pulse encodes the phases of carbon signals.

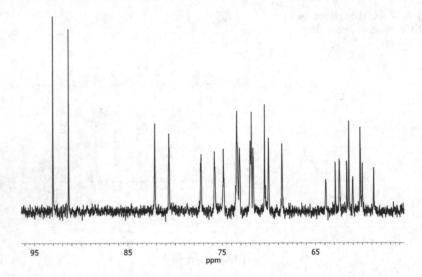

Fig. 4.15 An experimental proton-coupled ^{13}C spectrum of sucrose showing the correct multiplet structures obtained from INEPT$^+$ pulse sequence

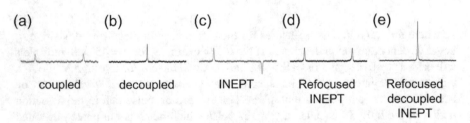

Fig. 4.16 An illustrative comparison of coupled, INEPT, refocused INEPT, and ^{1}H-decoupled ^{13}C spectra. The former two are recorded by direct ^{13}C observation, while the last one is acquired by ^{1}H decoupling during acquisition in the refocused INEPT experiment. The significant enhancement in the S/N ratio in the latter is clearly evident

The carbon resonances of CH, CH$_2$, and CH$_3$ groups vary with θ as $\sin\theta$, $2\sin\theta\cos\theta$, and $3\sin\theta\cos^2\theta$, respectively. Interestingly, this is identical to the patterns in the refocused INEPT scheme, except that here it is determined by the value of a flip angle as against the magnitude of the refocusing delay τ' in the refocused INEPT. DEPT is also more tolerant with regard to variations in J values. The flip angle-dependent sign encoding of the carbon signals in the DEPT experiment is shown in Fig. 4.18.

Clearly, the patterns seen in Fig. 4.18 suggest ways of observing particular types of carbons by choosing θ appropriately. For example, for $\theta = 90°$ only CH will be observed, and the other two will have zero intensity. For $\theta = 135°$ CH and CH$_3$ have positive intensities, while CH$_2$ has a negative intensity; while CH$_2$ and CH$_3$ will have their maximum intensities, CH will have less than its maximum intensity.

Fig. 4.17 A schematic of
DEPT NMR pulse sequence.
Changing the flip angle (θ) of
the final proton pulse (blue)
encodes the carbon signal
phases. For recording coupled
spectra, the decoupling during
detection should be
switched off

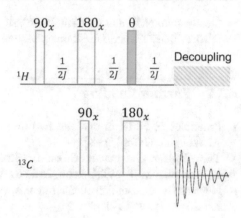

Fig. 4.18 Normalized signal
intensities of CH, CH_2, and
CH_3 carbons as a function of
the final proton pulse flip
angle

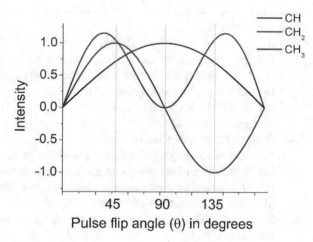

However, the DEPT experiment cannot be described using vector diagrams as was
done for the simplest two spin case of the INEPT scheme. Here, the magnetization
transfer pathway includes multiple-quantum transitions and their properties, and
such a complete analysis of the DEPT type of multipulse experiments requires
density matrix analysis of NMR phenomenon. These aspects will be introduced in
Chap. 5.

4.10 Summary

- The concept of magnetization transfer referred to as polarization transfer between
  nuclei in a given molecule is described. The nuclear Overhauser effect (NOE) is
  explained.
- The theoretical description in terms of population changes of the various energy
  levels is given.

- Steady-state NOE and transient NOE are presented.
- INEPT, INEPT$^+$, and DEPT pulse sequences are described.

4.11　Further Reading

- Principles of NMR in one and two dimensions, R. R. Ernst, G. Bodenhausen, A. Wokaun, Oxford, 1987
- The Nuclear Overhauser Effect in Structural and Conformational Analysis, D. Neuhaus, M. P. Williamson, 2nd ed., Wiley 2000
- Nuclear Overhauser Effect: Chemical Applications, J. H. Noggle, R. E. Schirmer, Academic Press, 1971.

4.12　Exercises

4.1. Polarization transfer between the nuclear spins is caused by
    (a)　J-coupling or dipolar coupling between the spins
    (b)　coupling with the magnetic field
    (c)　coupling with the RF
    (d)　none of the above

4.2. Positive NOE occurs because of
    (a)　domination of single-quantum transition probability
    (b)　domination of double-quantum transition probability
    (c)　domination of zero-quantum transition probability
    (d)　slow molecular motion

4.3. In an AX system while saturating the A spin, the polarization gain by the steady-state NOE is
    (a)　proportional to $\frac{\gamma_A}{\gamma_X}$
    (b)　proportional to $\frac{\gamma_X}{\gamma_A}$
    (c)　proportional to $\gamma_A \gamma_X$
    (d)　independent of γ_A and γ_X

4.4. In an AX system, the transfer of polarization from A spin to X spin in a selective population inversion (SPI) experiment requires
    (a)　selective inversion of the X spin transitions
    (b)　selective inversion of the A spin transitions
    (c)　selective inversion of one of the A spin transitions
    (d)　selective inversion of one of the X spin transitions

4.5. Polarization transfer in an SPI experiment results in
    A.　differential transfer of magnetization
    B.　net transfer of magnetization
    C.　positive and negative signals in the spectrum
    D.　all positive signals in the spectrum

Choose the correct answer.

(a) A and B

(b) B and D

(c) A and D

(d) A and C

4.6. The estimation of distance between spins A and X by a transient NOE experiment

(a) requires spin diffusion

(b) requires short mixing time

(c) requires large mixing time

(d) is independent of the mixing time

4.7. The spectral density function is

(a) power distribution as a function of frequency due to molecular motion

(b) dependent on the spectrometer frequency

(c) independent of the correlation time

(d) representation of the RF power

4.8. In an INEPT experiment,

(a) polarization is transferred from sensitive to insensitive nucleus

(b) polarization is transferred from insensitive to sensitive nucleus

(c) transfer efficiency is independent of the type of the nuclei

(d) there is no gain in the sensitivity

4.9. In a refocused INEPT experiment, for a C-H spin system with one bond coupling of 125 Hz, the refocusing delay should be

(a) 8 ms

(b) 4 ms

(c) 2 ms

(d) 1 ms

4.10. During the spin lock,

(a) the spins lose their identity and exchange magnetization

(b) spins precess with their characteristic frequencies

(c) individual spins do not interact with each other

(d) there is no exchange of magnetization

4.11. In C-H-refocused INEPT experiment, which of the following statement is correct?

(a) carbon and proton are decoupled during the spin echo elements.

(b) carbon and proton can be decoupled during the detection period.

(c) proton magnetization is detected.

(d) magnetization transfer occurs from carbon to proton.

4.12. The sensitivity enhancement in a NH-refocused INEPT experiment is by a factor of

(a) 4

(b) 10

(c) $\sqrt{1000}$

(d) 8

Density Matrix Description of NMR

5

Contents

5.1	Introduction	169
5.2	Density Matrix	170
5.3	Elements of Density Matrix	172
5.4	Time Evolution of Density Operator ρ	175
5.5	Matrix Representations of RF Pulses	179
5.6	Product Operator Formalism	184
	5.6.1 Basis Operator Sets	185
	5.6.2 Time Evolution of Cartesian Basis Operators	190
5.7	Summary	198
5.8	Further Reading	198
5.9	Exercises	198

Learning Objectives
- Fundamental aspects of spin dynamics
- Density matrix description
- Product operator formalism to understand NMR experiments

5.1 Introduction

In the Fourier transform NMR experiment described in Chap. 3, the data was collected after the application of an RF pulse to a spin system, which was in equilibrium. Thus, the information that is obtained is essentially steady-state information. Much more information about the spin system, energy level diagrams, cross-relaxation pathways, etc. can be obtained by monitoring transient effects following a perturbation to the spin system. For this, it is necessary to create a nonequilibrium

© The Author(s), under exclusive license to Springer Nature Switzerland AG 2022
R. V. Hosur, V. M. R. Kakita, *A Graduate Course in NMR Spectroscopy*,
https://doi.org/10.1007/978-3-030-88769-8_5

state prior to the application of the observe RF pulse so that the transient phenomena are reflected in the data that will be collected thereafter. Now, a nonequilibrium state of the spin system can be created in different ways, depending upon what the specific interest is. The design of proper experimental schemes suitable for the purpose in mind requires proper knowledge of the behavior of spins under the influence of various perturbing forces that may be used. Such an understanding can be most appropriately obtained by using the most fundamental "density matrix" formalism. It is impossible to obtain correct and predictable information from vector representations, as was done in the steady-state case. In this chapter, the focus will be on developing this "density matrix" formalism from the NMR point of view.

Sections 5.2, 5.3, 5.4, and 5.5 give a formal description of the theory of density matrix. This involves a fair amount of quantum mechanics and mathematical rigor. Section 5.6 onward, the product operator formalism provides a convenient tool for the evaluation of density matrices and density operators applicable to weakly coupled spin systems. Students who find the initial sections hard to grasp due to insufficient background can skip to Sect. 5.6 and continue to familiarize themselves with calculation of evolution of magnetization components through given pulse sequences. Chapter 6 makes use of these in an extensive manner.

5.2　Density Matrix

We have seen in Chap. 1 that the state of spin can be represented by a wave function which is of the form

$$\Psi(t) = \sum_m C_m(t) U_{m,I} \tag{5.1}$$

$U_{m, I}$ constitutes an orthonormal set of basis functions. We also know that in quantum mechanics, when we make a measurement of an observable of the spin system, we observe the time average or equivalently the ensemble average of its value, and this average value of the observable of the spin system is described by the *expectation value* of the corresponding operator. The expectation value of operator A is defined as

$$< A >=< \Psi \mid A \mid \Psi > \tag{5.2}$$

$$= \int \Psi^* A \ \Psi \ d\tau \tag{5.3}$$

For example, the expectation value of M_x, the operator for x-component of the magnetization, in terms of the functions $U_{m, I}$, is given by

$$< \Psi | M_x | \Psi >= \sum_m \sum_n C_m(t)^* C_n(t) < U_{m,I} | M_x | U_{n,I} > \tag{5.4}$$

Or, briefly,

$$< \Psi|M_x|\Psi >= \sum_m \sum_n C_m(t)^* C_n(t) < m|M_x|n > \qquad (5.5)$$

Since $<m|M_x|n>$ are constants, any variation of M_x results essentially from the changes in the coefficients. These products of coefficients, $C_m(t)^* C_n(t)$, can be conveniently arranged in the form of a matrix. It is useful to treat this matrix as made-up of matrix elements of a time-dependent operator, $P(t)$, operating on the basis set of functions.

$$< n|P(t)|m >= C_n(t)C_m(t)^* \qquad (5.6)$$

In this notation,

$$< \Psi|M_x|\Psi >= \sum_m \sum_n < n|P(t)|m >< m|M_x|n > \qquad (5.7)$$

Noting that, in general,

$$\sum_m |m >< m| = 1 \qquad (5.8)$$

Equation 5.7 reduces to

$$< \Psi|M_x|\Psi >= \sum_m \sum_n < n|P(t)M_x|n > \qquad (5.9)$$

$$= Tr\ \{PM_x\} \qquad (5.10)$$

In other words, the expectation value of M_x is given by the trace of the product of the matrix representations of P and M_x.

It is also easy to prove that P is a Hermitian operator:

$$< n|P(t)|m >=< m|P(t)|n>^* \qquad (5.11)$$

When we are dealing with an ensemble of spins, different spins will have different wave functions in the sense that the coefficients C_n's will be different for the individual spins. In such a case, one will have to take an ensemble average of these products to derive an average expectation value of the operator.

$$< \overline{M_x} >= \sum_m \sum_n \overline{C_m(t)^* C_n(t)} < m|M_x|n > \qquad (5.12)$$

The matrix formed by the ensemble averages of the products $C_m(t)^* C_n(t)$ is represented by another operator, ρ, which is defined as

$$< n\ |\ \rho(t)\ |\ m >= \overline{C_m(t)^* C_n(t)} \qquad (5.13)$$

This operator ρ is called as the "density operator."

5.3   Elements of Density Matrix

Matrix element ρ_{nm} is as given by Eq. 5.13.

$$\rho_{nm} = \overline{C_n(t)\, C_m(t)^*} \tag{5.14}$$

The coefficients C_n's are complex quantities, and hence an ensemble average can also be written as follows

$$\overline{C_n(t)\, C_m(t)^*} = \overline{\mid C_m \mid \mid C_n \mid e^{-i(\alpha_n - \alpha_m)}} \tag{5.15}$$

where the αs represent phases and the $\mid C_m \mid$ represents amplitudes.

At thermal equilibrium, by the hypothesis of random phases, all values of α in the range $0°$–$360°$ are equally probable, and hence the ensemble average vanishes for $m \neq n$, that is, all off-diagonal elements vanish. Nonvanishing of off-diagonal elements implies the existence of phase coherence between states. The diagonal elements $\mid C_m \mid^2$ represent the probabilities (populations) given by Boltzmann's distribution.

Thus,

$$\rho_{mn} = \frac{\delta_{mn} e^{\left(-\frac{E_n}{kT}\right)}}{Z} \tag{5.16}$$

where Z is the partition function given by

$$Z = \sum_{n=1}^{N} e^{-\frac{E_n}{kT}}$$

$$= \sum_{n=1}^{N} \left[1 - \frac{E_n}{kT} + \frac{1}{2!}\left(\frac{E_n}{kT}\right)^2 - \dots\dots\dots\dots \right] \tag{5.17}$$

where N is the number of states. Under high-temperature approximation, $\left(\frac{E_n}{kT}\right) \ll 1$, Z can be approximated ignoring the higher-order terms as

$$Z = \sum_{n=1}^{N} \left[1 - \frac{E_n}{kT} \right] \tag{5.18}$$

$$= \sum_{n=1}^{N} 1 - \frac{1}{kT} \sum_{n=1}^{N} E_n \tag{5.19}$$

For Zeeman interaction, $\sum_{n=1}^{N} E_n = 0$

Therefore,

$$Z = \sum_{n=1}^{N} 1 = N \tag{5.20}$$

If $\mathcal{H}$ is the Hamiltonian and $|i>$ are the eigenstates with eigenvalues λ_i

$$\mathcal{H}|i >= \lambda_i|i > \tag{5.21}$$

and

$$e^{\mathcal{H}}|i >= e^{\lambda_i}|i > \tag{5.22}$$

Therefore

$$<j \mid e^{\mathcal{H}}|i >=< j| e^{\lambda_i}|i >= \delta_{ij}e^{\lambda_i} \tag{5.23}$$

Thus, Eq. 5.16 can be rewritten as

$$\rho_{mn} = \frac{1}{N} < m \mid e^{-En/kT} \mid n >$$

$$= \frac{1}{N} < m \mid e^{-\frac{\mathcal{H}}{kT}} \mid n > \tag{5.24}$$

Thus,

$$\rho = \frac{1}{N} e^{-\frac{\mathcal{H}}{kT}} \tag{5.25}$$

Expanding this in power series, it becomes

$$\rho = \frac{1}{N} \left\{ 1 - \frac{\mathcal{H}}{kT} + \frac{1}{2!} \left(\frac{\mathcal{H}}{kT} \right)^2 - \cdots \cdots \cdots \right\} \tag{5.26}$$

Under high-temperature approximation

$$\rho = \frac{1}{N} \left\{ 1 - \frac{\mathcal{H}}{kT} \right\} \tag{5.27}$$

For one spin

$$\mathcal{H} = -\gamma \hbar H_0 I_z \tag{5.28}$$

For $I = 1/2$,

$$\rho = \frac{1}{2} \left(1 + \frac{\gamma \hbar H_0 I_z}{kT} \right) \tag{5.29}$$

Calculating the matrix element of the operator I_z, the matrix elements of ρ will be

$$\rho_{\alpha\alpha} = \frac{1}{2}\left(1 + \frac{\gamma\hbar H_0}{2kT}\right); \rho_{\beta\beta} = \frac{1}{2}\left(1 - \frac{\gamma\hbar H_0}{2kT}\right); \rho_{\alpha\beta} = \rho_{\beta\alpha} = 0 \qquad (5.30)$$

Thus,

$$\rho = \frac{1}{2}\begin{pmatrix} 1 & 0 \\ 0 & 1 \end{pmatrix} + \frac{\gamma\hbar H_0}{4kT}\begin{pmatrix} 1 & 0 \\ 0 & -1 \end{pmatrix} \qquad (5.31)$$

For multi-spin systems, the Hamiltonian will be

$$\mathcal{H} = \mathcal{H}_z + \mathcal{H}_J \qquad (5.32)$$

where $\mathcal{H}_z$ represents the Zeeman interaction and $\mathcal{H}_J$ represents the J-coupling interaction. Under high-field approximation, the contribution from $\mathcal{H}_J$ will be very small compared to that from $\mathcal{H}_z$, and then the J-coupling can be dropped for the evaluation of the elements of the density matrix.

Explicitly for the two-spin system AX,

$$\rho = \frac{1}{(2I+1)_A(2I+1)_X}\left(1 + \frac{\gamma\hbar H_0 I_z}{kT}\right) \qquad (5.33)$$

with $I_z = I_z(A) + I_z(X)$

The eigenstates of the spin system are $\alpha\alpha$, $\alpha\beta$, $\beta\alpha$, and $\beta\beta$. With these states, the matrix elements of ρ will be

$$\rho_{\alpha\alpha,\alpha\alpha} = \frac{1}{4}\left(1 + \frac{\gamma\hbar H_0}{kT}\right); \rho_{\beta\beta,\beta\beta} = \frac{1}{4}\left(1 - \frac{\gamma\hbar H_0}{kT}\right); \rho_{\alpha\beta,\alpha\beta} = \rho_{\beta\alpha,\beta\alpha} = \frac{1}{4} \qquad (5.34)$$

All the remaining elements will be zero.
Thus,

$$\rho = \frac{1}{4}\begin{bmatrix} 1 & 0 & 0 & 0 \\ 0 & 1 & 0 & 0 \\ 0 & 0 & 1 & 0 \\ 0 & 0 & 0 & 1 \end{bmatrix} + \frac{\gamma\hbar H_0}{4kT}\begin{bmatrix} 1 & 0 & 0 & 0 \\ 0 & 0 & 0 & 0 \\ 0 & 0 & 0 & 0 \\ 0 & 0 & 0 & -1 \end{bmatrix} \qquad (5.35)$$

In general,

$$\rho = \frac{1}{Z} + \left(\frac{K}{Z}\right)I_z \qquad (5.36)$$

where $K = \frac{\gamma\hbar H_0}{kT}$

5.4 Time Evolution of Density Operator ρ

An explicit understanding of the performance and characteristic features of an experiment is derived from the knowledge of the time evolution of the density operator through the experiment. To calculate this, we start from the relevant time-dependent Schrödinger equation:

$$\frac{-\hbar}{i} \frac{d\psi}{dt} = \mathcal{H}\psi \tag{5.37}$$

writing

$$\psi = \sum_n c_n(t) u_n \tag{5.38}$$

where $\{u_n\}$s constitute the orthonormal basis set of eigenstates.

Substituting Eq. 5.38 in Eq. 5.37, one obtains

$$\frac{-\hbar}{i} \sum_n \frac{dc_n(t)}{dt} u_n = \mathcal{H} \sum_n c_n(t) u_n \tag{5.39}$$

Taking the matrix elements with the state u_k, one obtains

$$\frac{-\hbar}{i} < k \left| \sum_n \frac{dc_n(t)}{dt} \right| n > = < k \mid \sum_n \mathcal{H} \mid c_n(t) \mid n > \tag{5.40}$$

$$= \sum_n c_n(t) < k|\mathcal{H}|n > \tag{5.41}$$

$$= \sum_n c_n(t) \mathcal{H}_{kn} \tag{5.42}$$

On the left-hand side, the only nonzero term will be $\frac{dc_k}{dt}$

Therefore,

$$\frac{-\hbar}{i} \frac{dc_k}{dt} = \sum_n c_n(t) \mathcal{H}_{kn} \tag{5.43}$$

Now,

$$\frac{d}{dt} < k|\rho|m > = \frac{d}{dt} \left(c_k c_m^* \right) \tag{5.44}$$

The ensemble average for the coefficients is implicit in this equation.

$$= c_k \frac{dc_m^*}{dt} + c_m^* \frac{dc_k}{dt} \tag{5.45}$$

From Eq. 5.43

$$\frac{dc_m^*}{dt} = \frac{i}{\hbar} \sum_n c_n^* \mathcal{H}_{nm} \qquad (5.46)$$

Thus, Eq. 5.45 reduces to

$$\frac{d}{dt} < k|\rho|m > = \frac{i}{\hbar} \sum_n c_k c_n^* \mathcal{H}_{nm} - \frac{i}{\hbar} \sum_n c_m^* c_n \mathcal{H}_{kn} \qquad (5.47)$$

$$= \frac{i}{\hbar} \sum_n \{< k|\rho|n >< n|\mathcal{H}|m > - < k|\mathcal{H}|n >< n|\rho|m >\} \qquad (5.48)$$

$$= \frac{i}{\hbar} < k|\rho\mathcal{H} - \mathcal{H}\rho|m > \qquad (5.49)$$

$$= \frac{i}{\hbar} < k | [\rho, \mathcal{H}] | m > \qquad (5.50)$$

Thus,

$$\frac{d\rho}{dt} = \frac{i}{\hbar} [\rho, \mathcal{H}] \qquad (5.51)$$

This is known as Liouville-von Neumann equation of motion for the density operator.

If the Hamiltonian is explicitly independent of time, then the solution of Eq. 5.51 is given as

$$\rho(t) = e^{-\frac{i}{\hbar}\mathcal{H}t} \rho(o) e^{\frac{i}{\hbar}\mathcal{H}t} \qquad (5.52)$$

This can be verified by explicit differentiation. Using Eq. 5.52, the off-diagonal elements of the density matrix can now be explicitly calculated.

$$< m|\rho(t)|n > = < m \left| e^{-\frac{i}{\hbar}\mathcal{H}t} \rho(o) e^{\frac{i}{\hbar}\mathcal{H}t} \right| n > \qquad (5.53)$$

$$\rho_{mn} = e^{\frac{i}{\hbar}(E_n - E_m)t} < m|\rho(o)|n > \qquad (5.54)$$

Substituting $E_m = h\nu_m$, $E_n = h\nu_n$, and $\omega_{mn} = 2\pi(\nu_m - \nu_n)$, we get

$$\rho_{mn} = e^{i\omega_{mn}t} < m|\rho(o)|n > \qquad (5.55)$$

Now, we also have from Eq. 5.14

$$< m|\rho|n > = \overline{c_m c_n^*} \qquad (5.56)$$

$$= \mid c_m \mid\mid c_n \mid e^{i(\alpha_m - \alpha_n)} \tag{5.57}$$

where the αs represent the phases and cs represent the amplitudes, which are independent of each other. Therefore, the no-vanishing of ρ_{mn} implies the existence of phase coherence between the spins in the states $|m>$ and $|n>$ in the ensemble. At thermal equilibrium, all phases occur with equal probability which implies that

$$\overline{c_m c_n^*} = 0 \tag{5.58}$$

Then comparing this with Eq. 5.54,

$$e^{\frac{i}{\hbar}(E_n - E_m)t} < m|\rho(o)|n> = 0 \tag{5.59}$$

Since the energy-dependent term which is oscillatory in time cannot be zero, it follows that

$$< m|\rho(o)|n> = 0 \tag{5.60}$$

Therefore, all off-diagonal elements of the density matrix vanish at all times. Any nonvanishing off-diagonal element implies a nonequilibrium state.

Summarizing, the density matrix in the most general case,

$$\rho = \begin{bmatrix} P_1 & c_{12}e^{i\omega_{12}t} & c_{13}e^{i\omega_{13}t} & \cdots & \cdots & \cdots & c_{1n}e^{i\omega_{1n}t} \\ c_{21}e^{i\omega_{21}t} & P_2 & c_{23}e^{i\omega_{23}t} & \cdots & \cdots & \cdots & c_{2n}e^{i\omega_{2n}t} \\ \vdots & \vdots & \vdots & \vdots & \vdots & \vdots & \vdots \\ \vdots & \vdots & \vdots & \vdots & \vdots & \vdots & \vdots \\ \vdots & \vdots & \vdots & \vdots & \vdots & \vdots & \vdots \\ \vdots & \vdots & \vdots & \vdots & \vdots & \vdots & \vdots \\ c_{n1}e^{i\omega_{n1}t} & c_{n2}e^{i\omega_{n2}t} & \cdots & \cdots & & \cdots & \cdots & P_n \end{bmatrix} \tag{5.61}$$

The measured signal in an NMR experiment is given by the expectation value of the relevant operator M_x, M_y, or $M_\pm$.

For example, for M_x,

$$< M_x > = Tr\,(\rho M_x) = Tr\,(M_x \rho) \tag{5.62}$$

For a single spin $\frac{1}{2}$ system, if ρ at the start of data collection has some phase coherence between the two-spin states α and β and the populations are not equilibrium populations, we can write

$$\rho(t) = \begin{bmatrix} P_1 & e^{i\omega_{12}t} \\ e^{-i\omega_{12}t} & P_2 \end{bmatrix} \tag{5.63}$$

Here we have assumed identical coefficients for the off-diagonal elements. Therefore,

$$Tr\left(M_x\rho\right) = Tr\left\{\frac{1}{2}\begin{bmatrix} 0 & 1 \\ 1 & 0 \end{bmatrix}\begin{bmatrix} P_1 & e^{i\omega_{12}t} \\ e^{-i\omega_{12}t} & P_2 \end{bmatrix}\right\} \tag{5.64}$$

$$= tr\left\{\frac{1}{2}\begin{bmatrix} e^{-i\omega_{12}t} & P_2 \\ P_1 & e^{i\omega_{12}t} \end{bmatrix}\right\} \tag{5.65}$$

$$= \cos\left(\omega_{12}t\right) \tag{5.66}$$

Including transverse relaxation, Eq. 5.66 will become

$$< M_x >= \cos\left(\omega_{12}t\right)e^{-t/T_2}$$

This oscillating function of time represents the frequency component of the time domain signal or the FID.

Extending to two spins,

$M_x = M_{1x} + M_{2x}$, and using the eigenstates $1 =| \alpha\alpha >, 2 =| \alpha\beta >, 3 =| \beta\alpha >,$ $4 =| \beta\beta >$

The matrix representation of M_x is

$$M_x = \frac{1}{2}\begin{bmatrix} 0 & 1 & 1 & 0 \\ 1 & 0 & 0 & 1 \\ 1 & 0 & 0 & 1 \\ 0 & 1 & 1 & 0 \end{bmatrix} \tag{5.67}$$

Assuming a nonequilibrium density operator of the form,

$$\rho(t) = \begin{bmatrix} P_1 & e^{i\omega_{12}t} & e^{i\omega_{13}t} & e^{i\omega_{14}t} \\ e^{-i\omega_{12}t} & P_2 & e^{i\omega_{23}t} & e^{i\omega_{24}t} \\ e^{-i\omega_{13}t} & e^{-i\omega_{23}t} & P_3 & e^{i\omega_{34}t} \\ e^{-i\omega_{14}t} & e^{-i\omega_{24}t} & e^{-i\omega_{34}t} & P_4 \end{bmatrix} \tag{5.68}$$

Here, ω_{12}, ω_{13}, ω_{24}, and ω_{34} represent the single-quantum coherences; ω_{14} and ω_{23} represent double-quantum and zero-quantum coherences, respectively.

The expectation value of M_x as per Eq. 5.62 is

$$< M_x >= Tr \, (\rho M_x)$$

$$= Tr \left\{ \frac{1}{2} \begin{bmatrix} 0 & 1 & 1 & 0 \\ 1 & 0 & 0 & 1 \\ 1 & 0 & 0 & 1 \\ 0 & 1 & 1 & 0 \end{bmatrix} \begin{bmatrix} P_1 & e^{i\omega_{12}t} & e^{i\omega_{13}t} & e^{i\omega_{14}t} \\ e^{-i\omega_{12}t} & P_2 & e^{i\omega_{23}t} & e^{i\omega_{24}t} \\ e^{-i\omega_{13}t} & e^{-i\omega_{23}t} & P_3 & e^{i\omega_{34}t} \\ e^{-i\omega_{14}t} & e^{-i\omega_{24}t} & e^{-i\omega_{34}t} & P_4 \end{bmatrix} \right\} \quad (5.69)$$

$$= \cos(\omega_{12}t) + \cos(\omega_{13}t) + \cos(\omega_{24}t) + \cos(\omega_{34}t) \quad (5.70)$$

Clearly, the off-diagonal elements representing single-quantum coherences are selected, and this constitutes the frequency component of the free induction decay— the detected signal. Of course, transverse relaxation causes decay of the signal. The double-quantum and zero-quantum coherences, even though they are present in the density operator, are not detected. These constitute a non-observable magnetization.

5.5 Matrix Representations of RF Pulses

We begin with the Liouville equation (5.51) with the Hamiltonian, including the radio frequency (RF) pulse explicitly:

$$\mathcal{H} = \mathcal{H}_0 + \mathcal{H}_1(t) \quad (5.71)$$

where $\mathcal{H}_0$ is the time-independent part of the Hamiltonian and $\mathcal{H}_1(t)$, which is time-dependent, represents the RF pulse.

Substituting Eq. 5.71 in Eq. 5.51, we get

$$\frac{d\rho}{dt} = \frac{i}{\hbar}[\rho, \mathcal{H}] = \frac{i}{\hbar}[\rho, \mathcal{H}_0 + \mathcal{H}_1(t)] \quad (5.72)$$

If $\mathcal{H}_1$ were nonexistent, the solution would have been

$$\rho(t) = e^{-\frac{i}{\hbar}\mathcal{H}_0 t}\rho(o)e^{\frac{i}{\hbar}\mathcal{H}_0 t} \quad (5.73)$$

Now we define and quantify ρ^* such that

$$\rho(t) = e^{-\frac{i}{\hbar}\mathcal{H}_0 t}\rho^*(t)e^{\frac{i}{\hbar}\mathcal{H}_0 t} \quad (5.74)$$

Such a solution satisfies the condition that at $t = 0$, ρ and ρ^* are identical. Differentiating equation (5.74) with respect to time, we get

$$\frac{d\rho}{dt} = -\frac{i}{\hbar}[\mathcal{H}_0, \rho] + e^{-\frac{i}{\hbar}\mathcal{H}_0 t}\frac{d\rho^*}{dt}e^{\frac{i}{\hbar}\mathcal{H}_0 t} \quad (5.75)$$

$$= \frac{i}{\hbar}[\rho, \mathcal{H}_0 + \mathcal{H}_1] \tag{5.76}$$

From this we get

$$\frac{d\rho^*}{dt} = \frac{i}{\hbar} e^{\frac{i}{\hbar}\mathcal{H}_0 t}[\rho, \mathcal{H}_1] e^{-\frac{i}{\hbar}\mathcal{H}_0 t} \tag{5.77}$$

$$= \frac{i}{\hbar} e^{\frac{i}{\hbar}\mathcal{H}_0 t}(\rho\mathcal{H}_1 - \mathcal{H}_1\rho)\, e^{-\frac{i}{\hbar}\mathcal{H}_0 t} \tag{5.78}$$

$$= \frac{i}{\hbar}\{ e^{\frac{i}{\hbar}\mathcal{H}_0 t}\rho e^{-\frac{i}{\hbar}\mathcal{H}_0 t} e^{\frac{i}{\hbar}\mathcal{H}_0 t}\mathcal{H}_1 e^{-\frac{i}{\hbar}\mathcal{H}_0 t} - e^{\frac{i}{\hbar}\mathcal{H}_0 t}\mathcal{H}_1 e^{-\frac{i}{\hbar}\mathcal{H}_0 t} e^{\frac{i}{\hbar}\mathcal{H}_0 t}\rho\, e^{-\frac{i}{\hbar}\mathcal{H}_0 t} \tag{5.79}$$

$$= \frac{i}{\hbar}[\rho^*, \mathcal{H}_1^*] \tag{5.80}$$

where

$$\mathcal{H}_1^* = e^{\frac{i}{\hbar}\mathcal{H}_0 t}\mathcal{H}_1 e^{-\frac{i}{\hbar}\mathcal{H}_0 t} \tag{5.81}$$

At $t = 0$, $\mathcal{H}_1^* = \mathcal{H}_1$

The transformation operator $e^{\frac{i}{\hbar}\mathcal{H}_0 t}$ represents the rotation about the static field axis and thus represents the transformation into the rotating frame. Such a representation is also called the interaction representation. Under resonance condition the evolution under $\mathcal{H}_0$ will be negligible. Thus, as we will show, during the high-power short-duration pulse, the Hamiltonian $\mathcal{H}_1^*$ will be identical to $\mathcal{H}_1$. Similarly, ρ^* will also become identical to ρ during the pulse.

We now calculate the matrix elements of $\mathcal{H}_1^*$:

$$< k|\mathcal{H}_1^*|m > = < k\,|\, e^{\frac{i}{\hbar}\mathcal{H}_0 t}\mathcal{H}_1 e^{-\frac{i}{\hbar}\mathcal{H}_0 t}\,|\, m > \tag{5.82}$$

$$= e^{\frac{i}{\hbar}(E_k - E_m)t} < k|\mathcal{H}_1|m > \tag{5.83}$$

If $\mathcal{H}_1 = \mathcal{H}_1(0)\, e^{-i\omega_{RF}t}$, which represents the RF pulse, then

$$< k|\mathcal{H}_1^*|m > = e^{\frac{i}{\hbar}(E_k - E_m - \hbar\omega_{RF})t} < k|\mathcal{H}_1(0)|m > \tag{5.84}$$

Now, $(E_k - E_m - \hbar\omega_{RF})$ is in the kHz range if "t" is in the μs range as in an RF pulse; the time-dependent term in (5.84) will be extremely slowly varying during the pulse and hence can be effectively considered to be constant. Thus, the matrix element $< k|\mathcal{H}_1^*|m >$ can be assumed to be independent of time; in fact, under resonance condition, $(E_k - E_m - \hbar\omega_{RF})$ will be zero, and there will be no time dependence at all. In other words, during the time of the pulse, $\mathcal{H}_1^*$ can be assumed to be time-independent and is equal to the amplitude of $\mathcal{H}_1$.

Under this condition, the solution of Eq. 5.80 can be written as

$$\rho^*(t) = e^{-\frac{i}{\hbar}\mathcal{H}_1 t}\rho^*(0)e^{\frac{i}{\hbar}\mathcal{H}_1 t} \tag{5.85}$$

And since $\rho^*(0) = \rho(0)$, Eq. 5.85 becomes

$$\rho^*(t) = e^{-\frac{i}{\hbar}\mathcal{H}_1 t}\rho(0)e^{\frac{i}{\hbar}\mathcal{H}_1 t} \tag{5.86a}$$

Following the discussion above, under resonance condition, note that for high-power pulse, resonance condition can be considered to be satisfied for all the frequencies in the spectrum at the same time; the effective field will be equal to the RF amplitude; the field along the z-axis will be zero; and thus evolution under the Hamiltonian $\mathcal{H}_0$ will be negligible. Thus, looking at Eq. 5.74, we can also replace $\rho^*(t)$ by $\rho(t)$ in Eq. 5.86a. Thus, the density operator transformation by the RF pulse can be described by

$$\rho(t) = e^{-\frac{i}{\hbar}\mathcal{H}_1 t}\rho(0)e^{\frac{i}{\hbar}\mathcal{H}_1 t} \tag{5.86b}$$

If the RF is applied along the x-axis,

$$\widehat{\mathcal{H}}_1 = \vec{\mu}.\vec{H}_1 = \gamma\hbar H_1\widehat{I}_x \tag{5.87}$$

The transformation operator $e^{-\frac{i}{\hbar}\mathcal{H}_1 t}$ thus becomes $e^{-i\beta\widehat{I}_x}$, where $\beta = \gamma H_1 t$ represents the rotation about the x-axis by angle β (flip-angle of the RF pulse). Thus, depending upon the length of the pulse, different rotation angles can be obtained.

For one spin, the I_q $(q = x, y, z)$ operator can be written as $\frac{1}{2}\sigma_q$, where σs are the Pauli spin matrices given as

$$\sigma_z = \begin{bmatrix} 1 & 0 \\ 0 & -1 \end{bmatrix}; \sigma_x = \begin{bmatrix} 0 & 1 \\ 1 & 0 \end{bmatrix}; \sigma_y = \begin{bmatrix} 0 & -i \\ i & 0 \end{bmatrix} \tag{5.88}$$

The Pauli matrices satisfy the condition:

$$\sigma_z^2 = \sigma_y^2 = \sigma_x^2 = 1 \tag{5.89}$$

Using this notation, the operator $e^{-i\beta\widehat{I}_x}$ can be expanded as a series:

$$e^{-i\beta\widehat{I}_x} = e^{-\frac{i\beta}{2}\sigma_x}$$

$$= 1 - \frac{i\beta}{2}\sigma_x + \frac{1}{2!}\left(\frac{i\beta}{2}\right)^2 - \frac{1}{3!}\left(\frac{i\beta}{2}\right)^3\sigma_x + \frac{1}{4!}\left(\frac{i\beta}{2}\right)^4$$

$$- \ldots\ldots\ldots\ldots \tag{5.90}$$

Regrouping the terms,

$$e^{-i\widehat{\beta I_x}} = \left(1 - \frac{1}{2!}\left(\frac{\beta}{2}\right)^2 + \frac{1}{4!}\left(\frac{\beta}{2}\right)^4 + \cdots\right) - i\left(\frac{\beta}{2} - \frac{1}{3!}\left(\frac{\beta}{2}\right)^3 - \cdots\right)\sigma_x \quad (5.91)$$

$$= \cos\left(\frac{\beta}{2}\right) - i\sigma_x \sin\left(\frac{\beta}{2}\right) \quad (5.92)$$

$$= \cos\left(\frac{\beta}{2}\right) - 2iI_x \sin\left(\frac{\beta}{2}\right)$$

Putting in matrix notation,

$$e^{-i\widehat{\beta I_x}} = \cos\left(\frac{\beta}{2}\right)\begin{bmatrix} 1 & 0 \\ 0 & 1 \end{bmatrix} - i\sin\left(\frac{\beta}{2}\right)\begin{bmatrix} 0 & 1 \\ 1 & 0 \end{bmatrix} \quad (5.93)$$

Thus, for one spin, a $90°$ $x-$pulse ($\beta = \frac{\pi}{2}$), the matrix representation becomes

$$R_x\left(\frac{\pi}{2}\right) = e^{-i\widehat{\frac{\pi}{2}I_x}} = \frac{1}{\sqrt{2}}\begin{bmatrix} 1 & -i \\ -i & 1 \end{bmatrix} \quad (5.94)$$

Similarly, for a $90°$ $y-$pulse, we get

$$R_y\left(\frac{\pi}{2}\right) = e^{-i\widehat{\frac{\pi}{2}I_y}} = \frac{1}{\sqrt{2}}\begin{bmatrix} 1 & -1 \\ 1 & 1 \end{bmatrix} \quad (5.95)$$

The matrices for π pulses turn out to be

$$R_x(\pi) = e^{-i\widehat{\pi I_x}} = \begin{bmatrix} 0 & -i \\ -i & 0 \end{bmatrix}; R_y(\pi) = e^{-i\widehat{\pi I_y}} = \begin{bmatrix} 0 & -1 \\ 1 & 0 \end{bmatrix} \quad (5.96)$$

The effect of these pulses on the density operator can be explicitly calculated using the matrix representations. For example, for a density operator represented by $\widehat{I_z}$, the transformation under $R_x\left(\frac{\pi}{2}\right)$ will be

$$\rho = R_x\left(\frac{\pi}{2}\right)\widehat{I_z}\,R_x^{-1}\left(\frac{\pi}{2}\right) \quad (5.97)$$

$$= \frac{1}{4}\begin{bmatrix} 1 & -i \\ -i & 1 \end{bmatrix}\begin{bmatrix} 1 & 0 \\ 0 & -1 \end{bmatrix}\begin{bmatrix} 1 & i \\ i & 1 \end{bmatrix} \quad (5.98)$$

$$= \frac{1}{2}\begin{bmatrix} 0 & i \\ -i & 0 \end{bmatrix} \quad (5.99)$$

$$= -\widehat{I_y} \quad (5.100)$$

So clearly, the z-magnetization is rotated onto the negative y-axis, when we apply a $\left(\frac{\pi}{2}\right)_x$ pulse.

Similarly, the transformation under $R_y\left(\frac{\pi}{2}\right)$ on density operator represented by $\widehat{I}_z$ is given in Box 5.1.

Box 5.1: Density Operator Transformation for the Effect of a $\left(\frac{\pi}{2}\right)_y$ Pulse on the $\widehat{I}_z$ Operator

For $R_y\left(\frac{\pi}{2}\right)$ pulse, the $\widehat{I}_z$ operator will transform as

$$\rho = R_y\left(\frac{\pi}{2}\right)\widehat{I}_z\, R_y^{-1}\left(\frac{\pi}{2}\right)$$

$$= \frac{1}{4}\begin{bmatrix} 1 & -1 \\ 1 & 1 \end{bmatrix}\begin{bmatrix} 1 & 0 \\ 0 & -1 \end{bmatrix}\begin{bmatrix} 1 & -1 \\ 1 & 1 \end{bmatrix} = \frac{1}{2}\begin{bmatrix} 0 & 1 \\ 1 & 0 \end{bmatrix}$$

$$= \widehat{I}_x$$

So clearly, the z-magnetization is rotated onto the positive x-axis, when we apply a $\left(\frac{\pi}{2}\right)_y$ pulse.

For a two-spin system, the matrix representations of the operators are calculated by direct products (Box 5.2).

$$R_x\left(\frac{\pi}{2}\right)(\text{non} - \text{selective}) = \frac{1}{2}\begin{bmatrix} 1 & -i \\ -i & 1 \end{bmatrix}\otimes\begin{bmatrix} 1 & -i \\ -i & 1 \end{bmatrix} \tag{5.101}$$

$$= \frac{1}{2}\begin{bmatrix} 1 & -i & -i & -1 \\ -i & 1 & -1 & -i \\ -i & -1 & 1 & -i \\ -1 & -i & -i & 1 \end{bmatrix} \tag{5.102}$$

Similarly,

$$R_y\left(\frac{\pi}{2}\right)(\text{non} - \text{selective}) = \frac{1}{2}\begin{bmatrix} 1 & -1 \\ 1 & 1 \end{bmatrix}\otimes\begin{bmatrix} 1 & -1 \\ 1 & 1 \end{bmatrix} \tag{5.103}$$

$$= \frac{1}{2} \begin{bmatrix} 1 & -1 & -1 & 1 \\ 1 & 1 & -1 & -1 \\ 1 & -1 & 1 & -1 \\ 1 & 1 & 1 & 1 \end{bmatrix} \tag{5.104}$$

Box 5.2: The Calculation of the Direct Product Between Two 2 × 2 Matrices

The direct product between two matrices P and Q can be represented as

$$P = \begin{bmatrix} a & b \\ c & d \end{bmatrix}, Q = \begin{bmatrix} A & B \\ C & D \end{bmatrix}$$

$$P \bigotimes Q = \begin{bmatrix} a & b \\ c & d \end{bmatrix} \bigotimes \begin{bmatrix} A & B \\ C & D \end{bmatrix}$$

$$= \begin{bmatrix} a \begin{bmatrix} A & B \\ C & D \end{bmatrix} & b \begin{bmatrix} A & B \\ C & D \end{bmatrix} \\ c \begin{bmatrix} A & B \\ C & D \end{bmatrix} & d \begin{bmatrix} A & B \\ C & D \end{bmatrix} \end{bmatrix}$$

Using these matrix representations for the pulses and the density operator, the evolution of the density operator through a multi-pulse experiment can be calculated.

5.6   Product Operator Formalism

In a generalized pulse sequence, as indicated in Fig. 5.1, the density operator evolution can be calculated as

$$\rho(t) = P_4 e^{-\frac{i}{\hbar}\mathcal{H}_3\tau_3} P_3 e^{-\frac{i}{\hbar}\mathcal{H}_2\tau_2} P_2 e^{-\frac{i}{\hbar}\mathcal{H}_1\tau_1} P_1 \rho(0) \, P_1^{-1} e^{\frac{i}{\hbar}\mathcal{H}_1\tau_1} P_2^{-1} e^{\frac{i}{\hbar}\mathcal{H}_2\tau_2} P_3^{-1} e^{\frac{i}{\hbar}\mathcal{H}_3\tau_3} P_4^{-1}$$

$$\tag{5.105}$$

This can be essentially broken into two types of transformations occurring successively.

$$\rho' = e^{-\frac{i}{\hbar}\mathcal{H}t} \rho \, e^{\frac{i}{\hbar}\mathcal{H}t} \text{ [for free evolution]} \tag{5.106}$$

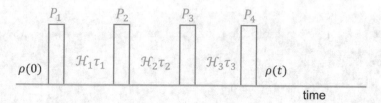

Fig. 5.1 A schematic of a multi-pulse sequence, which is used to calculate the density operator evolutions at different time points. P_s refer to the pulses, H_s refer to the Hamiltonians, and τ_s refer to the time for which the Hamiltonian is operative

and

$$\rho'' = P\rho P^{-1} \text{ [for pulses]} \tag{5.107}$$

To simplify this calculation, the product operator formalism has been developed for weakly coupled spin systems. The density operator is expressed as a linear combination of some basis operators, which constitute a complete set:

$$\rho(t) = \sum b_s(t) B_s \tag{5.108}$$

Thus,

$$\rho' = \sum b_s e^{-\frac{i}{\hbar}\mathcal{H}t} B_s \, e^{\frac{i}{\hbar}\mathcal{H}t} \text{ [for free evolution]} \tag{5.109}$$

$$\rho'' = \sum b_s P \, B_s \, P^{-1} \text{ [for pulses]} \tag{5.110}$$

In these two equations, ρ is the density operator at any particular instance in an experimental sequence. Thus, it is necessary to understand the transformational properties of individual B_s operators.

5.6.1 Basis Operator Sets

The basis operators can be defined in many ways: (i) Cartesian operators, (ii) single-element basis operators (polarization operators), and (iii) shift basis operators. The number of basis operator will depend on the number of coupled spins. For one spin, it will have four operators, which form a complete basis set. These are

Cartesian space; $\frac{E}{2}$, I_x, I_y, and I_z

Single-element operator space; I_α, I_β, I^+, I^-

Shift operator space; $\frac{E}{\sqrt{2}}$, I^+, I^-, I_0; $I_0 = \sqrt{2}I_z$

The corresponding matrix representations of various one-spin operators are given in Box 5.3.

Box 5.3: Matrix Representations of the Operators I_z, I_x, I_y, I^+, I^-, I_α, and I_β for the Case of One Spin $\frac{1}{2}$

$$I_z = \frac{1}{2}\begin{bmatrix} 1 & 0 \\ 0 & -1 \end{bmatrix}; I_x = \frac{1}{2}\begin{bmatrix} 0 & 1 \\ 1 & 0 \end{bmatrix}; I_y = \frac{1}{2}i\begin{bmatrix} 0 & -1 \\ 1 & 0 \end{bmatrix}; I^+ = \begin{bmatrix} 0 & 1 \\ 0 & 0 \end{bmatrix};$$

$$I^- = \begin{bmatrix} 0 & 0 \\ 1 & 0 \end{bmatrix}$$

$$I_\alpha = \begin{bmatrix} 1 & 0 \\ 0 & 0 \end{bmatrix}; I_\beta = \begin{bmatrix} 0 & 0 \\ 0 & 1 \end{bmatrix}$$

For n spins, in a coupled network, there will be 4^n elements in the basis operator sets. For example, for 2 spins, there will be a total of 16 operators. For the Cartesian space, these are

$$E$$

$$I_{1x}, I_{1y}, I_{1z}, I_{2x}, I_{2y}, I_{2z}$$

$$2I_{1x}I_{2x}, 2I_{1x}I_{2y}, 2I_{1x}I_{2z}$$

$$2I_{1y}I_{2x}, 2I_{1y}I_{2y}, 2I_{1y}I_{2z}$$

$$2I_{1z}I_{2x}, 2I_{1z}I_{2y}, 2I_{1z}I_{2z}$$

For three spins, labeled as AMQ, the Cartesian operator sets would be

$$E$$

$$I_{Ap}, I_{Mp}, I_{Qp} \quad p = x, y, z \quad \text{(a total of 9 operators)}$$

$$2I_{Ap}I_{Mr}, 2I_{Mp}I_{Qr}, 2I_{Ap}I_{Qr} \quad p, r = x, y, z \quad \text{(a total of 27 operators)}$$

$$4I_{Ap}I_{Mr}I_{Qs} \quad p, r, s = x, y, z \quad \text{(a total of 27 operators)}$$

Similar products can be written for other types of basis sets as well.

Matrix representations for all these operators can be derived, and these are explicitly listed in Table 5.1.

For one spin, the Cartesian space representations are

$$E = \frac{1}{2}\begin{bmatrix} 1 & 0 \\ 0 & 1 \end{bmatrix}; I_x = \frac{1}{2}\begin{bmatrix} 0 & 1 \\ 1 & 0 \end{bmatrix}; I_y = \frac{1}{2}\begin{bmatrix} 0 & -i \\ i & 0 \end{bmatrix}; I_z = \frac{1}{2}\begin{bmatrix} 1 & 0 \\ 0 & -1 \end{bmatrix} \quad (5.111)$$

For two spins, k and l,

$$E = \frac{1}{2}\begin{bmatrix} 1 & 0 \\ 0 & 1 \end{bmatrix} \otimes \begin{bmatrix} 1 & 0 \\ 0 & 1 \end{bmatrix}$$

Table 5.1 Matrix representations of product operators for a two-spin system

$I_E = \frac{1}{2}\begin{bmatrix}1&0&0&0\\0&1&0&0\\0&0&1&0\\0&0&0&1\end{bmatrix}$	$I_{kz} = \frac{1}{2}\begin{bmatrix}1&0&0&0\\0&1&0&0\\0&0&-1&0\\0&0&0&-1\end{bmatrix}$	$I_{lz} = \frac{1}{2}\begin{bmatrix}1&0&0&0\\0&-1&0&0\\0&0&1&0\\0&0&0&-1\end{bmatrix}$	$2I_{kz}I_{lz} = \frac{1}{2}\begin{bmatrix}1&0&0&0\\0&-1&0&0\\0&0&-1&0\\0&0&0&1\end{bmatrix}$
$I_{kx} = \frac{1}{2}\begin{bmatrix}0&0&1&0\\0&0&0&1\\1&0&0&0\\0&1&0&0\end{bmatrix}$	$I_{ky} = \frac{1}{2}\begin{bmatrix}0&0&-i&0\\0&0&0&-i\\i&0&0&0\\0&i&0&0\end{bmatrix}$	$2I_{kx}I_{lz} = \frac{1}{2}\begin{bmatrix}0&0&1&0\\0&0&0&-1\\1&0&0&0\\0&-1&0&0\end{bmatrix}$	$2I_{ky}I_{lz} = \frac{1}{2}\begin{bmatrix}0&0&-i&0\\0&0&0&i\\i&0&0&0\\0&-i&0&0\end{bmatrix}$
$I_{lx} = \frac{1}{2}\begin{bmatrix}0&1&0&0\\1&0&0&0\\0&0&0&1\\0&0&1&0\end{bmatrix}$	$I_{ly} = \frac{1}{2}\begin{bmatrix}0&-i&0&0\\i&0&0&0\\0&0&0&-i\\0&0&i&0\end{bmatrix}$	$2I_{kz}I_{lx} = \frac{1}{2}\begin{bmatrix}0&1&0&0\\1&0&0&0\\0&0&0&-1\\0&0&-1&0\end{bmatrix}$	$2I_{kz}I_{ly} = \frac{1}{2}\begin{bmatrix}0&-i&0&0\\i&0&0&0\\0&0&0&i\\0&0&-i&0\end{bmatrix}$
$2I_{kx}I_{lx} = \frac{1}{2}\begin{bmatrix}0&0&0&1\\0&0&1&0\\0&1&0&0\\1&0&0&0\end{bmatrix}$	$2I_{ky}I_{ly} = \frac{1}{2}\begin{bmatrix}0&0&0&-1\\0&0&1&0\\0&1&0&0\\-1&0&0&0\end{bmatrix}$	$2I_{kx}I_{ly} = \frac{1}{2}\begin{bmatrix}0&0&0&-i\\0&0&i&0\\0&-i&0&0\\i&0&0&0\end{bmatrix}$	$2I_{ky}I_{lx} = \frac{1}{2}\begin{bmatrix}0&0&0&-i\\0&0&-i&0\\0&i&0&0\\i&0&0&0\end{bmatrix}$

$$I_{kx} = \frac{1}{2} \begin{bmatrix} 0 & 1 \\ 1 & 0 \end{bmatrix} \otimes \begin{bmatrix} 1 & 0 \\ 0 & 1 \end{bmatrix}$$

$$I_{lx} = \frac{1}{2} \begin{bmatrix} 1 & 0 \\ 0 & 1 \end{bmatrix} \otimes \begin{bmatrix} 0 & 1 \\ 1 & 0 \end{bmatrix}$$

$$I_{ky} = \frac{1}{2} \begin{bmatrix} 0 & -i \\ i & 0 \end{bmatrix} \otimes \begin{bmatrix} 1 & 0 \\ 0 & 1 \end{bmatrix}$$

$$I_{ly} = \frac{1}{2} \begin{bmatrix} 1 & 0 \\ 0 & 1 \end{bmatrix} \otimes \begin{bmatrix} 0 & -i \\ i & 0 \end{bmatrix}$$

$$I_{kz} = \frac{1}{2} \begin{bmatrix} 1 & 0 \\ 0 & -1 \end{bmatrix} \otimes \begin{bmatrix} 1 & 0 \\ 0 & 1 \end{bmatrix}$$

$$I_{lz} = \frac{1}{2} \begin{bmatrix} 1 & 0 \\ 0 & 1 \end{bmatrix} \otimes \begin{bmatrix} 1 & 0 \\ 0 & -1 \end{bmatrix}$$

Similarly, for two spin products, for example, $2I_{kx}I_{ly}$, the matrix representation can be calculated as

$$2I_{kx}I_{ly} = \frac{1}{2} \begin{bmatrix} 0 & 1 \\ 1 & 0 \end{bmatrix} \otimes \begin{bmatrix} 0 & -i \\ i & 0 \end{bmatrix}$$

The complete list of matrix representations for two spins is given in Table 5.1. By examining the matrix representations, the following points become evident.

1. I_z operator represents the populations and the z-magnetizations.
2. I_x and I_y operators in a multi-spin system represent in-phase single-quantum coherences along the x- and y-axes, respectively.
3. $2I_{kx}I_{lz}$ and $2I_{ky}I_{lz}$ represent single-quantum coherences of k spin antiphase with respect to l along the x- and y-axes, respectively. Similar interpretations hold good for the l spin single-quantum coherences.
4. $2I_{kx}I_{ly}$, $2I_{ky}I_{lx}$, $2I_{kx}I_{lx}$, and $2I_{ky}I_{ly}$ represent mixtures of double-quantum and zero-quantum coherences, and suitable combinations of these represent pure double-quantum and single-quantum coherences.

$2I_{kx}I_{lx} + 2I_{ky}I_{ly}$ represents the x-component of zero-quantum coherence.
$2I_{kx}I_{ly} - 2I_{ky}I_{lx}$ represents the y-component of zero-quantum coherence.
$2I_{kx}I_{lx} - 2I_{ky}I_{ly}$ represents the x-component of double-quantum coherence
$2I_{kx}I_{ly} + 2I_{ky}I_{lx}$ represents the y-component of double-quantum coherence.

5. $2I_{kz}I_{lz}$ represents two-spin zz-order.

A pictorial representation of these coherences on the energy level diagram of a two-spin system is shown in Fig. 5.2.

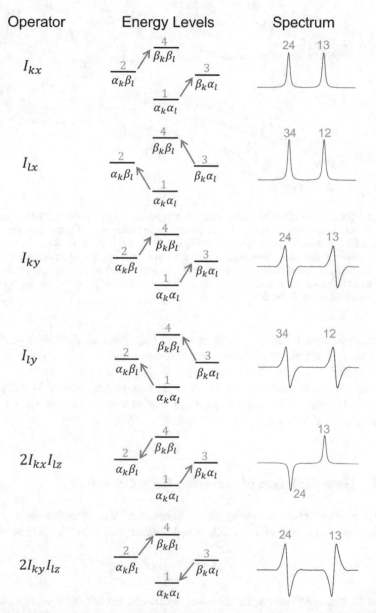

Fig. 5.2 Schematic drawings on the energy levels in a two-spin system (middle) to indicate the transitions represented by the individual operators on the left, and the corresponding spectra for different operators are shown on the right. Upward arrows indicate positive signals, and downward arrows indicate negative signals

Operator	Energy Levels	Spectrum

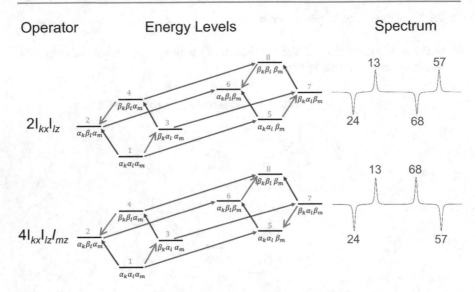

Fig. 5.3 Pictorial representations of the transitions represented by two-spin and three-spin product operators in a three-spin system (k, l, m) on the energy level diagram. In both cases, the operators represent the magnetization of k spin, and the spectrum on the right shows these four transitions. Upward arrows in the energy level diagram indicate positive signal, and downward arrows indicate negative signals. Different color codes are used to represent transitions belonging to the three spins. Note that arrows have been drawn for l and m spins as well for completeness, but the operators do not represent these transitions in any manner

Similar interpretations will hold good for two-spin and three-spin products in three-spin systems and other higher spin systems.

For example, a basis operator of type $4I_{Ax}I_{Mz}I_{Qz}$ represents a single-quantum coherence of A spin along the x-axis antiphase with respect to both M and Q spins.

Pictorial representations of a two-spin product in a three-spin system and a three-spin product in the three-spin system are shown in Fig. 5.3.

5.6.2   Time Evolution of Cartesian Basis Operators

5.6.2.1 Free Evolution Under the Influence of the Hamiltonian

The isotropic Hamiltonian for weakly coupled spin systems in liquids in units of $\hbar$ is

$$\mathcal{H} = \sum_k \omega_k I_{zk} + \sum_{k<l} 2\pi J_{kl} I_{zk} I_{zl} \qquad (5.112)$$

The first term represents the chemical shifts, and the second term represents the scalar couplings.

For a basis operator B_s, the evolution under the Hamiltonian is given by

$$B'_S = e^{-i\mathcal{H}t} B_s e^{i\mathcal{H}t} \tag{5.113}$$

$$= e^{-i\left(\sum_k \omega_k I_{zk} + \sum_{k<l} 2\pi J_{kl} I_{zk} I_{zl}\right)t} B_s e^{i\left(\sum_k \omega_k I_{zk} + \sum_{k<l} 2\pi J_{kl} I_{zk} I_{zl}\right)t} \tag{5.114}$$

Since the two parts of the Hamiltonian commute with each other, the terms in Eq. 5.114 can be shuffled without affecting the results.

$$B'_S = e^{-i\left(\sum_{k<l} 2\pi J_{kl} I_{zk} I_{zl}\right)t} \left(e^{-i\sum_k \omega_k I_{zk} t} B_s e^{i\sum_k \omega_k I_{zk} t} \right) e^{i\left(\sum_{k<l} 2\pi J_{kl} I_{zk} I_{zl}\right)t} \tag{5.115}$$

The central portion inside the bracket represents the evolution under chemical shift, and the outer terms represent the evolution under coupling. The two can be handled separately. One may also note that this order of evolutions can be interchanged because the two parts of the Hamiltonian commute with each other.

5.6.2.2 Chemical Shift Evolution

As an example, let us consider the evolution of the basis operator $B_s = I_{kx}$ representing the k spin magnetization.

So,

$$B'_S = e^{-i\omega_k I_{zk} t} I_{kx} e^{i\omega_k I_{zk} t} \tag{5.116}$$

From Eq. 5.92, this turns out to be

$$B'_S = \left\{ \cos\left(\frac{\omega_k t}{2}\right) - 2i\sin\left(\frac{\omega_k t}{2}\right) I_{kz} \right\} I_{kx} \left\{ \cos\left(\frac{\omega_k t}{2}\right) - 2i\sin\left(\frac{\omega_k t}{2}\right) I_{kz} \right\} \tag{5.117}$$

$$= \cos^2\left(\frac{\omega_k t}{2}\right) I_{kx} + 4\sin^2\left(\frac{\omega_k t}{2}\right) I_{kz} I_{kx} I_{kz} - i\sin(\omega_k t)[I_{kz}, I_{kx}] \tag{5.118}$$

The product $I_{kz} I_{kx} I_{kz}$ can be evaluated by individual matrix multiplication and turns out to be

$$I_{kz} I_{kx} I_{kz} = \frac{1}{8} \begin{bmatrix} 1 & 0 \\ 0 & -1 \end{bmatrix} \begin{bmatrix} 0 & 1 \\ 1 & 0 \end{bmatrix} \begin{bmatrix} 1 & 0 \\ 0 & -1 \end{bmatrix} = \frac{1}{8} \begin{bmatrix} 0 & -1 \\ -1 & 0 \end{bmatrix} = -\frac{1}{4} I_{kx} \tag{5.119}$$

Thus, Eq. 5.118 reduces to

$$B'_S = \cos^2\left(\frac{\omega_k t}{2}\right) I_{kx} - \sin^2\left(\frac{\omega_k t}{2}\right) I_{kx} + \sin(\omega_k t) I_{ky} \tag{5.120}$$

$$= \cos(\omega_k t) I_{kx} + \sin(\omega_k t) I_{ky} \tag{5.121}$$

5.6.2.3 Scalar Coupling Evolution

For the basis operator I_{kx}, the evolution can be written as

$$B_S'' = e^{-i2\pi J_{kl} I_{kz} I_{lz} t} I_{kx} \, e^{i2\pi J_{kl} I_{kz} I_{lz} t} \tag{5.122}$$

As shown in Box 5.4,

$$e^{-i2\pi J_{kl} I_{kz} I_{lz} t} = \cos\left(\frac{\pi J_{kl} t}{2}\right) - 4i \, \sin\left(\frac{\pi J_{kl} t}{2}\right) I_{kz} I_{lz} \tag{5.123}$$

Box 5.4: Explicit Derivation of Eq. 5.123

Let $I_{kz} I_{lz} = \frac{1}{4} A$ and $2\pi J_{kl} t = \beta$

$$A = \begin{bmatrix} 1 & 0 \\ 0 & -1 \end{bmatrix} \otimes \begin{bmatrix} 1 & 0 \\ 0 & -1 \end{bmatrix} = \begin{bmatrix} 1 & 0 & 0 & 0 \\ 0 & -1 & 0 & 0 \\ 0 & 0 & -1 & 0 \\ 0 & 0 & 0 & 1 \end{bmatrix}$$

$$A^2 = \begin{bmatrix} 1 & 0 & 0 & 0 \\ 0 & 1 & 0 & 0 \\ 0 & 0 & 1 & 0 \\ 0 & 0 & 0 & 1 \end{bmatrix}$$

$$e^{-i2\pi J_{kl}(A/4)t} = 1 - \frac{i\beta}{4} A + \left(\frac{i\beta}{4}\right)^2 \frac{A^2}{2!} - \left(\frac{i\beta}{4}\right)^3 \frac{A^3}{3!} + \cdots$$

$$= 1 - \frac{i\beta}{4} A + \left(\frac{i\beta}{4}\right)^2 \frac{1}{2!} - \left(\frac{i\beta}{4}\right)^3 \frac{A}{3!} + \cdots$$

$$= \cos\left(\frac{\pi J_{kl} t}{2}\right) - 4i \, \sin\left(\frac{\pi J_{kl} t}{2}\right) I_{kz} I_{lz}$$

Substituting Eq. 5.123 into Eq. 5.122,

$$B_S'' = \left\{ \cos\left(\frac{\pi J_{kl} t}{2}\right) - 4i \, \sin\left(\frac{\pi J_{kl} t}{2}\right) I_{kz} I_{lz} \right\} I_{kx} \left\{ \cos\left(\frac{\pi J_{kl} t}{2}\right) + 4i \, \sin\left(\frac{\pi J_{kl} t}{2}\right) I_{kz} I_{lz} \right\} \tag{5.124}$$

After some algebra (Box 5.5) similar to that in the calculation of shift evolution (5.119),

$$B_S'' = I_{kx} \cos(\pi J_{kl}t) + 2I_{ky}I_{lz} \sin(\pi J_{kl}t) \qquad (5.125)$$

Box 5.5: Explicit Derivation of Eq. 5.125

$$B_S'' = \left\{ \cos\left(\frac{\pi J_{kl}t}{2}\right) - 4i \sin\left(\frac{\pi J_{kl}t}{2}\right) I_{kz}I_{lz} \right\} I_{kx} \left\{ \cos\left(\frac{\pi J_{kl}t}{2}\right) + 4i \sin\left(\frac{\pi J_{kl}t}{2}\right) I_{kz}I_{lz} \right\}$$

$$= \left(\cos^2\left(\frac{\pi J_{kl}t}{2}\right) I_{kx} - \sin^2\left(\frac{\pi J_{kl}t}{2}\right) I_{kx} \right) - 2i \sin(\pi J_{kl}t)[I_{kz}, I_{kx}]I_{lz}$$

$$B_S'' = I_{kx} \cos(\pi J_{kl}t) + 2I_{ky}I_{lz} \sin(\pi J_{kl}t)$$

Similar calculations starting with other basis operators reveal that they form rotation groups, as indicated in Fig. 5.4. In Fig. 5.4a, operators I_x, I_y, and I_z form a group, which means they transform among themselves. For example, I_x and I_y interconvert under the influence of free evolution (I_z operator). In Fig. 5.4b, operator terms $2I_{kz}I_{lz}$, $2I_{ky}I_{lz}$, and I_{kx} form a rotation group under J-coupling evolution ($2I_{kz}I_{lz}$ operator). I_{kx} and $2I_{ky}I_{lz}$ interconvert among themselves under the influence of J-coupling evolution. Similarly, $2I_{kz}I_{lz}$, $2I_{kx}I_{lz}$, and I_{ky} form a rotation group under J-coupling evolution ($2I_{kz}I_{lz}$ operator). I_{ky} and $2I_{kx}I_{lz}$ interconvert among themselves under the influence of J-coupling evolution.

For example,

$$I_{kx} \xrightarrow{J-\text{coupling evolution}} I_{kx} \cos(\pi J_{kl}t) + 2I_{ky}I_{lz} \sin(\pi J_{kl}t)$$

$$I_{ky} \xrightarrow{J-\text{coupling evolution}} I_{ky} \cos(\pi J_{kl}t) - 2I_{kx}I_{lz} \sin(\pi J_{kl}t)$$

$$2I_{kx}I_{lz} \xrightarrow{J-\text{coupling evolution}} 2I_{kx}I_{lz} \cos(\pi J_{kl}t) + I_{ky} \sin(\pi J_{kl}t)$$

$$2I_{ky}I_{lz} \xrightarrow{J-\text{coupling evolution}} 2I_{ky}I_{lz} \cos(\pi J_{kl}t) - I_{kx} \sin(\pi J_{kl}t) \qquad (5.126)$$

5.6.2.4 Rotation by Pulses

This is represented by the transformation:

$$R_q B_s R_q^{-1} \quad q = x, y \qquad (5.127)$$

We describe here a few cases:

(i) $B_s = I_z$

(a)

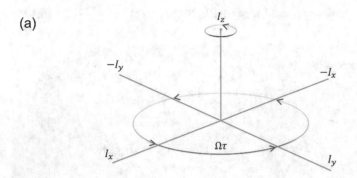

(b)

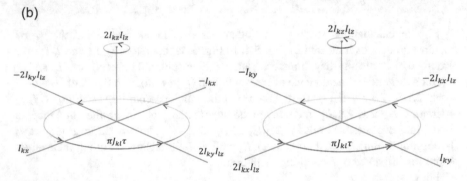

Fig. 5.4 (a) The free evolution of magnetization under Zeeman Hamiltonian (chemical shift evolution) and (b) scalar coupling evolutions. In either case, the Hamiltonian is represented along the z-axis, and the x- and y-axes represent the operators resulting from the respective evolutions. In each figure, the operators involved form rotation groups. See text for explicit transformations

For a 90_x pulse, the transformation will be

$$R_x\left(\frac{\pi}{2}\right) I_z R_x^{-1}\left(\frac{\pi}{2}\right) = \frac{1}{\sqrt{2}}\begin{bmatrix} 1 & -i \\ -i & 1 \end{bmatrix} \frac{1}{2}\begin{bmatrix} 1 & 0 \\ 0 & -1 \end{bmatrix} \frac{1}{\sqrt{2}}\begin{bmatrix} 1 & i \\ i & 1 \end{bmatrix} = \frac{1}{2}\begin{bmatrix} 0 & i \\ -i & 0 \end{bmatrix} = -I_y$$

Thus,

$$I_z \overset{90_x}{\rightarrow} -I_y \tag{5.128}$$

(ii) $B_s = I_y$

For a 90_x pulse, the transformation will be

$$R_x\left(\frac{\pi}{2}\right) I_y R_x^{-1}\left(\frac{\pi}{2}\right) = \frac{1}{\sqrt{2}}\begin{bmatrix} 1 & -i \\ -i & 1 \end{bmatrix} \frac{1}{2}\begin{bmatrix} 0 & -i \\ i & 0 \end{bmatrix} \frac{1}{\sqrt{2}}\begin{bmatrix} 1 & i \\ i & 1 \end{bmatrix} = \frac{1}{2}\begin{bmatrix} 1 & 0 \\ 0 & -1 \end{bmatrix} = I_z$$

Thus,

$$I_y \overset{90_x}{\rightarrow} I_z \qquad (5.129)$$

(iii) $B_s = I_x$

For a 90_x pulse, the transformation will be

$$R_x\left(\frac{\pi}{2}\right)I_x R_x^{-1}\left(\frac{\pi}{2}\right) = \frac{1}{\sqrt{2}}\begin{bmatrix} 1 & -i \\ -i & 1 \end{bmatrix}\frac{1}{2}\begin{bmatrix} 0 & 1 \\ 1 & 0 \end{bmatrix}\frac{1}{\sqrt{2}}\begin{bmatrix} 1 & i \\ i & 1 \end{bmatrix} = \frac{1}{2}\begin{bmatrix} 0 & 1 \\ 1 & 0 \end{bmatrix} = I_x$$

Thus, I_x is invariant under R_x pulse.

For multi-spin basis operators, the effects of pulses can be applied to individual spins.

For example,

$$2I_{kx}I_{lz} \overset{90_x(k)+90_x(l)}{\rightarrow} -2I_{kx}I_{ly} \qquad (5.130)$$

This represents the conversion of antiphase x-magnetization of k spin into a mixture of zero- and double-quantum coherences.

$$2I_{ky}I_{lz} \overset{90_x(k)+90_x(l)}{\rightarrow} -2I_{kz}I_{ly} \qquad (5.131)$$

This represents the conversion of antiphase y-magnetization of k spin into antiphase y-magnetization spin l. This is referred to as the coherence transfer from spin k to spin l. In general, it is seen that the application of RF pulses to antiphase magnetization in multi-spin systems causes coherence transfer among the spins. This forms the basis of many multi-pulse experiments in homo- and heteronuclear multi-spin systems.

The effects of various transformations under the influence of pulses are schematically shown in Fig. 5.5.

5.6.2.5 Calculation of the Spectrum of a J-Coupled Two-Spin System

In this section, we illustrate the calculation of the spectrum of a simple two-spin system, kl, in the standard FTNMR experiment (Fig. 5.6), using the product operator formalism.

To begin with the system is in equilibrium, and this is represented by the equilibrium density operator, ρ (see Eq. 5.36), which is proportional to I_z operator.

$$\rho \propto I_z = I_{kz} + I_{lz} \qquad (5.132)$$

This represents magnetization along the z-axis. On application of a 90_x pulse, the magnetization rotates to $-y$-axis (see Fig. 5.5).

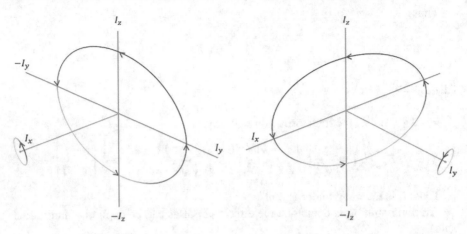

Fig. 5.5 The effect of $90°_x$ and $90°_y$ pulses on different magnetization components. The bigger circle indicates the rotation of magnetization components, while the smaller circle indicates the axis along which the pulse is applied

Fig. 5.6 One pulse FT-NMR experiment

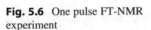

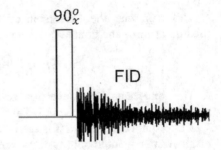

$$-I_y = -\left(I_{ky} + I_{ly}\right) \tag{5.133}$$

This will then evolve under chemical shift and J-coupling Hamiltonians. Both the spins evolve independently and can thus be treated independently. Considering the k spin, chemical shift evolution for time t leads to (ignoring the negative sign in the beginning) (see Fig. 5.4a)

$$I_{ky} \rightarrow I_{ky} \, \cos{(\omega_k t)} - I_{kx} \sin{(\omega_k t)} \tag{5.134}$$

Under J-coupling Hamiltonian, $2I_{kz}I_{lz}$ (see Fig. 5.4b and Eq. 5.126), the I_{ky} and I_{kx} operators evolve, leading to

$$\left\{ \left[\boldsymbol{I}_{ky} \cos \pi J_{kl}t - 2\boldsymbol{I}_{kx}\boldsymbol{I}_{lz} \sin \pi J_{kl}t \right] \cos \omega_k t - \left[\boldsymbol{I}_{kx} \cos \pi J_{kl}t + 2\boldsymbol{I}_{ky}\boldsymbol{I}_{lz} \sin \pi J_{kl}t \right] \sin \omega_k t \right\} \tag{5.135}$$

As discussed earlier only the first term and the third terms in Eq. 5.135 are observable and contribute to the spectrum. If we observe only the y-magnetization, then we need to consider the first term only. After taking the trace with I_{ky}, the signal (FID) will be represented by the time-dependent coefficients of this term. This is given by

$$\text{signal} = \cos(\omega_k t)\cos(\pi J_{kl}t) \tag{5.136}$$

Including transverse relaxation in the FID, the signal will be

$$\text{signal} = \cos(\omega_k t)\cos(\pi J_{kl}t)e^{-t/T_{2k}} \tag{5.137}$$

Substituting, $\omega_k = 2\pi v_k$

$$\text{signal} = \cos(2\pi v_k t)\cos(\pi J_{kl}t)e^{-t/T_{2k}} \tag{5.138}$$

$$\text{signal} = \frac{1}{2}\{\cos(2\pi v_k t + \pi J_{kl}t) + \cos(2\pi v_k t - \pi J_{kl}t)\}e^{-t/T_{2k}} \tag{5.139}$$

After the real (or cosine) Fourier transformation, this leads to absorptive spectral lines at $(v_k + \frac{J_{kl}}{2})$ and $(v_k - \frac{J_{kl}}{2})$.

Similarly, starting from the z-magnetization of the l spin, the final signal will be

$$\text{signal} = \cos(2\pi v_l t)\cos(\pi J_{kl}t)e^{-t/T_{2l}} \tag{5.140}$$

$$\text{signal} = \frac{1}{2}\{\cos(2\pi v_l t + \pi J_{kl}t) + \cos(2\pi v_l t - \pi J_{kl}t)\}e^{-t/T_{2l}} \tag{5.141}$$

Thus, for spin l, we will obtain absorptive signals at $(v_l + \frac{J_{kl}}{2})$ and $(v_l - \frac{J_{kl}}{2})$.
Thus, in the final spectrum (Fig. 5.7), we will get the doublets of k and l spins.

$$\textbf{Spectrallines}: \left(v_k + \frac{J_{kl}}{2}\right) \text{ and } \left(v_k + \frac{J_{kl}}{2}\right); \left(v_l + \frac{J_{kl}}{2}\right) \text{ and } \left(v_l - \frac{J_{kl}}{2}\right) \tag{5.142}$$

If we choose to observe the x-component of the signal in Eq. 5.135 and perform the same cosine transformation, we get the same four signals but with dispersive line shapes.

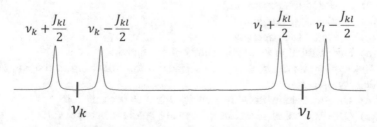

Fig. 5.7 A schematic of the J-coupled spectrum for a two-spin system, where v_k and v_l are the frequencies of k and l spins, respectively, and J_{kl} is the J-coupling between k and l spins

5.7   Summary

- The concept of density matrix description of NMR is described with some mathematical rigor.
- The product operator formalism which provides a simple and easy-to-handle description of density operator calculations for NMR pulse sequences is presented.
- A simple calculation for a two-spin system is presented as an illustration.

5.8   Further Reading

- Principles of Magnetic Resonance, C. P. Slichter, 3rd ed., Springer, 1990
- Principles of NMR in one and two dimensions, R. R. Ernst, G. Bodenhausen, A. Wokaun, Oxford, 1987
- Spin Dynamics, M. H. Levitt, 2nd ed., Wiley 2008
- Understanding NMR Spectroscopy, J. Keeler, Wiley, 2005
- Protein NMR Spectroscopy, J. Cavanagh, N. Skelton, W. Fairbrother, M. Rance, A, Palmer III, 2nd ed., Elsevier, 2006

5.9   Exercises

5.1 For a three-spin system ($I = 1/2$), the density operator has
   - (a) 9 elements
   - (b) 6 elements
   - (c) 64 elements
   - (d) 3 elements

5.2 If ρ is the density operator, the expectation value of M_x operator is given by
   - (a) $Tr(M_x)$
   - (b) $Tr(M_x)^2$
   - (c) $Tr(M_x\rho)$
   - (d) $Tr\{(M_x)^2\rho\}$

5.3 Equilibrium density operator
   - (a) is related to I_z operator
   - (b) is related to I_x operator
   - (c) is related to I_y operator
   - (d) has no relation to angular momentum operators

5.4 The hypothesis of random phases leads to the following in the equilibrium density operator.
   - (a) Diagonal elements in the density operator become zero.
   - (b) Off-diagonal elements in the density operator become zero.
   - (c) Both diagonal and off-diagonal elements become zero.
   - (d) It has no effect on the diagonal and off-diagonal elements of the density operator.

5.5 For a two-spin system ($I = 1/2$), which of the following is true?

(a) $I_z |\alpha\alpha> = |\alpha\alpha>$

(b) $I_x |\alpha\alpha> = |\alpha\alpha>$

(c) $I_y |\alpha\alpha> = |\alpha\alpha>$

(d) $I_z |\alpha\alpha> = |\alpha\beta>$

5.6 For a spin with $I = 5/2$, the partition function is

(a) 5/2

(b) 3/2

(c) 6

(d) 4

5.7 An FID arises from

(a) diagonal elements of a density operators

(b) single-quantum coherences in the density operators

(c) zero-quantum coherences in the density operators

(d) multiple-quantum coherences in the density operators

5.8 The off-diagonal elements of the density matrix represent

(a) the time evolution of isolated spins in the energy levels

(b) deviations from equilibrium populations

(c) the phase coherence of the spins in different energy levels

(d) the populations of the spins in individual energy levels

5.9 An RF pulse with a flip angle β applied along the x-axis is represented by

(a) βI_x

(b) $e^{-i\beta I_x}$

(c) $\beta(I_x)^2$

(d) $\beta^2(I_x)^2$

5.10 For a spin with precessional frequency ω_i, the field along z-axis in the rotating frame under resonance condition is

(a) H_0

(b) 0

(c) $\frac{\omega_i}{\gamma}$

(d) H_1

5.11 For a single spin ($I = 1/2$), the matrix representation of π pulse along the y-axis is given by

(a) $\begin{bmatrix} 0 & -1 \\ -i & 0 \end{bmatrix}$

(b) $\begin{bmatrix} 1 & -1 \\ -i & 1 \end{bmatrix}$

(c) $\begin{bmatrix} 1 & -1 \\ 1 & 1 \end{bmatrix}$

(d) $\begin{bmatrix} 0 & -1 \\ 1 & 0 \end{bmatrix}$

5.12 The basis operator $2I_{kx}I_{lz}$ represents
    (a) in-phase magnetization of l spin
    (b) x-magnetization of k spin anti phase with represent to l spin
    (c) in-phase magnetization of k spin
    (d) z-magnetization of l spin

5.13 $(2I_{kx}I_{ly} + 2I_{ky}I_{lx})$ represents
    (a) zero-quantum coherence of spin k and l
    (b) double-quantum coherence of spin k and l
    (c) mixture of double-quantum and zero-quantum coherences
    (d) total k spin magnetization

5.14 For a system of three spins ($I = 1/2$), the total number of basis operator is
    (a) 9
    (b) 27
    (c) 64
    (d) 81

5.15 In a three-spin system ($I = 1/2$), the operator term $I_{kx}I_{lz}I_{mz}$ represents
    (a) z-magnetization of l spin
    (b) z-magnetization of m spin
    (c) in-phase x-magnetization of k spin
    (d) x-magnetization of k spin antiphase to m and l spins

5.16 In a two-spin system k, l, the I_{kx} operator evolves under the J-coupling Hamiltonian for a time t to produce
    (a) y-magnetization of k spin
    (b) y-magnetization of k spin antiphase to l spin
    (c) x-magnetization of k spin antiphase to l spin
    (d) double-quantum coherence between k and l spin

5.17 Which combination of the operators form a rotation group?
    (a) I_{kx}, I_{ky}, $2I_{kx}I_{lz}$
    (b) I_{kx}, $2I_{ky}I_{lz}$, $2I_{kz}I_{lz}$
    (c) I_{kx}, $2I_{kx}I_{lz}$, $2I_{ky}I_{lz}$
    (d) I_{kx}, I_{lz}, $2I_{ky}I_{lz}$

5.18 The coherence transfer from k spin to l spin occurs due to
    (a) evolution under chemical shift
    (b) evolution under J-coupling
    (c) application of RF pulse along the y-axis to k spin
    (d) application of RF pulse to anti phase magnetization of k spin

5.19 An RF pulse applied along the x-axis causes
    (a) magnetization to align along the x-axis
    (b) rotation of the magnetization in the x-z plane
    (c) rotation of the magnetization in the y-z plane
    (d) rotation of the magnetization in the x-y plane

5.20 Which combination of the operators form a rotation group?
    (a) I_{kx}, I_{ky}, I_{kz}
    (b) I_{kx}, I_{ky}, $2I_{kz}I_{lz}$
    (c) I_{kz}, I_{lz}, I_{ky}
    (d) $2I_{kx}I_{lz}$, $2I_{ky}I_{lz}$, $2I_{kz}I_{lz}$

5.21 Which of the following statement is true?
   (a) $2I_{kx}I_{ly}$ represents a pure double-quantum coherence.
   (b) $2I_{kx}I_{lz}$ is an observable operator.
   (c) $2I_{ky}I_{lz}$ evolves under coupling to produce $2I_{kx}I_{lz}$.
   (d) I_{kx} and I_{ky} are observable operators.

5.22 A spin echo arises because of
   (a) refocusing of chemical shifts
   (b) refocusing of coupling constants
   (c) inhomogeneity in the main field
   (d) inaccuracy in RF pulses

5.23 In a spin echo experiment, refocusing of coupling evolution occurs when
   (a) the spin echo period is equal to $1/4J$
   (b) the spin echo period is equal to $1/2J$
   (c) the spin echo period is equal to $1/J$
   (d) the spin echo period is equal to $1/3J$

5.24 In the given pulse sequence, at the beginning of the detection, which of the
   following statement is true?

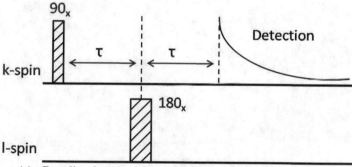

   (a) Coupling between k and l spins is effectively refocused.
   (b) Chemical shift evolution of l spin is refocused.
   (c) Chemical shift evolution of k spin is refocused.
   (d) Magnetization of k spin is inverted.

5.25 In a C-H INEPT experiment, magnetization is transferred from proton to
   carbon, which of the following operator transformation is valid?
   (a) $H_x \rightarrow H_z C_y$
   (b) $H_x \rightarrow H_x C_y$
   (c) $H_x \rightarrow H_z C_z$
   (d) $H_x \rightarrow H_x C_z$

5.26 Calculate the matrix representations of the operators, $2I_x S_y$ and $2I_z S_z$, in the
   eigenbasis of the weak coupling Hamiltonian.

5.27 Prove the commutator relationship: $\left[2I_\alpha S_{\alpha'}, 2I_\beta S_{\beta'}\right] = 0$, if $\alpha \neq \alpha'$ and $\beta \neq \beta'$
   simultaneously. α, α' and β, β' can be x, y, or z.

5.28 Calculate the effect of (a) $R_x(\pi)$ and (b) $R_y(\pi)$ pulses on the density operator
   represented by $\widehat{I}_z$ using matrix representations.

Multidimensional NMR Spectroscopy

6

Contents

6.1	Introduction	204
6.2	Two-Dimensional NMR	205
6.3	Two-Dimensional Fourier Transformation in NMR	207
6.4	Peak Shapes in Two-Dimensional Spectra	209
6.5	Quadrature Detection in Two-Dimensional NMR	211
6.6	Types of Two-Dimensional NMR Spectra	212
	6.6.1 Two-Dimensional Resolution/Separation Experiments	213
	6.6.2 Two-Dimensional Correlation Experiments	219
	6.6.3 Two-Dimensional Heteronuclear Correlation Experiments	238
	6.6.4 Combination of Mixing Sequences	248
6.7	Three-Dimensional NMR	249
	6.7.1 The CT-HNCA Experiment	252
	6.7.2 The HNN Experiment	257
	6.7.3 The Constant Time HN(CO)CA Experiment	263
	6.7.4 The HN(C)N Experiment	266
6.8	Summary	270
6.9	Further Reading	270
6.10	Exercises	271
Reference		276

The original version of the chapter has been revised. A Correction to this chapter can be found at
https://doi.org/10.1007/978-3-030-88769-8_8

Learning Objectives
- Introducing new dimensions in NMR
- Different types of two- and three-dimensional NMR spectra
- Benefits of multidimensional NMR spectra in terms of resolution enhancement and extractable information

6.1   Introduction

The most significant development in NMR after the discovery of FTNMR is undoubtedly multidimensional NMR spectroscopy, although one can say in the retrospective that FTNMR had already paved the way for its development. The essence of this statement in more explicit words is that multidimensional NMR exploits the fact that in FTNMR, the excitation of the spins and detection of their response are separated in time. The first ideas of extending the dimensionality of NMR to two from the conventional one-dimensional NMR was put forward by Jean Jeener in 1971. The technique has grown since then, in an explosive manner, and continues to develop unabatedly. The tremendous success of these experiments is due to the fact that they permit the display of pairwise interactions between spins in a given molecule in the form of cross-peaks in a plane. Quantitative interpretations of these correlations have revealed structural and dynamical information on such large molecules as proteins and nucleic acids—a hitherto unthinkable fact. With this, NMR entered the realm of biology, a subject with an ocean of unsolved problems both at macroscopic and microscopic levels.

This chapter begins by introducing the concepts in a pedagogic manner; progresses gradually in complexity and rigor, illustrating the explicit calculations in few cases; and quickly jumps into more complex experiments. In these complex experiments, used in biomolecular NMR or structural biology, explicit step-by-step calculations are not shown, but the final results which help to understand the performance of the experiments are presented. Certainly, the discussion is not exhaustive, but indicative. It will expose the students to the barrage of developments, so that those who would continue research in such advanced topics can pursue with the details at a later stage.

A generalized scheme of multidimensional NMR experiment is based on the idea of "segmentation of time axis," as shown in Fig. 6.1.

The experimental scheme begins with a "preparation period" during which the spin system is prepared in a suitable state. It can consist of a simple delay or a combination of pulses and delays or other kinds of perturbations as desired. For example, in a simple FTNMR experiment, the single-pulse excitation constitutes the preparation period during which x-y magnetization is created. A pair of 90° pulses separated by a constant evolution time constitutes the preparation period for multiple quantum spectroscopy, etc.

Fig. 6.1 Segmentation of the time axis

The "evolution periods" $t_1, t_2, \ldots t_n$ are variables and generate $(n+1)$ dimensional time domain data, which after $(n+1)$ dimensional Fourier transformation yields the $(n+1)$ dimensional spectrum. The evolution periods help to frequency label the individual spins or group of spins with their characteristic single-quantum or multiple-quantum frequencies. Various types of manipulations with the frequencies are possible during these periods.

Mns constitute the so-called mixing periods, the most important part of the experimental scheme. It is the "mixing" which establishes correlations between frequencies in adjacent evolution periods. Different kinds of correlations can be established by exploiting different types of interactions between the spins. The most common types of interactions exploited are J-coupling interactions and through-space dipolar interactions. Hundreds of pulse sequences have been published till date. In the following sections, we shall discuss at length the principle and developments in two-dimensional (2D) NMR, which laid the foundation for higher-dimensional experiments for specific purposes.

6.2 Two-Dimensional NMR

The details of performing a two-dimensional NMR experiment are shown schematically in Fig. 6.2.

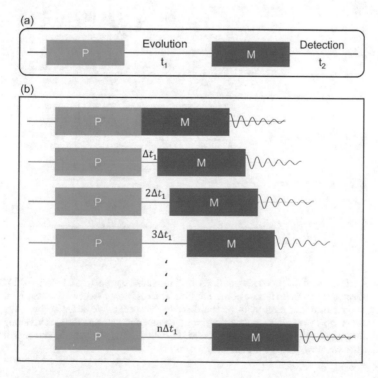

Fig. 6.2 (a) The segmentation of the time axis for the two-dimensional experiment. (b) Details of the experiment showing systematic incrementation of the evolution period, t_1

P and M are the preparation and mixing periods, respectively, and t_1 and t_2 are the evolution and detection periods, respectively. The experiment involves a collection of a number of free induction decays for systematically incremented values of t_1. The final data set will thus be a matrix, $S(t_1, t_2)$. Fourier transformation, with respect to t_2, results in a series of one-dimensional spectra in which the amplitudes and phases of the signals depend upon the value of t_1. Variations of these entities as a function of time carry the frequency information present during the t_1 period (Fig. 6.3), and thus Fourier transformation of these data along t_1 results in a two-dimensional frequency domain spectrum, $S(F_1, F_2)$.

$$S(t_1, t_2) \xrightarrow{\text{FT}} S(F_1, F_2) \tag{6.1}$$

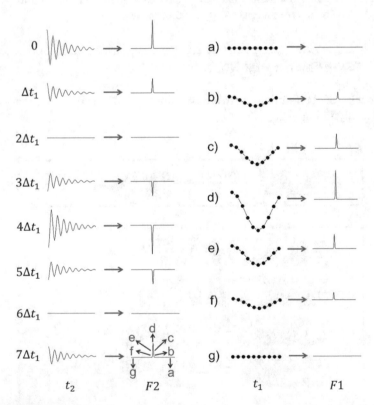

Fig. 6.3 A schematic of the processing of the two-dimensional data, $S(t_1, t_2)$. Individual FIDs (first column) collected for different t_1 time points are Fourier transformed (second column, $F2$ spectra). By taking the intensities at each point on the discrete $F2$ spectra, we arrive at the t_1-dependent profiles (column 3). These FIDs are then Fourier transformed to generate the spectra along the $F1$ axis. For illustration, only one line is considered. In both $F2$ and $F1$ spectra, one sees intensity modulations as we move through the frequency axes

Fig. 6.4 A schematic of a two-dimensional spectrum considering two spins k and l. ω_k and ω_l represent the resonance frequencies of the two spins

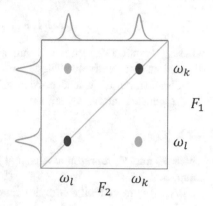

Consider a spin k whose x-y magnetization has been created by the preparation period, and this magnetization precesses with a frequency ω_k during the evolution period. At the end of the period t_1 the magnetization has components $M_k(0) \cos \omega_k t_1$ and $M_k(0) \sin \omega_k t_1$ along, say the y- and x-axes, respectively; $M_k(0)$ is the magnetization at the beginning of the evolution period. Let us now assume that the mixing period converts one of the above components (say $M_k(0) \sin \omega_k t_1$) into unobservable magnetization such as z-magnetization or multiple quantum coherence. From the remaining, part of the magnetization is transferred to say spin l, which has a characteristic frequency ω_l. The detected signal, as a function of t_2, will then have two contributions.

$$A = a\, M_k(0)\, \cos\, \omega_k t_1\, .\cos\, \omega_k t_2 \tag{6.2}$$

$$B = b\, M_k(0)\, \cos\, \omega_k t_1\, .\cos\, \omega_l t_2 \tag{6.3}$$

Here it is assumed that only the y-component of the magnetization is detected during t_2 period; a and b are some coefficients representing the relative contributions. These equations represent, of course, oversimplification made to convey the concepts clearly, and we will return to rigorous calculations later on. The component A, after two-dimensional Fourier transformation, results in a peak which has the same frequency ω_k along both $F1$ and $F2$ axis. The component B, results in a peak which has frequency ω_k along $F1$ and frequency ω_l along $F2$. The former is called the "diagonal peak" and the latter the "cross-peak" in the two-dimensional spectrum. A similar description applies to the magnetization originating from the l spin. A schematic of the resultant two-dimensional spectrum is shown in Fig. 6.4.

6.3   Two-Dimensional Fourier Transformation in NMR

A two-dimensional frequency spectrum, $S(F1, F2)$, will be generated from a two-dimensional time domain data set, $S(t_1, t_2)$, by the two-dimensional Fourier transformation. This is mathematically represented as

$$S(F1, F2) = \mathcal{F}^{(1)} \mathcal{F}^{(2)} S(t_1, t_2) \tag{6.4}$$

where $\mathcal{F}^{(1)}$ and $\mathcal{F}^{(2)}$ represent Fourier transformation operators along the t_1 and t_2 dimensions, respectively. These have to be carried out independently. Clearly, two-dimensional FT is a succession of one-dimensional FT.

$\mathcal{F}$ in general can be written as

$$\mathcal{F} = \mathcal{F}_c - i\mathcal{F}_s \tag{6.5}$$

where $\mathcal{F}_c$ and $\mathcal{F}_s$ represent cosine and sine transforms, respectively, as discussed in Chap. 3.

$S(t_1, t_2)$ is in general a complex function represented as

$$S(t_1, t_2) = \text{Re } \{S(t_1, t_2) + i \text{ Im } S(t_1, t_2)\} \tag{6.6}$$

$$= S_r(t_1, t_2) + i\, S_i(t_1, t_2) \tag{6.7}$$

Similarly,

$$S(F1, F2) = S_r(F_1, F_2) + i\, S_i(F_1, F_2) \tag{6.8}$$

Further,

$$S(F1, F2) = \left(\mathcal{F}_c{}^1 - i\mathcal{F}_s{}^1\right)\left(\mathcal{F}_c{}^2 - i\mathcal{F}_s{}^2\right)\{S_r(t_1, t_2) + i\, S_i(t_1, t_2)\} \tag{6.9}$$

From this, it follows

$$S_r(F_1, F_2) = \mathcal{F}^{cc}\{S_r(t_1, t_2)\} - \mathcal{F}^{ss}\{S_r(t_1, t_2)\} + \mathcal{F}^{cs}\{S_i(t_1, t_2)\} \\ + \mathcal{F}^{sc}\{S_i(t_1, t_2)\} \tag{6.10}$$

$$S_i(F_1, F_2) = \mathcal{F}^{cc}\{S_i(t_1, t_2)\} - \mathcal{F}^{ss}\{S_i(t_1, t_2)\} - \mathcal{F}^{cs}\{S_r(t_1, t_2)\} \\ - \mathcal{F}^{sc}\{S_r(t_1, t_2)\} \tag{6.11}$$

where

$$\mathcal{F}^{cc}\{S_r(t_1, t_2)\} = \int\limits_{-\infty}^{+\infty} dt_1 \cos \omega_1 t_1 \int\limits_{-\infty}^{+\infty} dt_2 \cos \omega_2 t_2\ S_r(t_1, t_2) \tag{6.12}$$

$$\mathcal{F}^{ss}\{S_r(t_1, t_2)\} = \int\limits_{-\infty}^{+\infty} dt_1 \sin \omega_1 t_1 \int\limits_{-\infty}^{+\infty} dt_2 \sin \omega_2 t_2\ S_r(t_1, t_2) \tag{6.13}$$

$$\mathcal{F}^{cs}\{S_r(t_1, t_2)\} = \int\limits_{-\infty}^{+\infty} dt_1 \cos\omega_1 t_1 \int\limits_{-\infty}^{+\infty} dt_2 \sin\omega_2 t_2 \ S_r(t_1, t_2) \qquad (6.14)$$

$$\mathcal{F}^{sc}\{S_r(t_1, t_2)\} = \int\limits_{-\infty}^{+\infty} dt_1 \sin\omega_1 t_1 \int\limits_{-\infty}^{+\infty} dt_2 \cos\omega_2 t_2 \ S_r(t_1, t_2) \qquad (6.15)$$

Similar equations hold good for $S_i(t_1, t_2)$ as well.

Since for t_1 and $t_2 < 0$, there is no signal, the transformations will have to be considered only for the range $0 < t < \infty$.

The general principles of Fourier transformation discussed in Chap. 3 are applicable here as well, along both axes, $F1$ and $F2$ of the two-dimensional spectrum. Sensitivity and resolutions along the two axes are governed by the same considerations of sampling rate, acquisition time, data size, zero filling, window multiplications, etc. The acquisition times along the t_1 and t_2 directions are generally represented as $t_1{}^{max}$ and $t_2{}^{max}$, respectively. While increasing $t_2{}^{max}$ can be simply accomplished by increasing the size of the FID data, increasing $t_1{}^{max}$ amounts to collecting more number of t_1 increments, and this contributes dearly to the total experimental time. Thus, for two-dimensional experiments, it is very essential to optimize the number of t_1 increments and the data need be collected only until that value of t_1 where the signal is present appreciably. The data is not actually collected during t_1, and this decision will have to be taken by calculation, by comparing $t_1{}^{max}$ value with the T_2 of the spin system, roughly observable from one-dimensional FIDs. Typically, $t_1{}^{max}$ is limited to 50–150 ms range depending upon the type of the experiment.

6.4 Peak Shapes in Two-Dimensional Spectra

The time domain signal ($S(t_1, t_2)$) is a superposition of many coherences. Considering a particular combination of coherences between levels $t \rightarrow u$ in t_1 domain and $r \rightarrow s$ in t_2 domain, the time domain signal for this pair will be

$$S_{tu,\ rs}(t_1, t_2) = S_{tu,rs}(0,0)e^{\{(-i\omega_{tu}-\lambda_{tu})t_1\}}e^{\{(-i\omega_{rs}-\lambda_{rs})t_2\}} \qquad (6.16)$$

where the λs represent the T_2 relaxation rates for the respective coherences.

Define

$$Z_{tu,rs} = S_{tu,rs}(0,0) \qquad (6.17)$$

Then,

$$S_{tu,rs}(\omega_1, \omega_2) = Z_{tu,rs} \left\{ \frac{1}{i\Delta\omega_{tu} + \lambda_{tu}} \right\} \left\{ \frac{1}{i\Delta\omega_{rs} + \lambda_{rs}} \right\} \tag{6.18}$$

where

$$\Delta\omega_{tu} = \omega_1 + \omega_{tu}, \Delta\omega_{rs} = \omega_2 + \omega_{rs}. \tag{6.19}$$

Equation 6.18 can be rewritten as

$$S_{tu,rs}(\omega_1, \omega_2) = Z_{tu,rs} \left\{ \frac{\lambda_{tu}}{(\Delta\omega_{tu})^2 + (\lambda_{tu})^2} - \frac{i\Delta\omega_{tu}}{(\Delta\omega_{tu})^2 + (\lambda_{tu})^2} \right\}$$

$$\times \left\{ \frac{\lambda_{rs}}{(\Delta\omega_{rs})^2 + (\lambda_{rs})^2} - \frac{i\Delta\omega_{rs}}{(\Delta\omega_{rs})^2 + (\lambda_{rs})^2} \right\} \tag{6.20}$$

In each of the angular brackets, the first term which is real represents an absorptive line shape (A), and the second term which is imaginary represents a dispersive line shape (D).

Thus,

$$S_{tu,rs}(F1, F2) = Z_{tu,rs}\{A_{tu}(F_1) - iD_{tu}(F_1)\}\{A_{rs}(F_2) - iD_{rs}(F_2)\} \tag{6.21}$$

$$= Z_{tu,rs}\{A_{rs}A_{tu} - D_{rs}D_{tu}\} - i\{D_{tu}A_{rs} + A_{tu}D_{rs}\} \tag{6.22}$$

This indicates that both the real and imaginary parts of the spectrum have mixed phases, along both the frequency axes. Figure 6.5 shows the appearances of the peaks for different peaks shapes along the $F1$ and $F2$ axes. Absorptive peak shapes produce the highest resolution in the spectra and thus are preferred.

The time domain signal can be classified into two categories:

(1) The evolution in t_1 modulates the phase of the detected signal (e.g., $e^{i\omega_{tu}t_1} \cdot f(t_2)$). This is called phase modulation.
(2) The evolution in t_1 modulates the amplitude of the detected signal (e.g., $\cos\omega_{tu}t_1 \cdot f(t_2)$). This is called amplitude modulation.

Several methods have been designed to obtain pure absorptive spectra, and the most common is to perform real Fourier transformation with respect to t_1. We show how absorptive lines can be obtained when the detected signal is amplitude modulated by evolution during t_1.

Consider

$$S_{tu,rs}(t_1, t_2) = \cos \omega_{tu}t_1 \cdot e^{-i\omega_{rs}t_2} \cdot e^{-\lambda_{tu}t_1 - \lambda_{rs}t_2} \tag{6.23}$$

Real cosine Fourier transformation of this data is given by

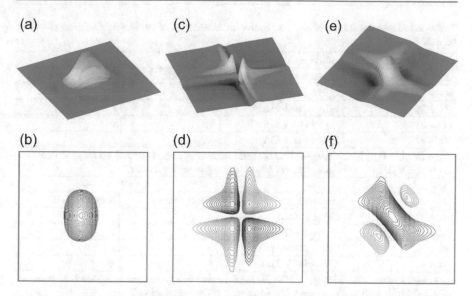

Fig. 6.5 Peak shapes in the two-dimensional spectra. Top row shows stacked plot representations, while the bottom row represents the corresponding contour representations of the same peak shapes. The left peak represents absorptive shape along both the frequency axes. The central picture represents dispersive line shapes along both the frequency axes, and the right picture represents absorptive along $F2$ and dispersive along $F1$ axes; the reverse is also possible. Such mixed line shapes are referred to as "mixed phases"

$$S_{tu,rs}(\omega_1, \omega_2) = \int_0^\infty \int_0^\infty S_{tu,rs}(t_1, t_2) \cos \omega_1 t_1 . e^{-i\omega_2 t_2} dt_1 dt_2 \qquad (6.24)$$

This leads to

$$S_{tu,rs}(\omega_1, \omega_2) = \frac{1}{2} \{A_{tu}(\omega_1) + A_{tu}(-\omega_1)\}\{A_{rs}(\omega_2) - iD_{rs}(\omega_2)\} \qquad (6.25)$$

If the real part of the spectrum along ω_2 is selected, one can obtain pure absorptive peak along both frequency axes.

This, however, results in the duplication of peaks at $\pm\omega_{tu}$, which is artificial. So, this can be avoided by doing quadrature detection along the t_1 axis, as discussed in the next section.

6.5 Quadrature Detection in Two-Dimensional NMR

Here we need to consider how positive and negative frequencies can be discriminated in both $F1$ and $F2$ dimensions of the two-dimensional experiment. As far as $F2$ dimension is concerned, it comes from the detected signal during the t_2 time period, and the procedures described in Chap. 3 are applicable here as well. However, along the F1 dimension, there is a difficulty because the data is not

Table 6.1 Protocols for data collection in the three methods of quadrature detection along the F1 axis

(a) TPPI (time proportional phase incrementation)

Experiment no.	Increment	Pulse phase	Receiver phase
$(4k + 1)$	$t_1(0) + (4k)\Delta$	x	x
$(4k + 2)$	$t_1(0) + (4k + 1)\Delta$	y	x
$(4k + 3)$	$t_1(0) + (4k + 2)\Delta$	$-x$	x
$(4k + 4)$	$t_1(0) + (4k + 3)\Delta$	$-y$	x

The index $k = 0, 1, 2,\ldots, (N/4)-1$; N is the total number of experiments along the t_1 dimension; $\Delta = 1/(2SW_1)$; $t_1(0) = $ ideally zero, but practically a few microseconds.

(b) States

Experiment no.	Increment	Pulse phase	Receiver phase
$(4k + 1)$	$t_1(0) + (4k)2\Delta$	x	x
$(4k + 2)$	$t_1(0) + (4k)2\Delta$	y	x
$(4k + 3)$	$t_1(0) + (4k + 1)2\Delta$	x	x
$(4k + 4)$	$t_1(0) + (4k + 1)2\Delta$	y	x

The index $k = 0, 1, 2,\ldots, (N/4)-1$; N is the total number of experiments along the t_1 dimension; $\Delta = 1/(2SW_1)$; $t_1(0) = $ ideally zero, but practically a few microseconds.

(c) States-TPPI

Experiment no.	Increment	Pulse phase	Receiver phase
$(4k + 1)$	$t_1(0) + (4k)2\Delta$	x	x
$(4k + 2)$	$t_1(0) + (4k)2\Delta$	y	x
$(4k + 3)$	$t_1(0) + (4k + 1)2\Delta$	$-x$	$-x$
$(4k + 4)$	$t_1(0) + (4k + 1)2\Delta$	$-y$	$-x$

The index $k = 0, 1, 2,\ldots, (N/4)-1$; N is the total number of experiments along the t_1 dimension; $\Delta = 1/(2SW_1)$; $t_1(0) = $ ideally zero, but practically a few microseconds.

actually collected during the "t_1" period. Different strategies are adopted for this purpose, by manipulating the way the increments in t_1 are adjusted along with the receiver phases. There are three methods which are known to achieve this, and these are described in Table. 6.1 (Cavanagh page 323, in this table, the pulse phase refers to the phase of the pulse immediately prior to the t_1 evolution period). For the TPPI method, the increment Δt_1 is half of that in STATES and TPPI-STATES methods.

6.6   Types of Two-Dimensional NMR Spectra

The known two-dimensional NMR experiments can be broadly classified into three categories:

(i) Resolution/separation experiments
(ii) Correlation experiments
(iii) Multiple-quantum experiments

Hybrid experiments have also been devised which use some of the ideas in two different classes of experiments.

6.6.1 Two-Dimensional Resolution/Separation Experiments

The primary aim in these experiments is to separate the different interactions in the Hamiltonian. In high-resolution NMR, this amounts to the separation of the Zeeman ($\mathcal{H}_z$) and the coupling Hamiltonians ($\mathcal{H}_J$).

$$\mathcal{H} = \mathcal{H}_z + \mathcal{H}_J \tag{6.26}$$

Different strategies can be defined depending upon the nature of the information required in the final spectrum.

6.6.1.1 Two-Dimensional Heteronuclear Separation Experiments

Figure 6.6 illustrates such a concept (pulse sequences a and b). In (a), the $F2$ axis of the final spectrum contains both ^{13}C chemical shift and ^{13}C-^{1}H coupling information, whereas the $F1$ axis contains only the ^{13}C chemical shift information. In (b), the reverse occurs, i.e., the $F1$ axis has both ^{13}C chemical shift and ^{13}C-^{1}H coupling constants, while the $F2$ axis has only ^{13}C chemical shift information. This was the first two-dimensional experiment ever recorded. Figure 6.7 shows an experimental spectrum corresponding to Fig. 6.6b.

Figure 6.8 illustrates another situation where the $F1$ axis has only scalar coupling constants and the $F2$ axis has the chemical shift information. This represents a complete separation of the coupling and chemical shift Hamiltonians along the two axes.

The product operator description of the experiment is explicitly given in the following paragraphs.

The density operator (ρ) terms at the time points 1 to 3 indicated in Fig. 6.8 are

$$\rho_1 = C_z \tag{6.27}$$

$$\rho_2 = -C_y \tag{6.28}$$

$$\rho_3 = -[C_y \cos \pi J_{HC} t_1 - 2C_x H_z \sin \pi J_{HC} t_1] \tag{6.29}$$

The terms C_x, C_y, and C_z refer to the x-, y-, and z-components of the ^{13}C magnetization. And $2C_x H_z$ represents x-magnetization of carbon antiphase with respect to proton. In Eq. 6.29, the second term does not lead to observable magnetization in t_2 because of proton decoupling. The first term evolves under chemical shift only during t_2.

Therefore,

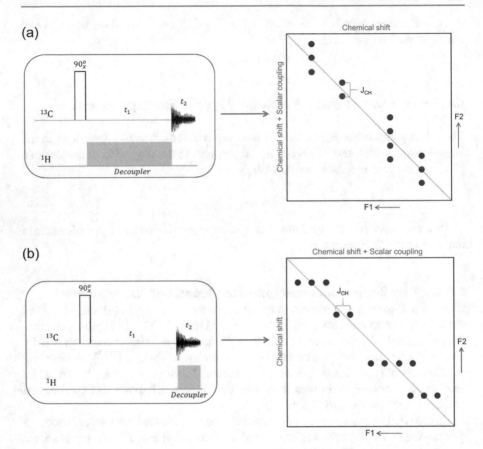

Fig. 6.6 Two-dimensional experiments, where the ^{13}C chemical shift and ^{13}C-^{1}H coupling constants are separated on the $F1$ and $F2$ dimensions. In (**a**) the $F2$ axis has both chemical shifts and coupling constants, while the $F1$ axis has only chemical shifts. The reverse is true in (**b**)

$$C_y \xrightarrow{\mathcal{H}_Z} [C_y \cos\omega_C t_2 - C_x \sin\omega_C t_2] \tag{6.30}$$

Then, assuming y-detection, the density operator at time point 4 in Fig. 6.8 is

$$\rho_4 = C_y \cos\omega_C t_2 \cos\pi J_{HC} t_1 \tag{6.31}$$

This leads to exclusively coupling information along t_1 and chemical shift information along t_2.

The experimental spectrum corresponding to this pulse scheme is shown in Fig. 6.9.

6.6.1.2 Two-Dimensional Homonuclear Separation Experiments

Figure 6.10 shows a pulse scheme for obtaining the separation of interactions in homonuclear systems. This is often referred to as two-dimensional J-resolved

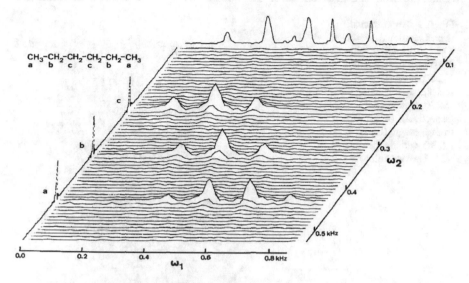

Fig. 6.7 Experimental spectrum demonstrating the scheme in Fig. 6.6b. (Reproduced from J. Chem. Phys. 63, 5490 (1975), with the permission of AIP Publishing)

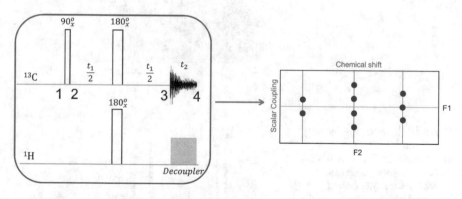

Fig. 6.8 Schematic of ^{13}C chemical shift and ^{13}C-^{1}H scalar coupling separation in the two-dimensional spectrum. The $F2$ axis displays ^{13}C chemical shifts, while the $F1$ axis displays the multiplicity at each carbon site. See text for more details

(JRES) experiment. The pulse sequence can be analyzed using the product operator formalism.

For a weakly coupled two-spin system (k and l, $I = 1/2$), the density operator terms at different time points along the pulse sequence are

$$\rho_1 = I_{kz} + I_{lz} \tag{6.32}$$

Fig. 6.9 Experimental two-dimensional J-resolved NMR spectrum of cholesterol displaying the separation of ^{13}C chemical shift and ^{13}C-^{1}H scalar coupling, along the $F2$ and $F1$ dimensions, respectively. (Reproduced from J. Magn. Reson. 29, 587 (1978), with the permission from Elsevier Publishing)

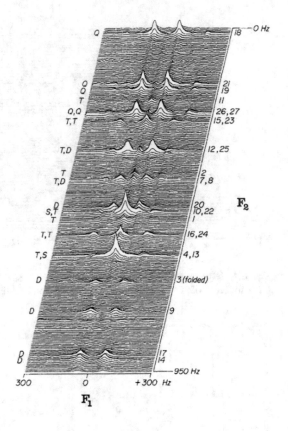

Fig. 6.10 Schematic of the two-dimensional J-resolved pulse sequence. Numbers 1–4 indicate the time points at which the density operators are calculated (see text)

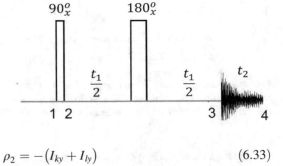

$$\rho_2 = -\left(I_{ky} + I_{ly}\right) \tag{6.33}$$

During the next spin echo period chemical shifts are refocused, and thus spins evolve under the scalar coupling Hamiltonian (H_J) only. Explicitly the evolutions of product operators are shown for spin k only. Similar calculations apply for the spin l as well. Now, ρ_2 evolves under scalar coupling during the spin echo (t_1) and during the detection period t_2, thus for the total time period t_1+t_2.

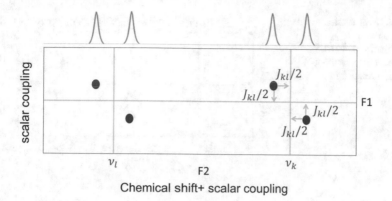

Chemical shift+ scalar coupling

Fig. 6.11 Schematic two-dimensional JRES for a two-spin system. Projection shown on the top represents the one-dimensional spectrum

$$-I_{ky} \xrightarrow{\mathcal{H}_J} -I_{ky}\cos \pi J_{kl}(t_1 + t_2) + 2I_{kx}I_{lz}\sin \pi J_{kl}(t_1 + t_2) \tag{6.34}$$

In this equation the second term which represents antiphase magnetization is not observable during t_2 period. So, considering the chemical shift evolution of the first term during t_2, one gets

$$-\cos \pi J_{kl}(t_1 + t_2)I_{ky} \xrightarrow{\mathcal{H}_z} -\cos \pi J_{kl}(t_1 + t_2)\{I_{ky}\cos \omega_k t_2 - I_{kx}\sin \omega_k t_2\} \tag{6.35}$$

Assuming y-detection, the signal is proportional to

$$-\cos \pi J_{kl}(t_1 + t_2)\cos \omega_k t_2 \tag{6.36}$$

$$= -\{\cos \pi J_{kl}t_1 \cos \pi J_{kl}t_2 - \sin \pi J_{kl}t_1 \sin \pi J_{kl}t_2\}\cos \omega_k t_2 \tag{6.37}$$

We see that along the t_2 axis, there are both chemical shifts and coupling constants, whereas along the t_1 axis, there is only scalar coupling information. This results in a spectrum of the type shown in Fig. 6.11. The peaks align themselves at an angle of 45° with respect to the $F2$ axis. The detected signal has both cosine and sine modulations along both t_1 and t_2 axes. The cosine modulation results in an absorptive shape, while the sine modulation results in dispersive line shape, after Fourier transformation. Thus, the peaks will have mixed phases. This requires a magnitude mode calculation of the spectra. Such calculation can be extended to multi-spin systems as well. An experimental spectrum demonstrating these features is shown in Fig. 6.12.

In Figs. 6.11 and 6.12, we notice that the coupling information is present along both axes, and it would be desirable to have a complete separation of the chemical shift and coupling information. This can be achieved by performing a shearing transformation on the peaks as indicated schematically in Fig. 6.13.

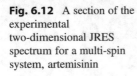

Fig. 6.12 A section of the experimental two-dimensional JRES spectrum for a multi-spin system, artemisinin

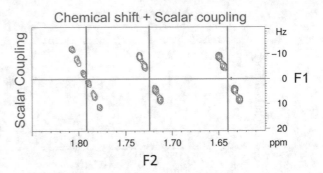

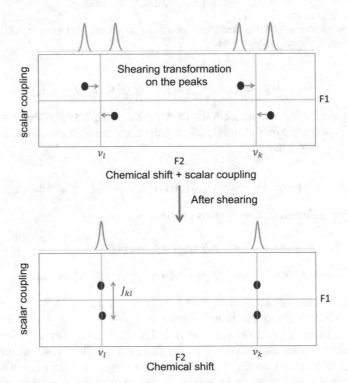

Fig. 6.13 Shearing transformation on the peaks in the two-dimensional J-resolved spectrum. The shearing transformation eliminates the coupling information along the *F*2 (horizontal axis, bottom picture)

Figure 6.14 shows the result of a shearing transformation in an experimental spectrum of a multi-spin system demonstrating the complete separation of the chemical shift and scalar coupling information along the *F*2 and *F*1 axes, respectively.

Fig. 6.14 A section of the experimental spectrum of artemisinin demonstrating the effect of shearing transformation

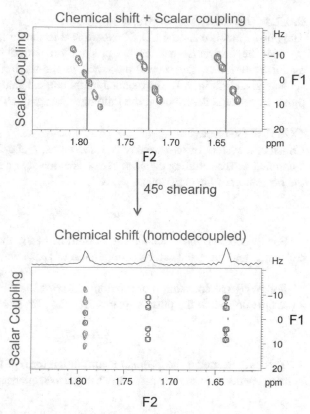

Interestingly, the projection of the spectrum on the $F2$ axis shows a completely homodecoupled spectrum of the spin system. This is an interesting way of achieving broadband homonuclear decoupling in complex spin systems. Along the $F1$ axis, the resolution is sufficiently high because of the small spectral width; therefore, the coupling constants can be measured very precisely. Further, because of the spin echo in the t_1 period, the external magnetic field inhomogeneity effects of line broadening are eliminated, which enhances the resolution along the $F1$ dimension. This technique has been extremely useful in separating out the multiplets in complex spin systems and measure accurately the various coupling constants.

6.6.2   Two-Dimensional Correlation Experiments

These experiments are designed to correlate two frequencies in a given one-dimensional spectrum with regard to various interactions between the spin systems in a molecule under consideration. The very first experiment in this context was proposed by Jean Jeener. This experiment has been popularly known as correlated spectroscopy or COSY.

6.6.2.1 The COSY Experiment

The pulse sequence for the COSY experiment is shown in Fig. 6.15.

Here the first pulse acts as the preparation period which is followed by the evolution period t_1. The second pulse acts as the mixing period of the generalized scheme given in Fig. 6.1. The detailed mathematical analysis of the working of this pulse sequence is described in the following paragraphs.

COSY of Two Spins

Consider a system of two weakly coupled spins, k and l (with $I = 1/2$). They are J-coupled with a coupling constant of J_{kl}. The density operator of the spin system at the beginning of the experiment, ρ_1, is

$$\rho_1 = I_{kz} + I_{lz} \tag{6.38}$$

For illustration, we calculate the evolution of I_{kz} through the pulse sequence explicitly, and the same can be extrapolated to I_{lz}, as well.

Following the convention of rotations described in Chap. 5, the density operator ρ_2 at time point 2 in the pulse sequence, for the spin k, is

$$\rho_2 = -I_{ky} \tag{6.39}$$

This evolves under the Zeeman Hamiltonian ($\omega_k I_{kz}$), for a period t_1 yielding the density operator ρ_3, at time point 3 in the pulse sequence.

$$\rho_3 = -\left(I_{ky} \cos \omega_k t_1 - I_{kx} \sin \omega_k t_1\right) \tag{6.40}$$

Next, considering evolution under the J−coupling Hamiltonian ($2\pi J_{kl} I_{kz} I_{lz}$), the density operator will be $\rho_3{}'$:

Fig. 6.15 Schematic pulse sequence of the COSY experiment. Numbers 1–5 indicate the time points at which the density operator calculations are reported (see text)

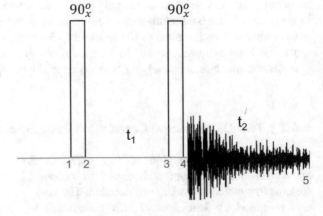

$$\rho_3' = -\{[I_{ky}\cos\pi J_{kl}t_1 - 2I_{kx}I_{lz}\sin\pi J_{kl}t_1]\cos\omega_k t_1$$
$$-[I_{kx}\cos\pi J_{kl}t_1 + 2I_{ky}I_{lz}\sin\pi J_{kl}t_1]\sin\omega_k t_1\} \qquad (6.41)$$

The last pulse transforms these operators to yield a density operator ρ_4, at time point 4 of the pulse sequence.

$$\rho_4 = -\{[I_{kz}\cos\pi J_{kl}t_1 + 2I_{kx}I_{ly}\sin\pi J_{kl}t_1]\cos\omega_k t_1$$
$$-[I_{kx}\cos\pi J_{kl}t_1 - 2I_{kz}I_{ly}\sin\pi J_{kl}t_1]\sin\omega_k t_1\} \qquad (6.42)$$

Since the data is collected soon after, one needs to look at only those terms in the density operator which are observable as per definitions given in Chap. 5 ($Tr[I_x B(-s)] \neq 0$). Thus, the observable part of ρ_4 is ρ_4^{obs}:

$$\rho_4^{obs} = [I_{kx}\cos\pi J_{kl}t_1 - 2I_{kz}I_{ly}\sin\pi J_{kl}t_1]\sin\omega_k t_1\} \qquad (6.43)$$

The first term in Eq. 6.43 which represents x-magnetization of the k spin evolves during the t_2 period with frequencies characteristic of k spin. Therefore, this will produce a diagonal peak ($F1 = F2 = \omega_k$) in the final two-dimensional spectrum. The second term which represents y-magnetization of l spin evolves during the t_2 period with frequencies characteristic of l spin. Therefore, this term will produce a "cross-peak" ($F1 = \omega_k$; $F2 = \omega_l$). Both these peaks will have fine structures, which contain the coupling information.

We now calculate the evolution of the terms in Eq. 6.43 during the t_2 time period. Here again both chemical shift and coupling evolutions have to be considered explicitly.

The first (diagonal peak) term in Eq. 6.43:

Chemical shift evolution leads to the density operator ρ_{5d} given by

$$\rho_{5d} = [I_{kx}\cos\omega_k t_2 + I_{ky}\sin\omega_k t_2]f_d(t_1) \qquad (6.44)$$

where $f_d(t_1) = \cos\pi J_{kl}t_1\sin\omega_k t_1$.
Evolution under coupling generates the density operator ρ_{5d}' given by

$$\rho_{5d}' = \{[I_{kx}\cos\pi J_{kl}t_2 + 2I_{ky}I_{lz}\sin\pi J_{kl}t_2]\cos\omega_k t_2$$
$$+[I_{ky}\cos\pi J_{kl}t_2 - 2I_{kx}I_{lz}\sin\pi J_{kl}t_2]\sin\omega_k t_2\}f_d(t_1) \qquad (6.45)$$

Assuming that we measure the y-magnetization, the observable signal is given by $Tr[\rho_{5d}'I_{ky}]$.

$$Tr[\rho_{5d}'I_{ky}] = \cos\pi J_{kl}t_2\sin\omega_k t_2\, f_d(t_1)$$

$$= \cos \pi J_{kl} t_2 \, \sin \, \omega_k t_2 \, \cos \pi J_{kl} t_1 \, \sin \omega_k t_1 \qquad (6.46)$$

Explicitly this will lead to the following terms:

$$Tr\left[\rho_{5d}'I_{ky}\right] = \frac{1}{4} \{ \sin \, (\omega_k + \pi J_{kl})t_2 + \, \sin \, (\omega_k - \pi J_{kl})t_2 \} \{ \, \sin \, (\omega_k + \pi J_{kl})t_1$$
$$+ \sin \, (\omega_k - \pi J_{kl})t_1 \} \qquad (6.47)$$

This contributes to the detected FID.

The two-dimensional real Fourier transformation along the t_1 and t_2 dimensions leads to four peaks with a dispersive line shapes at the following frequencies (Hz).

$$(F1, F2) = \left[\left(v_k + \frac{J_{kl}}{2}\right), \left(v_k + \frac{J_{kl}}{2}\right)\right]; \text{positive, dispersive}$$

$$\left[\left(v_k + \frac{J_{kl}}{2}\right), \left(v_k - \frac{J_{kl}}{2}\right)\right]; \text{positive, dispersive}$$

$$\left[\left(v_k - \frac{J_{kl}}{2}\right), \left(v_k + \frac{J_{kl}}{2}\right)\right]; \text{positive, dispersive}$$

$$\left[\left(v_k - \frac{J_{kl}}{2}\right), \left(v_k - \frac{J_{kl}}{2}\right)\right]; \text{positive, dispersive} \qquad (6.48)$$

This results in a fine structure for the diagonal peak as indicated in Fig. 6.16. *The second (cross-peak) term in Eq. 6.43:*

Here, let us consider the J-evolution first. This leads to the density operator ρ_{5c}:

$$\rho_{5c} = \left[2I_{kz}I_{ly} \cos \pi J_{kl} t_2 - I_{lx} \sin \pi J_{kl} t_2\right] f_c(t_1) \qquad (6.49)$$

$$f_c(t_1) = \sin \pi J_{kl} t_1 \, \sin \, \omega_k t_1 \qquad (6.50)$$

Next, considering the shift evolution, we get ρ_{5c}' as

$$\rho_{5c}' = \{2I_{kz}\left[I_{ly} \cos \omega_l t_2 - I_{lx} \sin \omega_l t_2\right] \, \cos \, \pi J_{kl} t_2$$
$$- \left[I_{lx} \cos \omega_l t_2 + I_{ly} \sin \omega_l t_2\right] \, \sin \, \pi J_{kl} t_2 \} f_c(t_1) \qquad (6.51)$$

Again, assuming that we measure the y-magnetization, the observable signal is given by $Tr[\rho_{5c}'I_{ly}]$:

$$Tr\left[\rho_{5c}'I_{ly}\right] = \, \sin \omega_l t_2 \sin \pi J_{kl} t_2 \sin \omega_k t_1 \sin \pi J_{kl} t_1 \qquad (6.52)$$

This leads to four absorptive peaks at the following coordinates.

Fig. 6.16 Typical fine structure of the diagonal peaks in the COSY spectrum. They have in-phase dispersive line shapes along both frequency axes

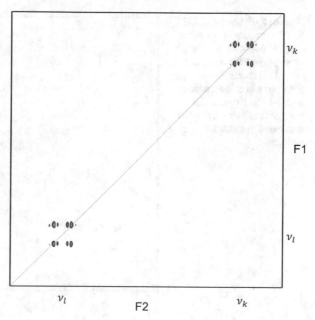

$$(F1, F2) = \left[\left(v_k + \frac{J_{kl}}{2} \right), \left(v_l + \frac{J_{kl}}{2} \right) \right] ; \text{positive, absorptive}$$

$$\left[\left(v_k + \frac{J_{kl}}{2} \right), \left(v_l - \frac{J_{kl}}{2} \right) \right] ; \text{negative, absorptive}$$

$$\left[\left(v_k - \frac{J_{kl}}{2} \right), \left(v_l - \frac{J_{kl}}{2} \right) \right] ; \text{positive, absorptive}$$

$$\left[\left(v_k - \frac{J_{kl}}{2} \right), \left(v_l + \frac{J_{kl}}{2} \right) \right] \text{negative, absorptive} \tag{6.53}$$

Similar expressions can be derived to obtain the peak list starting from the l spin magnetization. Thus, the overall two-dimensional spectrum for the $k - l$ spin system will look as shown in Fig. 6.17.

Figure 6.18 shows the phase-sensitive experimental spectrum of an AX sub-spin system of curcumin dissolved in $CDCl_3$.

COSY of Three Spins

The detailed calculation shown for the two-spin system can be extrapolated to three-spin systems as well. The following considerations will help in arriving at the appropriate fine structures of the peaks.

(i) The spectrum will have cross-peaks displaying the nature of the coupling network. For example, for a linear AMX system, there will be cross-peaks from A to M and M to X on one-side of the diagonal and M to A and X to M

Fig. 6.17 Schematic COSY
NMR spectrum of a weakly
coupled two-spin system. The
cross-peaks have antiphase
(+ and −) character and
absorptive line shapes along
both F2 and F1 axes. The
diagonal peaks have in-phase
dispersive line shapes along
both axes

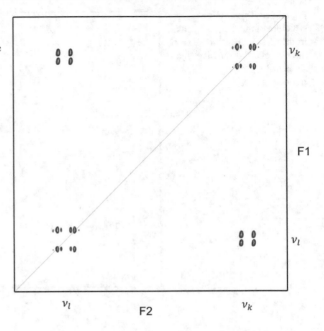

on the other side of the diagonal. All the three diagonal peaks will be present.
However, there will be no A to X and X to A cross-peaks, as there is no coupling
between A and X spins. Likewise, for a triangular coupling network, where all
the three coupling constants are nonzero, there will be A to M, A to X, M to X,
M to A, X to A, and X to M cross-peaks. Figure 6.19 shows the expected COSY
spectra for linear and triangular coupling networks.

(ii) Each cross-peak in the COSY spectrum arises as a result of the evolution under
one particular coupling constant. For example, in an AMX spin system, the A to
M (or M to A) cross-peak will result from the coupling J_{AM}. This coupling
constant leads to a splitting where lines will have positive and negative signs,
and this is called *active coupling*. The other coupling, for example, A to X, if it is
nonzero, leads to in-phase splitting and is called the *passive coupling*. Accordingly, the fine structures in the cross-peaks will depend upon the relative
magnitudes of the *active* and *passive* couplings. Figure 6.19 shows the fine
structures for the A($F2$) to M($F1$) cross-peak for two different cases of J_{AM} and
J_{AX} coupling constants in the linear AMX spin system (Fig. 6.20).
Continuing along the same lines, the fine structure in the A($F1$) to M($F2$) cross-
peak can be calculated, and this is shown in Fig. 6.21.
    For the triangular coupling network of the three spins A, M, and X, the fine
structures can be calculated for the individual cross-peaks following the same
procedure described. This is explicitly shown for the A to M cross-peak in
Fig. 6.22 for a particular choice of magnitudes of coupling constants. Notice
once again that in this cross-peak, J_{AM} is the active coupling, while J_{MX} and J_{AX}
are passive couplings.

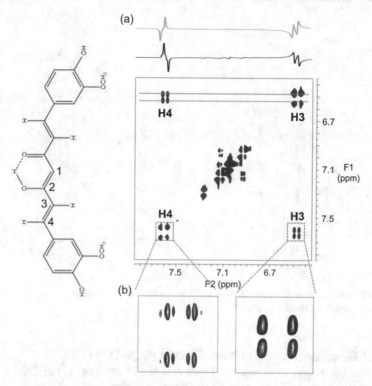

Fig. 6.18 (**a**) Phase-sensitive COSY spectrum of curcumin (AX sub-spin system) and (**b**) blowups of the cross- (right) and diagonal peaks (left). Horizontal cross-sections through the peaks at the gray- and green-colored lines are shown on the top

Disadvantages of COSY

The COSY experiment has the following disadvantages.

(i) The dispersive line shapes in the diagonal peaks produce long tails which hamper the resolution in the spectra. The cross-peaks which lie close to the diagonal would get masked out.

(ii) The diagonal peak has in-phase components, while the cross-peak has antiphase components. The resolution in the F1 dimension is determined by the number of increments one can acquire along the t_1 dimension, and this will be limited by the machine time. In that scenario, because of poor resolution along the F1 dimension, the peak intensities cancel because of the positive/negative character of the components in the cross-peak. On the other hand, the components in the diagonal peaks coadd because of the in-phase character. This results in huge diagonal peaks and tiny cross-peaks in the event of insufficient resolution in the spectrum.

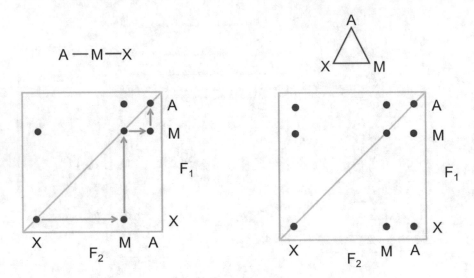

Fig. 6.19 Schematic appearance of the COSY spectrum for a system of three spins, AMX. The coupling patterns are shown on the top

6.6.2.2 Double-Quantum-Filtered COSY (DQF-COSY)

This experiment was designed to circumvent the limitations of the COSY experiment. The pulse sequence for the DQF-COSY is shown in Fig. 6.23.

The pulse sequence is similar to that of COSY up to the second pulse but for the fact that the phases (ϕ) of these two pulses need to be cycled and the data coadded or subtracted as discussed in the following. The scheme involves acquiring four transients with the pulse phase (ϕ) and the receiver phase (θ) incremented with every transient. This is indicated in Table 6.2.

The experiment can be analyzed in the same manner as was done for COSY. Considering a system of two weakly coupled spins (k, l), the density operator calculation follows the same steps as for COSY, and we rewrite the density operator at time point 4 of the pulse sequence:

$$\rho_4(\phi = x) = \left[-I_{kz}\cos \pi J_{kl}t_1 - 2I_{kx}I_{ly}\sin \pi J_{kl}t_1\right]\cos \omega_k t_1$$
$$+ \left[I_{kx}\cos \pi J_{kl}t_1 - 2I_{kz}I_{ly}\sin \pi J_{kl}t_1\right]\sin \omega_k t_1 \qquad (6.54)$$

Here the first two pulses are considered to be applied along the x axis ($\phi = x$). Repeating such an exercise with $\phi = y$ leads to the following density operator $\rho_4(y)$:

$$\rho_4(\phi = y) = \left[-I_{kz}\cos \pi J_{kl}t_1 + 2I_{ky}I_{lx}\sin \pi J_{kl}t_1\right]\cos \omega_k t_1$$
$$+ \left[I_{ky}\cos \pi J_{kl}t_1 + 2I_{kz}I_{lx}\sin \pi J_{kl}t_1\right]\sin \omega_k t_1 \qquad (6.55)$$

Similar calculations with $\phi = -x$ and $\phi = -y$ lead to the following density operators.

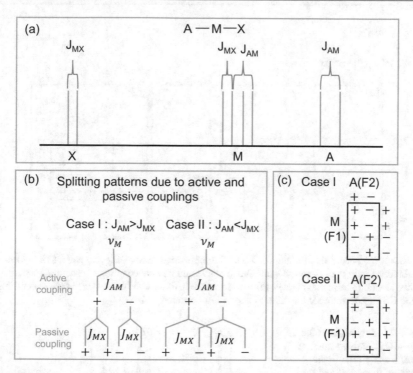

Fig. 6.20 (a) Fine structure in the one-dimensional spectrum for a linear AMX system. (b) Splitting patterns in the A-M cross-peak due to active and passive couplings for the M spin for two different cases of relative magnitudes of active and passive couplings. (c) The final fine structure in the cross-peak A($F2$) to M($F1$) in the COSY spectrum for the two cases considered in (b). Note that this peak lies in the lower triangle in the two-dimensional spectrum (Fig. 6.19)

Case I : $J_{AM} > J_{MX}$ Case II : $J_{AM} < J_{MX}$

```
      +    +    −    −              +    −    +    −
  ┌────────────────────┐       ┌────────────────────┐
 −│  −    −    +    +  │      −│  −    +    −    +  │
  │                    │ A(F1) │                    │ A(F1)
 +│  +    +    −    −  │      +│  +    −    +    −  │
  └────────────────────┘       └────────────────────┘
        M(F2)                        M(F2)
```

Fig. 6.21 Schematic fine structure in the A($F1$) to M($F2$) cross-peak. Note that this peak lies in the upper triangle in the two-dimensional spectrum (Fig. 6.19)

$$\rho_4(\phi = -x) = \left[-I_{kz}\cos \pi J_{kl}t_1 - 2I_{kx}I_{ly}\sin \pi J_{kl}t_1\right]\cos \omega_k t_1$$
$$+ \left[-I_{kx}\cos \pi J_{kl}t_1 + 2I_{kz}I_{ly}\sin \pi J_{kl}t_1\right]\sin \omega_k t_1 \qquad (6.56)$$

$$\rho_4(\phi = -y) = \left[-I_{kz}\cos \pi J_{kl}t_1 + 2I_{ky}I_{lx}\sin \pi J_{kl}t_1\right]\cos \omega_k t_1$$
$$+ \left[-I_{ky}\cos \pi J_{kl}t_1 - 2I_{kz}I_{lx}\sin \pi J_{kl}t_1\right]\sin \omega_k t_1 \qquad (6.57)$$

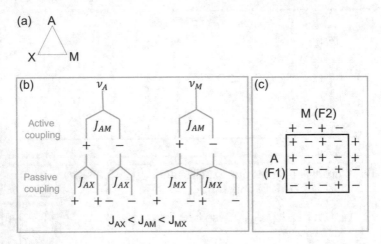

Fig. 6.22 (a) Schematic of triangular J-coupling network in the weakly coupled AMX system. (b) The splitting pattern of A and M spins due to active and passive coupling constants for a particular choice of their relative magnitudes. (c) Fine structure in the A to M cross-peak in the lower triangle of the COSY spectrum for the choice of coupling constants as in (b)

Fig. 6.23 Schematic of the DQF-COSY pulse sequence. Here, ϕ and θ refer to the phases of the pulses and the receiver, respectively. See text for more details. Numbers 1–5 indicate the time points at which the density operators are reported in the text

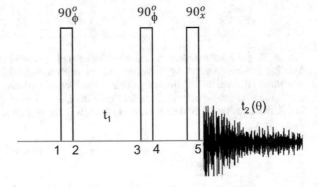

Table 6.2 Phase cycling for DQF-COSY pulse sequence

Scan no.	Pulse phase (ϕ)	Receiver phase (θ)
1	x	+
2	y	−
3	−x	+
4	−y	−

As per the receiver phase cycling the data is added or subtracted and then, the resultant density operator will be

$$\rho_4 = \rho_4(\phi = x) - \rho_4(\phi = y) + \rho_4(\phi = -x) - \rho_4(\phi = -y) \tag{6.58}$$

$$= -4\left(I_{ky}I_{lx} + I_{kx}I_{ly}\right)\sin \pi J_{kl}t_1 \cos \omega_k t_1 \tag{6.59}$$

The operators in Eq. 6.59 represent pure double-quantum coherences.

The third 90x pulse in the pulse sequence converts these terms into observable single-quantum coherences. The corresponding density operator ρ_5 will be

$$\rho_5 = -4\left(I_{kz}I_{lx} + I_{kx}I_{lz}\right)\sin \pi J_{kl}t_1 \cos \omega_k t_1 \tag{6.60}$$

This now consists of antiphase magnetizations of both k and l spins with the same phase. They will evolve during the t_2 period into in-phase magnetizations of k and l spins, whereby it becomes observable. Rewriting Eq. 6.60,

$$\rho_5 = -4\left(I_{kz}I_{lx} + I_{kx}I_{lz}\right)\mathrm{f}(t_1) \tag{6.61}$$

$$\mathrm{f}(t_1) = \sin \pi J_{kl}t_1 \cos \omega_k t_1 \tag{6.62}$$

Now evolve ρ_5 under $\mathcal{H}_J$ and $\mathcal{H}_z$ sequentially during t_2,

$$\rho_5 \xrightarrow{\mathcal{H}_J} -2\left\{\left[2I_{kz}I_{lx}\cos\left(\pi J_{kl}t_2\right) + I_{ly}\sin\left(\pi J_{kl}t_2\right)\right]\right.$$
$$\left. + \left[2I_{kx}I_{lz}\cos\left(\pi J_{kl}t_2\right) + I_{ky}\sin\left(\pi J_{kl}t_2\right)\right]\right\}\mathrm{f}(t_1) \tag{6.63}$$

The antiphase terms $2I_{kz}I_{lx}$ and $2I_{kx}I_{lz}$ are not observable and hence will not be considered further. The other terms will be evolved under the $\mathcal{H}_z$.

$$\xrightarrow{\mathcal{H}_z} -2\left\{\left[I_{ly}\cos\omega_l t_2 - I_{lx}\sin\omega_l t_2\right]\sin\left(\pi J_{kl}t_2\right)\right.$$
$$\left. + \left[I_{ky}\cos\omega_k t_2 - I_{kx}\sin\omega_k t_2\right]\sin\left(\pi J_{kl}t_2\right)\right\}f(t_1) \tag{6.64}$$

Assuming y-detection, we have the following signal:

$$\text{Signal (S)} = -2\left[\cos\omega_l t_2 \sin\left(\pi J_{kl}t_2\right) + \cos\omega_k t_2 \sin\left(\pi J_{kl}t_2\right)\right]\sin\pi J_{kl}t_1 \cos\omega_k t_1 \tag{6.65}$$

In this expression, the first term leads to the cross-peak, while the second term leads to the diagonal peak in the spectrum.

(a) Cross-peak

$$\cos\omega_l t_2 \sin\left(\pi J_{kl}t_2\right)\sin\pi J_{kl}t_1 \cos\omega_k t_1$$

$$= \frac{1}{4}\left[\sin\left(\omega_k + \pi J_{kl}\right)t_1 - \sin\left(\omega_k - \pi J_{kl}\right)t_1\right].$$
$$\left[\sin\left(\omega_l + \pi J_{kl}\right)t_2 - \sin\left(\omega_l - \pi J_{kl}\right)t_2\right] \tag{6.66}$$

$$= +\sin\left(\omega_k + \pi J_{kl}\right)t_1 \sin\left(\omega_l + \pi J_{kl}\right)t_2$$

$$- \sin\left(\omega_k + \pi J_{kl}\right)t_1 \sin\left(\omega_l - \pi J_{kl}\right)t_2$$

$$- \sin (\omega_k - \pi J_{kl})t_1 \sin (\omega_l + \pi J_{kl})t_2$$

$$+ \sin (\omega_k - \pi J_{kl})t_1 \sin (\omega_l - \pi J_{kl})t_2 \qquad (6.67)$$

This leads to the following peaks.

$$(F1, F2) = \left[\left(v_k + \frac{J_{kl}}{2} \right), \left(v_l + \frac{J_{kl}}{2} \right) \right]; \text{positive, dispersive}$$

$$\left[\left(v_k + \frac{J_{kl}}{2} \right), \left(v_l - \frac{J_{kl}}{2} \right) \right]; \text{negative, dispersive}$$

$$\left[\left(v_k - \frac{J_{kl}}{2} \right), \left(v_l + \frac{J_{kl}}{2} \right) \right]; \text{negative, dispersive}$$

$$\left[\left(v_k - \frac{J_{kl}}{2} \right), \left(v_l - \frac{J_{kl}}{2} \right) \right]; \text{positive, dispersive} \qquad (6.68)$$

A 90° phase shift will produce absorptive line shape for all the four peaks.

(b) Diagonal peak

$$\cos \omega_k t_2 \sin (\pi J_{kl} t_2) \sin \pi J_{kl} t_1 \cos \omega_k t_1$$

$$= \frac{1}{4} [\sin (\omega_k + \pi J_{kl})t_1 - \sin (\omega_k - \pi J_{kl})t_1].$$
$$[\sin (\omega_k + \pi J_{kl})t_2 - \sin (\omega_k - \pi J_{kl})t_2] \qquad (6.69)$$

$$= + \sin (\omega_k + \pi J_{kl})t_1 \ \sin (\omega_k + \pi J_{kl})t_2$$

$$- \sin (\omega_k + \pi J_{kl})t_1 \sin (\omega_k - \pi J_{kl})t_2$$

$$- \sin (\omega_k - \pi J_{kl})t_1 \sin (\omega_k + \pi J_{kl})t_2$$

$$+ \sin (\omega_k - \pi J_{kl})t_1 \sin (\omega_k - \pi J_{kl})t_2 \qquad (6.70)$$

This leads to the following peaks:

$$(F1, F2) = \left[\left(v_k + \frac{J_{kl}}{2} \right), \left(v_k + \frac{J_{kl}}{2} \right) \right]; \text{positive, dispersive}$$

$$\left[\left(v_k + \frac{J_{kl}}{2} \right), \left(v_k - \frac{J_{kl}}{2} \right) \right]; \text{negative, dispersive}$$

$$\left[\left(v_k - \frac{J_{kl}}{2} \right), \left(v_k + \frac{J_{kl}}{2} \right) \right] \text{negative, dispersive}$$

$$\left[\left(v_k - \frac{J_{kl}}{2} \right), \left(v_k - \frac{J_{kl}}{2} \right) \right]; \text{positive, dispersive} \qquad (6.71)$$

A 90° phase shift will produce absorptive line shape for all the four peaks.

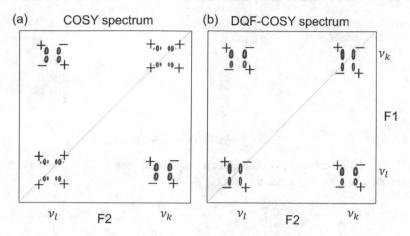

Fig. 6.24 Comparison of COSY and DQF-COSY spectra for a two-spin system. + and − indicate positive and negative signs of the peak components, respectively

Thus, the final spectrum for the two-spin system will appear as shown in Fig. 6.24. Clearly, the DQF-COSY spectrum shows better resolution than the COSY spectrum and is pretty clean in both the diagonal and cross-peaks.

6.6.2.3 Total Correlation Spectroscopy (TOCSY)

The COSY and the DQF-COSY resulted in fine structures in diagonal as well as cross-peaks. The DQF-COSY circumvented the shortcomings of COSY with respect to the diagonal. However, the fine structures still retain the antiphase nature of the components in the cross-peaks. In essence, this amounts to a differential transfer of magnetization between the spins. The antiphase character results in the cancellation of component intensities in the cross-peaks in the absence of sufficient resolution. This problem is circumvented by total correlation spectroscopy (TOCSY) which results in in-phase components and thus achieves net transfer of magnetization between the spins. The pulse sequence for the TOCSY experiment is shown in Fig. 6.25.

The pulse sequence starts with a 90° pulse, which creates transverse magnetization which then evolves during the period t_1 with characteristic frequencies. The so-called mixing here consists of a strong RF field or a train of pulses (often referred to as composite pulses) during which time the spins are locked in the rotating frame in the transverse plane along the x or the y-axis. During the spin lock transfer of coherence occurs among the J-coupled spins. For a two-spin system, k and l (spin 1/2), the effective Hamiltonian during the mixing period consists only of the J-coupling Hamiltonian, and the Zeeman interactions are eliminated. This Hamiltonian is given by

$$\mathcal{H}_e = 2\pi J I_k.I_l \tag{6.72}$$

Fig. 6.25 Pulse sequence for
the TOCSY experiment

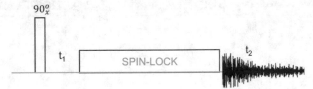

$$= 2\pi J\left(I_{kz}I_{lz} + I_{kx}I_{lx} + I_{ky}I_{ly}\right) \tag{6.73}$$

The evolution of the magnetization components under the influence of this
Hamiltonian is given by the following equation:

$$I_{kx} \overset{\mathcal{H}_{e}t}{\rightarrow} I_{kx}\left(\frac{1+\cos 2\pi Jt}{2}\right) + I_{lx}\left(\frac{1-\cos 2\pi Jt}{2}\right) + \left(I_{ky}I_{lz} - I_{ly}I_{kz}\right)\sin 2\pi Jt$$

$$\tag{6.74}$$

Complete transfer of magnetization will occur for time $t = 1/2J$. This is in contrast
to INEPT transfer of coherence where a complete transfer requires a time $t = 1/J$; in
the INEPT, the transfer occurs in two steps: the first step involving a spin-echo of
period $1/2J$ causes antiphase transfer, and in the second step, a second spin echo of
period $1/2J$ causes refocusing to generate in-phase magnetization ($I_{kx} \rightarrow 2I_{ly}I_{kz} \rightarrow I_{lx}$).

A similar equation can be written for the evolution of I_{lx}:

$$I_{lx} \overset{\mathcal{H}_{e}t}{\rightarrow} I_{lx}\left(\frac{1+\cos 2\pi Jt}{2}\right) + I_{kx}\left(\frac{1-\cos 2\pi Jt}{2}\right) + \left(I_{ly}I_{kz} - I_{ky}I_{lz}\right)\sin 2\pi Jt \tag{6.75}$$

The addition of Eqs. 6.74 and 6.75 leads to the following:

$$\left(I_{kx} + I_{lx}\right) \overset{\mathcal{H}_{e}t}{\rightarrow} \left(I_{kx} + I_{lx}\right) \tag{6.76}$$

This implies that the total x-magnetization is conserved through the mixing
sequence, and there is in-phase transfer ($I_{kx} \rightarrow I_{lx}$ and vice versa), retaining the
phase of the magnetization, i.e., $I_{kx} \rightarrow I_{lx}$, $I_{ky} \rightarrow I_{ly}$, and $I_{kz} \rightarrow I_{lz}$. Therefore, this
mixing is termed as isotropic mixing, and the Hamiltonian is termed as isotropic
Hamiltonian. After the mixing the magnetization components are detected in the t_2
time period. Two-dimensional Fourier transformation of the collected signal results
in a two-dimensional spectrum.

Detailed calculations for multi-spin systems show that the general conclusions
derived for the two-spin systems are valid for multi-spin systems as well. However,
an interesting feature of this experiment is the following. Considering a linear three-
spin system (AMX), where there is no coupling between the spins A and X, it turns
out that there will be a cross-peak between the spins A and X, provided both AM

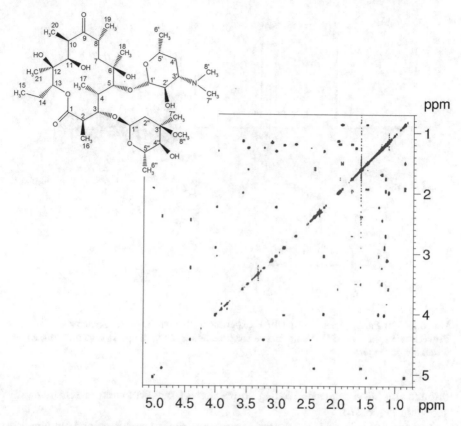

Fig. 6.26 Experimental TOCSY spectrum of erythromycin-A

coupling and MX coupling are nonzero. Thus, the TOCSY experiments relays magnetization through the network of coupled spins, providing valuable information about the coupling network in a given molecule. An experimental TOCSY spectrum is shown in Fig. 6.26.

Such a spectrum will enable to distinguish between a linear three-spin system, AMX, and a mixture of two two-spin systems AM and M'X with accidental degeneracy of the M and M' chemical shifts. In the latter case, there will be no cross-peak between A and X spins in the TOCSY spectrum, whereas the COSY or DQF-COSY will not be able to distinguish between these two situations.

6.6.2.4 Two-Dimensional Nuclear Overhauser Effect Spectroscopy (2D-NOESY)

This experiment represents an extension of the one-dimensional transient NOE to two dimensions. The pulse sequence for this is given in Fig. 6.27a. τ_m here is called the mixing time during which transfer of magnetization happens through dipolar interactions or the NOE effect. For uncoupled spin systems, the spin dynamics

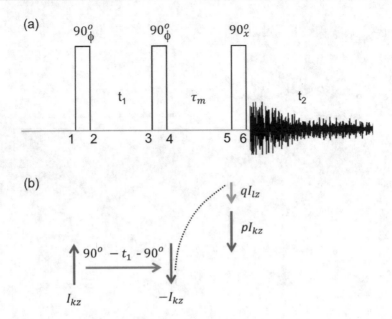

Fig. 6.27 (a) Pulse sequence of NOESY experiment. (b) Schematic of magnetization transfer between two spins (k and l). Numbers 1–6 are time points along the pulse sequence to facilitate discussion in the text

through the pulse sequence leading to transfer of magnetization is schematically shown in Fig. 6.27b.

For coupled spin systems, the pulse sequence can be analyzed by following the product operator formalism, as in the case of COSY. For a two-spin system, k and l (spin $= 1/2$), the density operator at time point 4 in the pulse sequence is given by

$$\rho_4 = \left[-I_{kz} \cos \pi J_{kl} t_1 - 2I_{kx} I_{ly} \sin \pi J_{kl} t_1 \right] \cos \omega_k t_1 \\ + \left[I_{kx} \cos \pi J_{kl} t_1 - 2I_{kz} I_{ly} \sin \pi J_{kl} t_1 \right] \sin \omega_k t_1 \qquad (6.77)$$

As demonstrated in the case of DQF-COSY, a phase cycling scheme is utilized to retain only the first term in Eq. 6.33. This is indicated in Table 6.3.

During the following period τ_m, transfer of z-magnetization occurs from spin k to spin l, as per the dipolar coupling-mediated relaxation of the spins (refer to Chap. 4). The final 90° pulse converts the z-magnetization into transverse magnetization for detection.

Since the transfer of magnetization during the mixing time is never complete, there will be magnetization components of both k and l spins (coupled or uncoupled) evolving during the detection period. These result in the diagonal and cross-peaks, respectively. Both diagonal and cross-peaks will have fine structure if the spins are J-coupled and the components will have in-phase character.

In multi-spin systems, transfer of magnetization will be governed by the relaxation matrix, as discussed in Chap. 4. There will be cross-peaks between every two

Scan	ϕ	Receiver
1	x	+
2	y	+
3	$-x$	+
4	$-y$	+

Table 6.3 Phase cycling for NOESY pulse sequence given in Fig. 6.27a

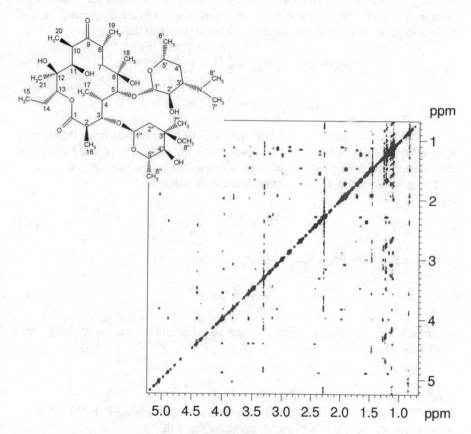

Fig. 6.28 Experimental NOESY spectrum of erythromycin-A

spins which have dipolar coupling contributing to their relaxation. Thus, the NOESY spectrum represents the network of dipolar-coupled spins in a given molecule. The cross-peak intensities will depend upon respective cross-relaxation rates for short mixing times compared to the spin-lattice relaxation time (T_1). These, in turn, are proportional to the inverse sixth power of the internuclear distances; in a sense the cross-peak intensities reflect the NOEs observed in a transient NOE experiment (see Chap. 4). Thus, the NOESY spectrum reflects the distance matrix representing the three-dimensional structure of a given molecule. An illustrative experimental spectrum is shown in Fig. 6.28.

The NOESY pulse sequence also reflects transfer of magnetization through chemical exchange mechanism. During the mixing time, transfer of z-magnetization can also happen via chemical exchange process, wherever it is present. Thus, in such situations, the cross-peak and diagonal peak intensities can also be monitored as a function of the mixing time to derive the exchange rates. For a symmetrical two-site exchange $A \leftrightarrow B$, with equal populations at the two sites, equal spin-lattice relaxation rates, and equal transverse relaxation rates, the intensities of the diagonal (a_{AA}, a_{BB}) and cross-peaks (a_{AB}, a_{BA}) are given by the following equations:

$$a_{AA}(\tau_m) = a_{BB}(\tau_m) = \frac{1}{2}\left[1 + e^{-2k\tau_m}\right]e^{-\tau_m/T_1} \tag{6.78}$$

$$a_{AB}(\tau_m) = a_{BA}(\tau_m) = \frac{1}{2}\left[1 - e^{-2k\tau_m}\right]e^{-\tau_m/T_1} \tag{6.79}$$

where k is the exchange rate and T_1 is the spin-lattice relaxation time. Equilibrium magnetization at the two sites is assumed to be the same. Figure 6.29 shows the dependence of the diagonal and cross-peak intensities on the mixing time.

The ratio of diagonal-to-cross-peak intensities will be

$$\frac{a_{AA}}{a_{AB}} = \frac{1 + e^{-2k\tau_m}}{1 - e^{-2k\tau_m}} \tag{6.80}$$

For short mixing times ($k\tau_m \ll 1$), Eq. 6.80 reduces to

$$\frac{a_{AA}}{a_{AB}} = \frac{1 - k\tau_m}{k\tau_m} \tag{6.81}$$

Thus, by monitoring the intensity ratios as a function of τ_m, the exchange rates can be calculated.

6.6.2.5 Two-Dimensional ROESY

ROESY represents the Overhauser experiment in the rotating frame (ROE). The pulse sequence for this experiment is given in Fig. 6.30.

Here, the mixing process and the consequent magnetization transfer is brought about by low-power spin lock on the transverse magnetization. The magnetization transfer ($I_{kx} \rightarrow I_{lx}$) occurs via transverse cross-relaxation, and the evolution of the magnetization components during the mixing time (τ_m) can be shown to be as follows.

$$I_{kx}(\tau_m) = \left(1 - \frac{\tau_m}{T_2}\right)\sin(\omega_k t_1)I_{kz}^0 - \sigma\tau_m \sin(\omega_l t_1)I_{lz}^0 \tag{6.82}$$

$$I_{lx}(\tau_m) = \left(1 - \frac{\tau_m}{T_2}\right)\sin(\omega_l t_1)I_{lz}^0 - \sigma\tau_m \sin(\omega_k t_1)I_{kz}^0 \tag{6.83}$$

Fig. 6.29 Diagonal (brown line) and cross-peak (green line) intensities in the presence of chemical exchange as a function of mixing time

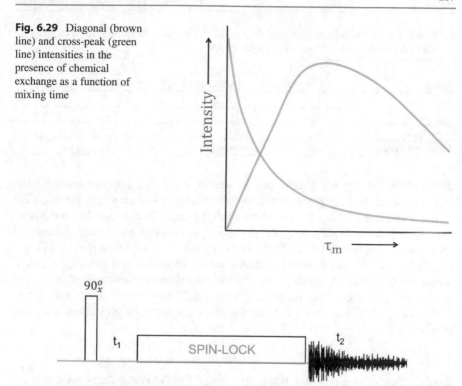

Fig. 6.30 Pulse sequence for two-dimensional ROESY experiment

In both these equations, the first term produces the diagonal peak in the end, and the second term produces the cross-peak. In the initial rate limit, i.e., $\frac{\tau_m}{T_2} \ll 1$, the diagonal peaks are positive, and the cross-peaks will be negative.

Table 6.4 shows the results of detailed calculations of the intensities as a function of spectrometer frequency (ω_o), correlation times (τ_c) of molecular tumbling, and chemical exchange rates (k) for ROESY and NOESY spectra. It is seen that the ROESY spectrum allows the discrimination of ROE and chemical exchange peaks, whereas NOESY will have ambiguities.

The ROESY experiment has some additional advantages in comparison to the NOESY, especially for molecules with $\omega_0\tau_c\sim1$. In such situations, the NOESY spectrum does not show magnetization transfer.

6.6.2.6 Application of Two-Dimensional Homonuclear Experiments in Structural Analysis of Small Organic Molecules: A Case Study of Artemisinin

The combined utilization of two-dimensional homonuclear NMR spectra, viz., DQF-COSY, TOCSY, and NOESY, helps in solving structures of molecules. Figure 6.31 represents the homonuclear two-dimensional NMR spectra recorded on artemisinin molecule (Fig. 6.31d). The chemical shift correlations obtained from

Table 6.4 Comparison of cross-peak and diagonal peak signs in NOESY and ROESY spectra for different molecular tumbling rates and chemical exchange

	NOESY		ROESY	
Condition	Diagonal peak	Cross-peak	Diagonal peak	Cross-peak
$\omega_0\tau_c \ll 1$	+	−	+	−
$\omega_0\tau_c \sim 1$	+	0	+	−
$\omega_0\tau_c \gg 1$	+	+	+	−
Chemical exchange (k)	+	+	+	+

the DQF-COSY (Fig. 6.31a) are useful for identifying the scalarly coupled spin pairs of artemisinin; for example, correlations with the spin 9 allow to get the chemical shift assignments of 8 and for one of the methyl groups. In contrast, the correlations from spin 9 in the TOCSY (Fig. 6.31b) spectrum facilitate monitoring the relayed spin network up to 4–5 bonds. In the present case, from spin 9 to spin 7, TOCSY correlations are observed. Besides, spatial information obtained from the NOESY spectrum (Fig. 6.31c) enables to obtain the three-dimensional structure of artemisinin molecule. The observed NOE correlations between the spin pairs, 12-5″, 12-6, 12-8′, and 8a-5a, confirm the given structure for artemisinin molecule (Fig. 6.31d).

6.6.3 Two-Dimensional Heteronuclear Correlation Experiments

Coherence transfer can also be effected between two different types of nuclear species, say I and S. Such experiments are referred to as heteronuclear correlation experiments. A variety of heteronuclear experiments can be designed, since the RF pulses can be applied selectively to either of the species and heteronuclear broadband decoupling can be incorporated without any constraints. Heteronuclear experiments have particular advantages:

(i) Increased sensitivity of indirect detection as evidenced in the INEPT pulse sequence.

(ii) The possibility of unraveling overlapping I resonances by exploiting the chemical shifts of the S spins and vice versa.

(iii) The correlation of chemical shifts of different nuclear species would facilitate assignments in complex systems.

In most cases, one of the two nuclear species is a rare nucleus (S) such as ^{13}C, ^{15}N, etc., while the other nucleus is usually a more sensitive species (I) such as ^{1}H, ^{19}F, etc.

6.6.3.1 Heteronuclear COSY

The simplest I-S correlation experiment (considering $I = {}^{1}H$ and $S = {}^{13}C$) is depicted in Fig. 6.32. This pulse sequence is very similar to the homonuclear COSY, except

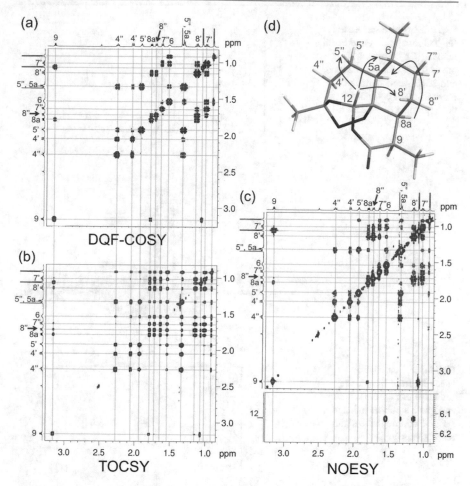

Fig. 6.31 Two-dimensional homonuclear correlation spectra DQF-COSY (**a**), TOCSY (**b**), and NOESY (**c**) recorded on artemisinin molecule dissolved in DMSO-D$_6$ solvent. The combined utilization of all these spectra resulted in the given structure for artemisinin molecule (**d**). The green-colored lines are useful to track the chemical shift correlations, whereas the red-colored arrows are the NOE correlations

the first 90° pulse is selective to only *I* spin. During the t_1 period, therefore, there are only *I* spin coherences. The pair of pulses on *I* and *S*, at the end of t_1 transfers coherence partially to the *S* spin. The magnetization is finally detected on the *S* spin. Thus, the two-dimensional spectrum will have only *I-S* correlation peaks, which retain the fine structure as in the COSY spectrum. Such a spectrum for two spins is schematically shown in Fig. 6.33. It has the antiphase property along both dimensions, and the separation between the components is equal to the coupling constant.

Fig. 6.32 Pulse sequence for
^{1}H-^{13}C correlation experiment
with carbon detection. H_z is
the starting ^{1}H magnetization,
and $2C_yH_z$ represents the ^{13}C
magnetization component at
the beginning of detection

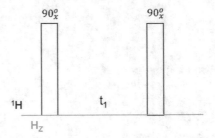

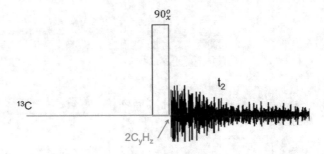

Fig. 6.33 Schematic
spectrum from the pulse
sequence in Fig. 6.31. Orange
and green symbols indicate
positive and negative signs,
respectively

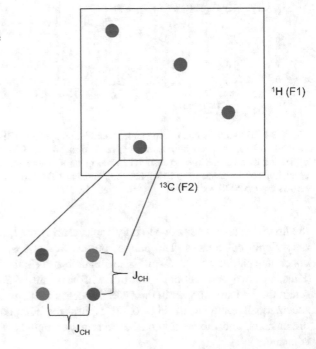

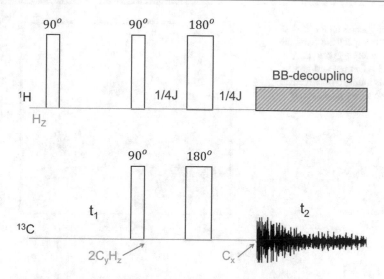

Fig. 6.34 Pulse sequence for ^{1}H-^{13}C correlation with carbon detection and proton decoupling during acquisition. Magnetization components at few time points along the sequence are indicated in cyan. BB implies broadband and J is the one-bond ^{1}H-^{13}C coupling constant

In most of these experiments, the correlation is established between nuclei, which are directly bonded. These one-bond coupling constants are usually very high, for example, J^1H$-^{13}$C ~ 120–160 Hz and J^1H$-^{15}$N ~ 90–100 Hz. While this enables a very efficient transfer of coherence, the large overall width of the cross-peak hampers the resolution in the spectra. Since the one-bond coupling constant does not add too much value for structural information of the molecules, it would be desirable to remove this coupling constant information from the spectrum. This is partially achieved by the pulse sequence shown in Fig. 6.34. In the pulse sequence, an additional spin echo block is added to refocus the S spin antiphase magnetization so that during detection of S magnetization, the I spins can be decoupled. This results in the collapse of the fine structure along the detection axis ($F2$), which is shown in Fig. 6.34. The components here will have twice the intensity as compared to Fig. 6.32.

A further improvement can be achieved by eliminating the coupling information altogether. This can be achieved in more than one ways (Fig. 6.35).

(A) *The HETCOR Pulse Sequence*

The pulse sequence for the HETCOR experiment (considering $I = {}^1$H and $S = {}^{13}$C) is shown in Fig. 6.36. It begins with the excitation of the I spin magnetization by a nonselective 90° pulse. Then this magnetization evolves during the t_1 period during which the I-S coupling is removed by the application of 180° pulse to the S spin in the middle of the t_1 period. Thus, during t_1, the I spins are labelled by their characteristic frequencies. Following the t_1 period, a spin echo block [τ_1 –

Fig. 6.35 Schematic ^{1}H-^{13}C
correlation spectrum from the
pulse sequence given in
Fig. 6.34. Orange and green
symbols indicate positive and
negative signs, respectively

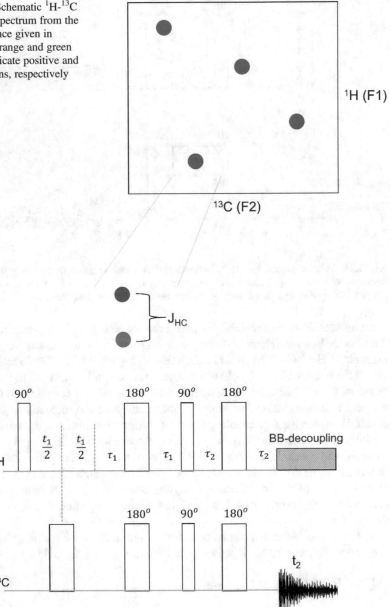

Fig. 6.36 Pulse sequence for the HETCOR experiment, which incorporates ^{1}H-^{13}C decoupling
along both F1 and F2 axes

180 $(I, S) - \tau_1$] is introduced during which I spin magnetization evolves under the I-
S coupling and generates antiphase I spin magnetization. The pair of 90° (I, S) pulses
at the end of the spin echo causes coherence transfer to the S spin, resulting in

Fig. 6.37 Schematic
HETCOR spectrum. The
peaks do not have any fine
structure

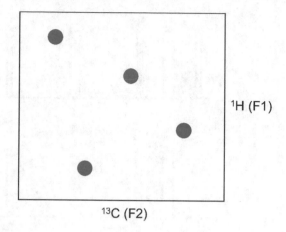

¹H (F1)

¹³C (F2)

antiphase S spin magnetization. The delay τ_1 ($\tau_1 = 1/4J_{IS}$) can be adjusted to cause near-complete transfer to the S spin. Then the antiphase S magnetization evolves during the following spin echo period, $[\tau_2-180\ (I,\ S)-\tau_2]$, to produce in-phase S spin magnetization. This magnetization is detected during t_2, while I spins are simultaneously decoupled. Thus, the resulting spectrum has only one peak for an I-S pair, as shown in Fig. 6.37. The signal-to-noise ratio (SNR) in this experiment is

$$\text{SNR} \propto \gamma_I (\gamma_S)^{\frac{3}{2}} \tag{6.84}$$

(B) *The HSQC Pulse Sequence*

This experiment improves upon the HETCOR experiment. The pulse sequence for the HSQC (heteronuclear single-quantum correlation) is depicted in Fig. 6.38. The experiment starts with an INEPT (refer to Sect. 4.7 in Chap. 4) block which achieves the transfer of I spin magnetization to S spin ($I_z \rightarrow 2I_zS_y$). This S spin magnetization is antiphase in character with respect to the coupled I spin and evolves during the following t_1 period under chemical shift Hamiltonian. Evolution under the I-S coupling is eliminated because of the 180° pulse applied to the I spin in the middle of the t_1 period. Thus, during the t_1 period, the S spins are labeled by their characteristic frequencies. The subsequent pair of 90° pulses on I and S transfers the magnetization back to the I spin as antiphase magnetization ($2I_zS_y \rightarrow 2I_yS_z$). This antiphase I magnetization is then refocused during the next spin echo block, $[\tau_2-180\ (I,\ S)\ -\tau_2]$, to generate in-phase I magnetization (I_x). This in-phase magnetization then evolves during t_2 with characteristic I spin frequencies, while the I-S coupling is removed by broadband S spin decoupling. Thus, the resultant spectrum has a single peak for an I-S pair, as shown Fig. 6.39.

The differences between the experiments A (HETCOR) and B (HSQC) are the following.

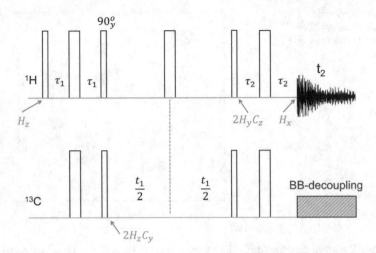

Fig. 6.38 Pulse sequence for the HSQC experiment. Narrow and wide rectangles indicate 90° and 180° pulses, respectively. Unless mentioned all the pulses are applied along the x-axis. Relevant magnetization components at certain time points are indicated in cyan

Fig. 6.39 Schematic appearance of the HSQC spectrum. The peaks do not have any fine structure

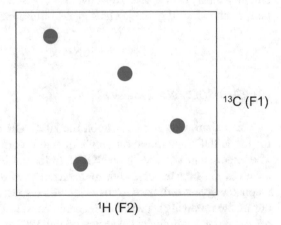

(i) In A, S spin magnetization is detected, whereas in B, I spin magnetization is detected. This has an impact on sensitivity, since the latter is proportional to $\gamma^{3/2}$ of the detected nucleus. Therefore, if I spin is ^{1}H and S spin is ^{13}C, then the HSQC experiment has a sensitivity gain of $\left(\frac{\gamma_H}{\gamma_C}\right)^{3/2}$. This is a factor of 8, which is a substantial gain in terms of the signal-to-noise ratio, which in turn amounts a gain by factor of 64 in terms of the experimental time. Similarly, for $S = {}^{15}$N and $I = {}^{1}$H, the gain will be a factor of ~1000, in terms of experimental time.

(ii) In A, the detected signal will have S frequencies, whereas in B, the detected signal will have I frequencies. The spectral range of S spin (^{13}C, ~140 ppm) is much larger compared to that of the I spin (^{1}H, ~10 ppm). Therefore, the

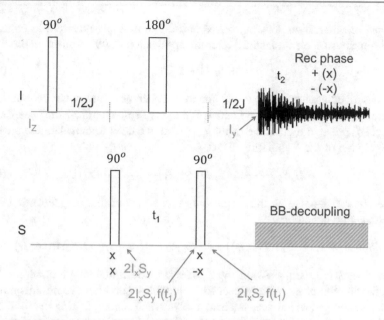

Fig. 6.40 Pulse sequence for the HMQC experiment. Relevant magnetization components at certain time points are indicated in cyan. The *I* spin pulses are applied along x-axis. The pulses on the *S* spin and the receiver are phase cycled as indicated

chemical shift dispersion along the detection axis will be higher in A as compared to that in B, even considering that the γ_H is four times γ_C.

(C) *The HMQC Pulse Sequence*

This experiment achieves coherence transfer from spin *I* to spin *S* via multiple quantum coherences; hence, this is termed as heteronuclear multiple-quantum coherence (HMQC) transfer experiment. The pulse sequence for HMQC is shown in Fig. 6.40.

In Fig. 6.40, the spin *I* is taken to represent the abundant species, and the spin *S* is taken to represent the rare heteronucleus (^{13}C (1.1% abundant)/^{15}N (0.37% abundant)). In the case of protons which are coupled ^{12}C or ^{14}N, the signals coming from these have to be eliminated. This is achieved by phase cycling the receiver ($+x, -x$) in consecutive scans, while the phase of the first 90° pulse on *I* spin remains as $+x$.

The experiment can be analyzed using the product operator formalism, and the flow of the magnetization can be described for a *I-S* two-spin system. The first 90° pulse along the x-axis creates transverse magnetization of the *I* spin:

$$I_z \rightarrow -I_y \tag{6.85}$$

The chemical shift evolution of the *I* magnetization is refocused by the 180° pulse kept at the middle of the entire evolution period before the start of the detection. Hence, this evolution need not be calculated. When the τ period is set equal to $1/2J_{IS}$, the I spin magnetization gets transferred entirely to multiple-quantum IS coherence (double-quantum + zero-quantum) after the application of 90° x-pulse on the S spin.

$$-I_y \rightarrow -2I_xS_y \tag{6.86}$$

As described earlier, this multiple quantum coherence does not evolve under the influence of J-coupling between I and S. Since the I spin chemical shift is refocused by the 180° pulse in the middle of the t_1 period, we need to calculate the chemical shift evolution of the S spin only. Thus,

$$-2I_xS_y \rightarrow -2I_x \left[S_y \cos (\omega_s t_1) - S_x \sin (\omega_s t_1) \right] \tag{6.87}$$

The last 90° x-pulse on the S spin converts a part of this magnetization into a single-quantum coherence.

$$-2I_x \left[S_y \cos (\omega_s t_1) - S_x \sin (\omega_s t_1) \right] \rightarrow -2I_xS_z \cos (\omega_s t_1) + 2I_xS_x \sin (\omega_s t_1) \tag{6.88}$$

The first term in Eq. 6.88 on the right-hand side is the single-quantum I spin magnetization antiphase with respect to S, and the second term represents multiple quantum coherence which does not lead to observable signal. During the following τ period, the antiphase I magnetization gets refocused into in-phase magnetization.

$$-2I_xS_z \cos (\omega_s t_1) \rightarrow -I_y \cos (\omega_s t_1) \tag{6.89}$$

During the detection period t_2, the I spin is decoupled from S spin and thus will only have chemical shift evolution. Thus, we will only have one cross-peak between I and S, as shown in Fig. 6.41. From explicit product operator calculations, the intensity of the cross-peaks in the final spectrum turns out to be proportional to [sin $(\pi J_{IS}\tau)]^2$.

In the examples shown, we have considered one I spin and one S spin. However, in real systems, there will be situations where an S spin is J-coupled to more than one I spin species, which may be scalarly coupled among themselves. Assuming I and

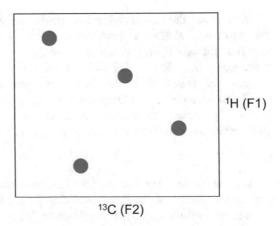

Fig. 6.41 Schematic HMQC spectrum for I-S spin systems

^{1}H (F1)

^{13}C (F2)

K are two such spins of the same nuclear species, a similar calculation will lead to the following observable operator at the start of the detection period.

$$\rho'(t_1) = -I_y \cos(\omega_S t_1) \cos(\pi J_{IK} t_1) + 2I_x K_z \cos(\omega_S t_1) \sin(\pi J_{IK} t_1) \qquad (6.90)$$

This indicates the following:
(i) There will be splitting along the indirect dimension (F1) due to J_{IK}.
(ii) The sum of cosine-cosine and cosine-sine products in Eq. 6.90 results in the superposition of in-phase absorptive and antiphase dispersive line shapes along the F1 axis; this results in mixed phases.

After considering the evolution of I spin magnetization terms in Eq. 6.90 during the following t_2 period, the observable part of the density operator will be

$$\rho''(t_2) = f'(t_1)\{I_y \cos(\omega_I t_2) \cos(\pi J_{IK} t_2) - I_x \sin(\omega_I t_2) \cos(\pi J_{IK} t_2)\}$$
$$+ f''(t_1)\{I_y \cos(\omega_I t_2) \sin(\pi J_{IK} t_2) - I_x \sin(\omega_I t_2) \sin(\pi J_{IK} t_2)\} \qquad (6.91)$$

where

$$f'(t_1) = \cos(\omega_S t_1) \cos(\pi J_{IK} t_1) \text{ and } f''(t_1) = \cos(\omega_S t_1) \sin(\pi J_{IK} t_1)$$

If we assume the detection of y-magnetization, the resultant signal will be

$$\text{Signal} = \cos(\omega_I t_2)[\cos(\pi J_{IK} t_2) f'(t_1) + \sin(\pi J_{IK} t_2) f''(t_1)]$$

$$= \{[\cos(\omega_I + \pi J_{IK})t_2 + \cos(\omega_I - \pi J_{IK})t_2] f'(t_1) + [\sin(\omega_I + J_{IK})t_2 - \sin(\omega_I - \pi J_{IK})t_2] f''(t_1)\}$$
$$(6.92)$$

From this it follows that there will be splitting along the direct dimension ($F2$) due to J_{IK}. Further, there will be superposition of in-phase absorptive and antiphase dispersive line shapes along the detection axis ($F2$), which results in mixed phases.

The dispersive component of the signal can be purged by inserting a 90° y-pulse on I spin prior to the detection period. The antiphase I spin operator is converted into antiphase k spin operator ($2I_x K_z \rightarrow 2I_z K_x$). This results in a cross-peak between K and S spins. If the k spin is not coupled to S spin, this can lead to a confusion with regard to the correlations, though it will have antiphase components as against in-phase components in the I-S cross-peak. These complications do not occur in the HSQC spectra.

6.6.3.2 Heteronuclear Multiple Bond Correlation (HMBC)

The heteronuclear correlation experiments, HSQC, HETCOR and HMQC, rely on the transfer of coherences based on one-bond coupling constants. Often it is necessary to establish correlations via multiple bond coupling constants for unambiguous resonance assignments and structure elucidations. If one has to optimize the coherence transfer to reflect these correlations, the delay τ has to be chosen to be $\frac{1}{2J_{IS}^{\text{long}}}$,

Fig. 6.42 Peak intensities in HMQC experiment as a function of the long-range coupling constant while assuming the T_2 relaxation value equal to 1 s

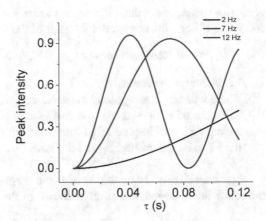

where I and S are spins separated by multiple bonds. These coupling constants are much smaller compared to one-bond couplings, and there is also a large variation in these couplings (1–15 Hz). Simultaneous optimization with respect to all these couplings is not possible. As discussed earlier for HMQC, the intensity of the cross-peaks in the final spectrum turns out to be proportional to $\left[\sin\left(\pi J_{IS}^{\text{long}}\tau\right)\right]^2$. Figure 6.42 shows the variation in the intensities with τ for three different values of long-range couplings (2 Hz, 7 Hz, and, 12 Hz).

From this, it can be seen that the smaller the coupling, the larger is the required delay. In such a situation, relaxation also plays an important role in determining the intensity of the cross-peak. Figure 6.43 shows a pulse sequence designed to circumvent some of these problems. This experiment is referred to as heteronuclear multiple bond correlation (HMBC). It differs from the HMQC pulse sequence in only one sense; i.e., the last refocusing τ period (Fig. 6.40) is eliminated, and accordingly, the decoupling of S spin has also been removed, while this saves on the relaxation loss and the intensity will be proportional to $\left[\sin\left(\pi J_{IS}^{\text{long}}\tau\right)\right]$ (Fig. 6.44), which is better than $\left[\sin\left(\pi J_{IS}^{\text{long}}\tau\right)\right]^2$ dependence. This results in an antiphase splitting of the cross-peak along the detection dimension.

6.6.4   Combination of Mixing Sequences

Depending upon the desired information in the two-dimensional spectrum, it is possible to design pulse sequences, which have a mix of different types of coherence transfer steps discussed in the previous sections. For example, HSQC can be combined with TOCSY or COSY or NOESY transfer, HMQC can be combined with TOCSY or COSY or NOESY, etc. Some typical pulse sequences to achieve these features are shown in Fig. 6.45. The corresponding spectra are shown in Fig. 6.46.

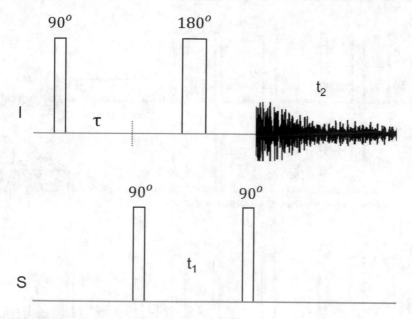

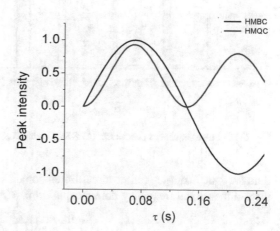

Fig. 6.43 Pulse sequence for the HMBC experiment

Fig. 6.44 A comparison of intensities in HMQC and HMBC spectra for a small J-value (7 Hz) while assuming the T_2 relaxation value equal to 1 s

6.7   Three-Dimensional NMR

The ideas discussed in the context of two-dimensional NMR can be extended to include another dimension resulting in a three-dimensional spectrum. A schematic of such an experiment is indicated in Fig. 6.47.

This consists of a *preparation period*, two *evolution periods* (t_1 and t_2), two *mixing periods* (M1 and M2), and a *direct detection period* (t_3). The resulting time

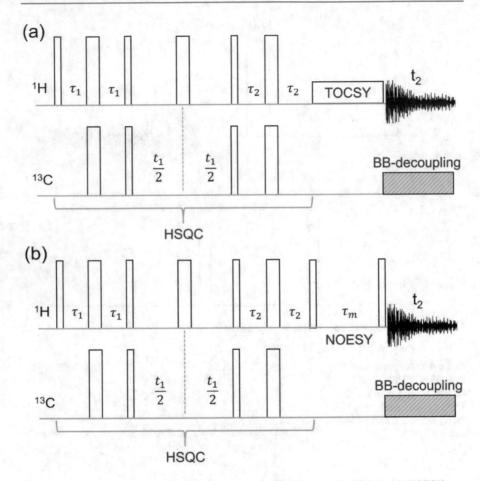

Fig. 6.45 Pulse sequences combining (**a**) HSQC with TOCSY and (**b**) HSQC with NOESY

domain data, $S(t_1, t_2, t_3)$, after three-dimensional Fourier transformation produces a three-dimensional frequency domain spectrum, $S(F1, F2, F3)$.

$$S(t_1, t_2, t_3) \overset{3D-FT}{\rightarrow} S(F1, F2, F3) \tag{6.93}$$

A variety of three-dimensional spectra can be generated by choosing appropriate mixing sequences, M1 and M2. For example, if M1 is chosen to result in a HSQC type of the transfer of coherence with its independent evolution period t_1, and M2 is chosen to result in a TOCSY type of transfer with the evolution period t_2, then in the end, we generate a three-dimensional HSQC-TOCSY spectrum, schematically shown in Fig. 6.48.

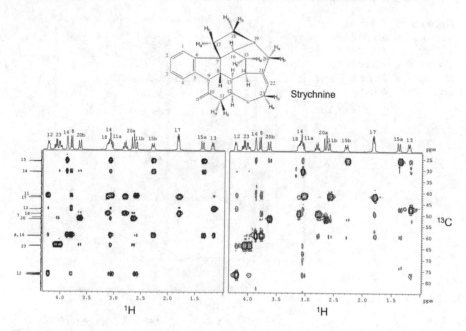

Fig. 6.46 Spectra of (**a**) HSQC-TOCSY and (**b**) HSQC-NOESY recorded on strychnine. It is adapted from the Bruker website

Fig. 6.47 A schematic of a three-dimensional experiment

Similar combinations can be made with COSY, TOCSY, NOESY, HSQC, HMQC, HMBC, etc., to generate a variety of three-dimensional spectra.

A large variety of three-dimensional spectra have been designed for biomolecular applications, especially proteins. These differ in magnetization transfer pathways along the protein chain. While these are covered in several elegant monographs, some of these are indicated in Fig. 6.49, wherein the pathways of magnetization transfer through the chain are indicated.

These rely on the transfer of magnetization via evolution under the influence of one- and two-bond couplings along the polypeptide chain. These coupling constants are independent of the amino acid sequence in the chain, and their typical values are shown in Fig. 6.50.

In the following, we describe briefly a few experiments to demonstrate the analysis of these pulse sequences, in general. All these experiments require proteins uniformly enriched in ^{13}C and ^{15}N isotopes. These are routinely achieved by standard techniques in recombinant protein production. These experiments also require spectrometers equipped with three independent channels, ^{1}H, ^{13}C, and ^{15}N.

Fig. 6.48 Schematic of three-dimensional HSQC-TOCSY spectrum. The TOCSY relay is seen along the F2 axis in the three-dimensional spectrum

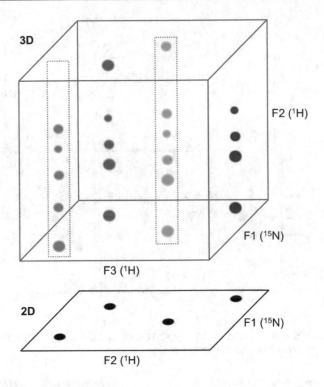

6.7.1   The CT-HNCA Experiment

Figure 6.51 shows the pulse sequence for the constant time (CT)-HNCA experiment.

The flow of magnetization through the pulse sequence is schematically shown in Eq. 6.94:

$$H_i^N \rightarrow N_i(t_1) \rightarrow C^\alpha(i, i-1)(t_2) \rightarrow N_i \rightarrow H_i^N(t_3) \qquad (6.94)$$

It starts with the ^{1}H magnetization, H_z, and the evolution of this magnetization through the pulse sequence can be calculated using the product operator formalism. At time point "a," after the first INEPT transfer from amide proton (H_i^N) to N_i along the backbone, the density operator is given by

$$\rho_a = -2H_{iz}^N N_{iy} \qquad (6.95)$$

H_{iz}^N refers to z-magnetization of the amide proton (H_i^N) of the ith residue along the polypeptide chain, and N_{iy} refers to the y-component of the backbone ^{15}N spin of the ith residue. Thus, this operator represents antiphase ^{15}N magnetization with respect to the amide proton (H_i^N).

Following this, the ^{15}N magnetization evolves for a constant time period T under different Hamiltonians:

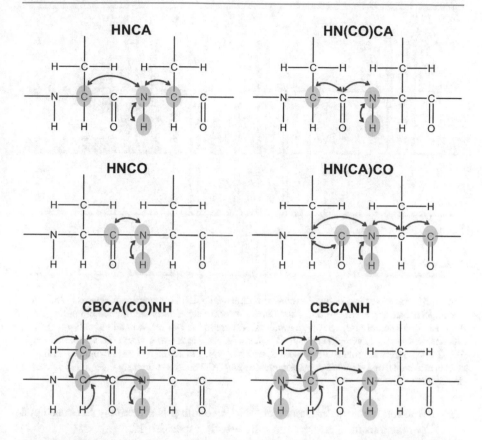

Fig. 6.49 Schematic of magnetization transfer pathways in HNCA, HN(CO)CA, HNCO, HN (CA)CO, CBCA(CO)NH, and CBCANH experiments. Red arrows identify magnetization transfers during the pulse sequence, and the atoms enclosed in cyan circles are the nuclei participating the transfer process

Fig. 6.50 One- and two-bond coupling constants relevant for magnetization transfers shown in Fig. 6.48

(i)   Under the influence of ^{15}N chemical shifts for a period t_1.

$$-2H^N_{iz}N_{iy} \rightarrow -2H^N_{iz}(N_{iy}\cos(\omega_{Ni}t_1) - N_{ix}\sin(\omega_{Ni}t_1))\qquad(6.96)$$

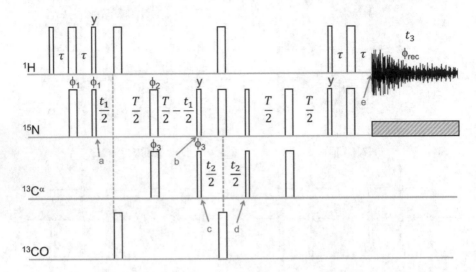

Fig. 6.51 Pulse sequence for the constant time (CT)-HNCA experiment. Narrow and wide rectangular bars represent 90° and 180° pulses, respectively. Pulses are applied along the x-axis unless indicated otherwise. The phase cycles $\phi1$–$\phi3$ are as follows: $\phi1 = x$ and $-x$; $\phi2 = 4(x)$, $4(y)$, $4(-x)$, and $4(-y)$; and $\phi3 = 2(x)$ and $2(-x)$. ϕrec $= x, -x, -x, x, -x, x, x$, and $-x$. The period $\tau = 1/4J_{NH}$. T represents the constant time period, which is typically 22–25 ms. Broadband decoupling of ^{15}N is achieved using standard composite decoupling. Alphabets a–e identify time points discussed in the text

(ii) Evolution under the influence of ^{15}N-^{1}H coupling is effectively refocused and ^{15}N magnetization remains anti-phase with respect to ^{1}H.

(iii) Under the influence of one-bond ($N_i - C_i^\alpha$) and two-bond ($N_i - C_{(i-1)}^\alpha$) couplings for the period T.

(iv) Evolution under ^{15}N-^{13}CO coupling is removed by the application of 180° band-selective pulse to the ^{13}CO spins and also to ^{15}N.

During the period $\frac{t_1}{2} - \pi(CO) - \frac{t_1}{2}$, decoupling happens due to the 180° pulse on the carbonyl spins.

During the next period, $\left[\left(\frac{T}{2} - \frac{t_1}{2}\right) - \pi\left(^{15}N\right) - \left(\frac{T}{2} - \frac{t_1}{2}\right)\right]$, decoupling happens due to the 180° pulse on ^{15}N.

Thus, at time point "b," the density operator is given by

$$\rho_b = \left\{ 4H_{iz}^N N_{ix} C_{iz}^\alpha \Gamma_1(T) + 4H_{iz}^N N_{ix} C_{(i-1)z}^\alpha \Gamma_2(T) \right\} \cos\left(\omega_{N_i} t_1\right) \tag{6.97}$$

where

$$\Gamma_1(T) = \sin\left(\pi^1 J_{C^\alpha N} T\right) \cos\left(\pi^2 J_{C^\alpha N} T\right) \tag{6.98}$$

$$\Gamma_2(T) = \cos\left(\pi^1 J_{C^\alpha N}T\right)\sin\left(\pi^2 J_{C^\alpha N}T\right) \tag{6.99}$$

$\Gamma_1(T)$ and $\Gamma_2(T)$ represent transfer efficiencies, which are seen to be dependent on the constant time period T and the magnitudes of the coupling constants. In this calculation relaxation has been ignored. However, relaxation will be occurring which causes an exponential decay ($\exp(-R_N T)$, where R_N is the transverse relaxation rate of ^{15}N magnetization. Therefore, the constant time period T has to be properly optimized for the efficient transfer of magnetization without losing too much signal.

The first term in Eq. 6.97 represents ^{15}N magnetization of the ith residue, antiphase with respect to H^N and C^α of the ith residue. The second term represents ^{15}N magnetization of the ith residue antiphase with respect to H^N of the ith residue and C^α of the $(i-1)^{th}$ residue. Thus, a sequential correlation between i and $(i-1)$ residues is created. Following the application of a pair of 90° pulses to ^{15}N and C^α spins at the end of the T period results in the density operator at time point "c":

$$\rho_c = \left\{4H^N_{iz}N_{iz}C^\alpha_{iy}\Gamma_1(T) + 4H^N_{iz}N_{iz}C^\alpha_{(i-1)y}\Gamma_2(T)\right\}\cos(\omega_{N_i}t_1) \tag{6.100}$$

Now the magnetization is on C^α of "i" (first term in 6.100) and $(i-1)$ (second term in 6.100) residues. This magnetization evolves for the t_2 period under the influence of C^α chemical shifts. All the coupling evolutions (except $C^\alpha - C^\beta$ coupling) are eliminated by simultaneous 180° pulses on CO, ^{15}N, and ^1H channels. At the end of the t_2 evolution (i.e., at time point "d"), the relevant density operator is given by

$$\rho_d = \left\{4H^N_{iz}N_{iz}C^\alpha_{iy}\cos\left(\omega_{C^\alpha_i}t_2\right)\Gamma_1(T) + 4H^N_{iz}N_{iz}C^\alpha_{(i-1)y}\cos\left(\omega_{C^\alpha_{i-1}}t_2\right)\Gamma_2(T)\right\}\cos(\omega_{N_i}t_1)\cdot\cos\left(\pi J_{C^\alpha C^\beta}t_2\right) \tag{6.101}$$

This magnetization is then transferred back to the coupled ^{15}N spins by the simultaneous application of 90° pulses on C^α and ^{15}N. The ^{15}N magnetization which is antiphase with C^α and also ^1H then evolves for the constant time period T to refocus the antiphase character with respect to ^{15}N. At the end of the T period, we have ^{15}N magnetization which is antiphase with respect to coupled H^N spins. A pair of 90° pulse on ^{15}N and ^1H's at this point transfers the magnetization to amide protons. This proton magnetization is antiphase with respect to the ^{15}N, and during the next INEPT block gets refocused to produce in-phase amide proton magnetization. The relevant density operator at this point "e" is given by

$$\rho_e = H^N_{ix}\left\{\cos\left(\omega_{C^\alpha_i}t_2\right)\Gamma_1(T) + \cos\left(\omega_{C^\alpha_{i-1}}t_2\right)\Gamma_2(T)\right\}\cos(\omega_{N_i}t_1)\cdot\cos(\pi J_{C^\alpha C^\beta}t_2) \tag{6.102}$$

This proton magnetization is then detected during the time t_3, while ^{15}N is decoupled in a broadband fashion. Thus, the resulting three-dimensional spectrum can be represented by

$$S(t_1, t_2, t_3) \xrightarrow{3D-FT} S[F1(^{15}N), F2(C^\alpha), F3(H^N)] \qquad (6.103)$$

This is schematically shown in Fig. 6.52a. The $F2$–$F3$ cross-section plane at particular $^{15}N_i$ along $F1$ through this three-dimensional spectrum is shown in Fig. 6.52b. It is clearly seen that this experiment allows establishing correlations between two neighboring amino acid residues, which allows sequential walk along the polypeptide chain as indicated in Fig. 6.53. Each strip shows correlations between the amide protons of a particular residue, say i, to the ^{15}N of the same residue i (self-peak) and to the ^{15}N of the previous residue $i-1$ (sequential peak). Typically, the self-peak has slightly higher intensity than the sequential peak.

In practical terms one has to scan through the ^{15}N planes along the $F1$ axis to find H^N–C^α correlation peaks at the appropriate chemical shifts to establish such connectivities. While this works elegantly when the chemical shift dispersions are very good, difficulties arise when there are degeneracies in the chemical shifts. This happens particularly for C^α chemical shifts in disordered and flexible regions of proteins, and sequential connectivities become ambiguous. Several other three-dimensional experiments have been designed to circumvent such problems, and these have been described in great details in many other books (see, e.g., Cavanagh et al., protein NMR spectroscopy). We describe one particular development which has not been covered in any book. Even here, we restrict to the very basic technique (Sanjay et al. 2001); several additions, improvements, and enhancements have been published in the literature.

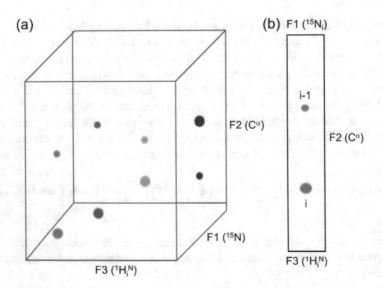

Fig. 6.52 (a) Schematic spectrum of three-dimensional CT-HNCA. (b) The $F2$–$F3$ cross-section at a particular ^{15}N chemical shift along $F1$. Different colors are used to distinguish between the residues, and larger and smaller circles indicate self- and sequential correlations

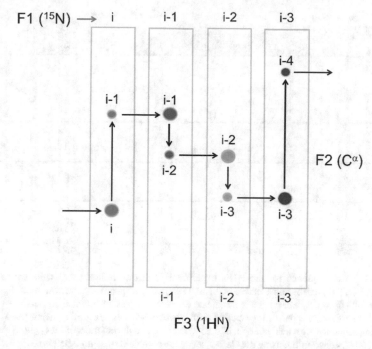

Fig. 6.53 Sequential walk through the polypeptide chain from residues i to $i-4$. Selected $F2$–$F3$ strips at $F1$ chemical shifts indicated on the top are aligned to show the sequential connectivities through the polypeptide sequence

6.7.2   The HNN Experiment

This experiment is derived by simple modification of the HNCA, and the pulse sequence is shown in Fig. 6.54. It follows the magnetization transfer pathway shown in Fig. 6.55.

The basic differences with respect to the HNCA are: (i) Both the $F1$ and $F2$ axes have ^{15}N chemical shifts, whereas in HNCA $F1$ has ^{15}N and $F2$ has C^{α} chemical shifts, and (ii) an additional coherence transfer step is included to transfer the magnetization to the neighboring residues ($i \rightarrow i-1$, $i \rightarrow i+1$). The periods $2T_N$ and $2\tau_{CN}$ in Fig. 6.54 are constant time periods during which magnetization transfers take place. In the first $2T_N$ period, the transfer happens from ^{15}N of residue "i" to C^{α} spins of residues "i" and ($i-1$). During the $2\tau_{CN}$ constant time period, magnetization transfer occurs from C^{α} of "i" residue to ^{15}N of i and $i+1$ residues; likewise, the transfer also occurs from C^{α} of ($i-1$) the residue to ^{15}N of i and ($i-1$) residues. Thus, a sequential correlation gets established between three consecutive residues, $i-1$, i, and $i+1$. The constant time periods, $2T_N$ and $2\tau_{CN}$, are adjusted to be around 22–30 ms. The z-field gradients used in the pulse sequence destroy the unwanted transverse components of the magnetization at different stages. Just before the

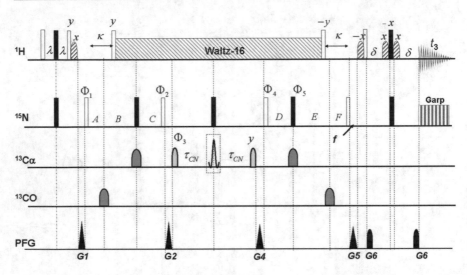

Fig. 6.54 Pulse sequence for the HNN experiment. Narrow (hollow) and wide (filled black) rectangular bars represent nonselective 90° and 180° pulse, respectively. Narrow lobe (light blue) and wide lobe (gray) on carbon channel indicate selective 90° and 180° pulse, respectively. Unless indicated otherwise, the pulses are applied with phase x. Proton decoupling using the Waltz-16 decoupling sequence with field strength of 6.25 kHz is applied during most of the t_1 and t_2 evolution periods, and ^{15}N decoupling using the Garp-1 sequence with field strength 0.9 kHz is applied during acquisition. The ^{13}C carrier frequency for pulses, respectively, on $^{13}C^\alpha$ and ^{13}CO channels are set at 54.0 ppm and 172.5 ppm. The strengths of the $^{13}C^\alpha$ pulses (standard Gaussian cascade Q3 (180°) and Q5 (90°) pulses) are adjusted so that they cause minimal excitation of carbonyl carbons. The 180° ^{13}CO-shaped pulse (width 200 μs) had a standard Gaussian cascade Q3 pulse profile with minimal excitation of $^{13}C^\alpha$. The delays are set to $\lambda = 2.7$ ms, $\kappa = 5.4$ ms, $\delta = 2.7$ ms. The delay τ_{CN} used for the evolution of one-bond and two-bond $^{13}C^\alpha - {}^{15}N$coupling is around 12–16 ms and must be optimized. The values for the individual periods containing t_1 are $A = t_1/2$, $B = T_N$, and $C = T_N-t_1/2$. The values for the individual period containing t_2 are $D = T_N-t_1/2$, $E = T_N$, and $F = t_1/2$. The delay $2T_N$ is set to 24–28 ms. Phase cycling for the experiment is $\Phi_1 = 2(x), 2(-x)$; $\Phi_2 = \Phi_3 = x, -x$; $\Phi_4 = x$; and $\Phi_5 = 4(x), 4(-x)$ and receiver $= 2(x), 4(-x)$, and $2(x)$. Frequency discrimination in t_1 and t_2 is achieved using states-TPPI phase cycling of Φ_1 and Φ_4, respectively, along with the receiver phase. The gradient (sine-bell shaped; 1 ms) levels are optimized between 30% and 80% of the maximum strength of 53 G/cm in the z-direction. These destroy the unwanted transverse magnetization components

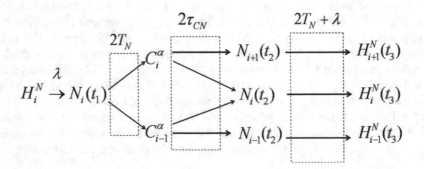

Fig. 6.55 A schematic of magnetization transfer pathway through the HNN pulse sequence

detection, the Watergate pulse block is used to achieve an efficient water suppression.

The experiment can be analyzed in detail using the product operator formalism. Considering a chain of four residues, $i-2$ to $i+1$ along the polypeptide chain, the intensities of the diagonal $\left(I_i^d; F1 = F2 = N_i, \quad F3 = H_i^N\right)$ and cross-peaks $\left(\left(I_{i-1}^c; F1 = N_i, \quad F2 = N_{i-1}, F3 = H_{i-1}^N\right)\right.$ and $\left(I_{i+1}^c; F1 = N_i, \quad F2 = N_{i+1}, \quad F3 = H_{i+1}^N\right)\left.\right)$ in the $F2$–$F3$ plane of the HNN spectrum turn out to be

$$I^d = -\left(E_1^2 E_3 E_9 K_{i1}^d + E_2^2 E_5 E_{10} K_{i2}^d\right) \tag{6.104}$$

$$I_{i-1}^c = E_1 E_4 E_7 E_9 K_{i-1}^c$$

$$I_{i+1}^c = E_2 E_6 E_8 E_{10} K_{i+1}^c \tag{6.105}$$

where

$$E_1 = \cos p_i T_N \sin q_{i-1} T_N$$

$$E_2 = \sin p_i T_N \cos q_{i-1} T_N$$

$$E_3 = \cos p_{i-1} \tau_{CN} \cos q_{i-1} \tau_{CN}$$

$$E_4 = \sin p_{i-1} \tau_{CN} \sin q_{i-1} \tau_{CN}$$

$$E_5 = \cos p_i \tau_{CN} \cos q_i \tau_{CN}$$

$$E_6 = \sin p_i \tau_{CN} \sin q_i \tau_{CN}$$

$$E_7 = \sin p_{i-1} T_N \cos q_{i-2} T_N$$

$$E_8 = \cos p_{i+1} T_N \sin q_i T_N$$

$$E_9 = \cos n_{i-1} \tau_{CN}$$

$$E_{10} = \cos n_i \tau_{CN} \tag{6.106}$$

and

$$p_i = 2\pi^1 J\left(C_i^\alpha - N_i\right); q_i = 2\pi^2 J\left(C_i^\alpha - N_{i+1}\right); n_i = 2\pi^1 J\left(C_i^\alpha - C_i^\beta\right) \tag{6.107}$$

$$K_{i1}^d = \exp\left(-4T_N R_{2i}^N - 2\tau_{CN} R_{2,i-1}^\alpha\right)$$

$$K_{i2}^d = \exp\left(-4T_N R_{2i}^N - 2\tau_{CN} R_{2i}^\alpha\right)$$

$$K_{i-1}^c = \exp\left(-2T_N\left(R_{2i}^N + R_{2,i-1}^N\right) - 2\tau_{CN} R_{2,i-1}^\alpha\right)$$

$$K_{i+1}^c = \exp\left(-2T_N\left(R_{2i}^N + R_{2,i+1}^N\right) - 2\tau_{CN}R_{2i}^\alpha\right) \tag{6.108}$$

^1J and ^2J represent one-bond and two-bond N-C$^\alpha$ coupling constants and R_2^N and R_2^α's are the various ^{15}N and C$^\alpha$ transverse relaxation rates, respectively.

This data after three-dimensional Fourier transformation yields the three-dimensional NMR spectrum.

$$S(t_1, t_2, t_3) \overset{3D-FT}{\rightarrow} S\left[F1\left(^{15}N\right), F2\left(^{15}N\right), F3\left(H^N\right)\right] \tag{6.109}$$

Equation 6.104 (I^d) gives rise to the diagonal peak ($F1 = F2 = N_i$, $F3 = H_i^N$) in the three-dimensional spectrum. The first term in Eq. 6.105 yields the cross-peak (I_{i-1}^c; $F1 = N_i$, $F2 = N_{i-1}$, $F3 = H_{i-1}^N$). The second term in Eq. 6.105 yields another cross-peak $\left(I_{i+1}^c; F1 = N_i, F2 = N_{i+1}, F3 = H_{i+1}^N\right)$. A schematic representation of the three-dimensional spectrum is shown in Fig. 6.56a, and in Fig. 6.56b, c are shown, respectively, the $F1 - F3$ plane at $F2 = N_i$ and $F2 - F3$ plane at $F1 = N_i$.

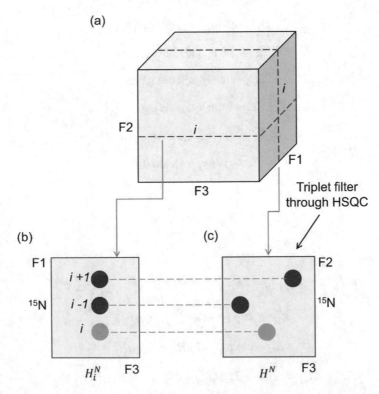

Fig. 6.56 (a) Schematic representation of the three-dimensional spectrum of HNN. (b) Schematic representation of $F1 - F3$ plane at a particular $F2 = N_i$. (c) Schematic representation of $F2 - F3$ plane at $F1 = N_i$. Cyan peaks are self-peaks ($F2 = F1 = N_i$), and the red peaks are sequential peaks

Clearly, the $F2 - F3$ plane at a particular $F1 = N_i$ is a triplet filter through the HSQC spectrum displaying exclusively the peaks of three consecutive residues ($i-1$, i, $i+1$) along the polypeptide chain. The orthogonal $F1 - F3$ plane at a particular $F2 = N_i$ shows correlations from the amide proton of residue "i" to the ^{15}N of residues $i-1$, i, and $i+1$. This feature eliminates the need to scan through the ^{15}N planes as in the HNCA experiment to establish sequential correlations.

Figure 6.57 shows the coherence transfer efficiencies and consequent intensities of the diagonal and cross-peaks with and without including relaxation. These curves indicate the optimum value to be chosen for the $2\tau_{CN}$ period. As mentioned earlier, a value of 22–30 ms turns out to be the optimum choice, which gives reasonable intensities for both the diagonal and the cross-peaks.

The HNN experiment has an additional interesting feature in the patterns of peaks. The diagonal and cross-peaks will have different combinations of positive and negative signs depending upon the nature of the residues in the triplet sequence represented by the chosen plane. This feature arises because of the fact that during the $2\tau_{CN}$ period, the magnetization on C^α evolves under the influence of $C^\alpha - C^\beta$ coupling; the coefficients E9 and E10 which reflect this coupling evolution contribute to the change in sign patterns of the diagonal and cross-peaks. Since the glycine

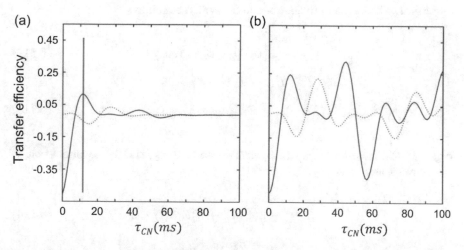

Fig. 6.57 Plots of the HNN coherence transfer efficiencies. The transfer functions for the diagonal peak I^d and the cross-peaks I^c_{i-1}, I^c_{i+1}. Here, (**a**) is for the transfer efficiencies calculated with relaxation terms, while (**b**) is for calculations without the relaxation terms. The transfer efficiency is plotted as a function of τ_{CN}. The plots were calculated by using, $J_{C^\alpha C^\beta}$, $J_{C^\alpha CO}$, and J_{NCO} values of 35, 55, and 15 Hz, respectively. The $^1J_{C^\alpha N}$, and $^2J_{C^\alpha N}$ values have been chosen to be 10.5 and 8.5 Hz, respectively. The value of T_N used in the transfer functions for HNN was 14.0 ms. Thick and dotted lines represent diagonal and sequential peaks, respectively. The vertical red line indicates the optimum choice for the τ_{CN} value. For this choice, the diagonal and cross-peaks have opposite signs. (Reproduced from Journal of Magnetic Resonance. 181, 21 (2006), with the permission of Elsevier Publishing)

residues do not have a C^β carbon, they appear distinctly and generate different peak patterns depending upon the position of the glycine in the triplet sequence. Four different cases of triplet amino acid sequences can be considered: (i) XGZ, (ii) GYZ, (iii) G'GZ, and (iv) ZXY, where X, Y, and Z can be any amino acid residue other than glycine and proline and G is glycine. The cases (i)–(iii) are special cases containing glycine in the triplet sequence, and case (iv) is a general case. For the three special cases (i) to (iii), the relevant density operators before the start of detection are given by the following:

(i)   -XGZ-: In this case the $C^\alpha - C^\beta$ couplings are absent and hence $E_{10} = 1$. Thus, the transfer efficiencies are as follows:

$$I^d = -\left(E_1^2 E_3 E_9 K_{i1}^d + E_2^2 E_5 K_{i2}^d\right) \tag{6.110}$$

$$I_{i-1}^c = E_1 E_4 E_7 E_9 K_{i-1}^c$$

$$I_{i+1}^c = E_2 E_6 E_8 K_{i+1}^c \tag{6.111}$$

(ii) -GYZ-: In this case $C^\alpha - C^\beta$ coupling of the $(i-1)$th residue vanishes and hence E9=1. The transfer efficiencies can be written as follows:

$$I^d = -\left(E_1^2 E_3 K_{i1}^d + E_2^2 E_5 E_{10} K_{i2}^d\right) \tag{6.112}$$

$$I_{i-1}^c = E_1 E_4 E_7 K_{i-1}^c$$

$$I_{i+1}^c = E_2 E_6 E_8 E_{10} K_{i+1}^c \tag{6.113}$$

(iii) -G'GZ-: Here both E_{10} and E_9 terms become unity, and the equations can be written as:

$$I^d = -\left(E_1^2 E_3 K_{i1}^d + E_2^2 E_5 K_{i2}^d\right) \tag{6.114}$$

$$I_{i-1}^c = E_1 E_4 E_7 K_{i-1}^c$$

$$I_{i+1}^c = E_2 E_6 E_8 K_{i+1}^c \tag{6.115}$$

The calculated peak patterns in $F1$–$F3$ planes for various combinations of triplets of sequences involving a glycine residue at different positions in the triplet are shown in Fig. 6.58. If there is a proline residue at either $(i-1)$ or $(i+1)$ position in the triplet, the corresponding peak will not appear in the spectrum.

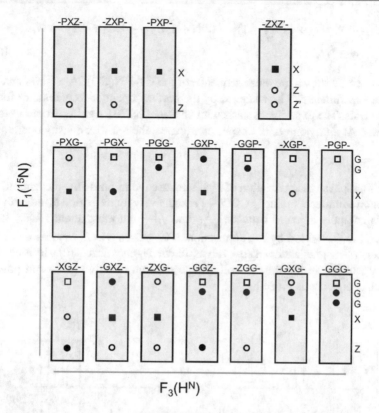

Fig. 6.58 Schematic patterns in the $F1$–$F3$ planes at the $F2$ chemical shift of the central residue in the triplets mentioned on the top of each panel, in the HNN spectra for various special triplet sequences. X, Z, and Z' are any residue other than glycine (G) and proline (P). Squares are the diagonal peaks and circles are the sequential peaks. Filled and open symbols represent positive and negative signals, respectively. In all cases, the peaks are aligned at the $F3$ (H^N) chemical shift of the central residue

6.7.3 The Constant Time HN(CO)CA Experiment

The HNCA experiment described earlier establishes the correlation between residue "i" and ($i-1$) along the polypeptide chain. A particular cross-section plane along the ^{15}N axis shows peaks between amide protons of residue "i" and C^α carbons of residues i and $i-1$. However, a priori, it is not possible to identify the i and $i-1$ peak individually, unambiguously. The HN(CO)CA experiment has been designed to circumvent this problem by adopting a different magnetization transfer pathway, which allows the flow of magnetization through the pulse sequence in one direction along the polypeptide chain. This is as indicated in Eq. 6.116.

$$H_i^N \rightarrow N_i(t_1) \rightarrow CO(i-1) \rightarrow C^\alpha(i-1)(t_2) \rightarrow CO(i-1) \rightarrow N_i$$
$$\rightarrow H_i^N(t_3) \tag{6.116}$$

Figure 6.59 shows the pulse sequence of the CT-HN(CO)CA experiment.

The experiment can be analyzed by the product operator formalism as for other experiments. The experiment starts with an initial INEPT transfer from H_i^N to N_i of residue i. At time point a, in the pulse sequence, the relevant density operator is

$$\rho_a = -2H_{iz}^N N_{iy} \tag{6.117}$$

This antiphase magnetization of N_i is refocused to in-phase magnetization, which then evolves under coupling to $CO(i-1)$ exclusively for the period $\delta_1 + \delta_2 + \delta_3 = 2\delta_1$. This is normally adjusted between $\frac{1}{2J_{NCO}}$ and $\frac{1}{3J_{NCO}}$ and most often it is set to $\frac{1}{3J_{NCO}}$. Note, C^α is decoupled by the application 180° pulses on C^α channel and the ^{15}N channel. N_i-magnetization also evolves under N_i-chemical shifts leading to frequency labeling in the time period t_1. The relevant density operator at point b in the pulse sequence is given by

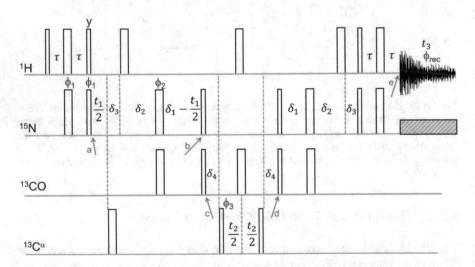

Fig. 6.59 Pulse sequence for the CT-HN(CO)CA experiment. Wide and narrow rectangles indicate 180° and 90° pulses, respectively. Typically, the delays are $2\delta_1 \approx 22$ ms $\left[\approx \frac{1}{3J_{NCO}}\right]$, $2\delta_3 \approx \frac{1}{2J_{NH}}$, $\delta_2 = (\delta_1 - \delta_3)$, and δ_4 in the range $\frac{1}{3J_{C^\alpha CO}}$ to $\frac{1}{2J_{C^\alpha CO}}$. Unless mentioned, the pulse phases are along the x-axis. The phase cycles mentioned are $\phi1 = x, -x$; $\phi2 = 4(x), 4(y), 4(-x), 4(-y)$; $\phi3 = 2(x), 2(-x)$. and $\phi rec = x, -x, -x, x, -x, x, x, -x$. Quadrature detection in the t_1 and t_2 dimensions is achieved by incrementing independently the phases $\phi1$ and $\phi3$, respectively, along with the receiver phase, as in a states-TPPI manner

$$\rho_b = 2\, N_{iy} CO_{(i-1)z} \cos(\omega_{N_i} t_1) \sin(2\pi J_{NCO}\delta_1) \sin(2\pi J_{NH}\delta_3) \qquad (6.118)$$

The simultaneous application of 90° pulses on CO and ^{15}N channels at this point causes transfer into antiphase CO magnetization, and the relevant density operator at time point c is given by

$$\rho_c = -2\, N_{iz} CO_{(i-1)y} \cos(\omega_{N_i} t_1) \sin(2\pi J_{NCO}\delta_1) \sin(2\pi J_{NH}\delta_3) \qquad (6.119)$$

This magnetization then evolves under C^α coupling for the period δ_4, and it is transferred to $C^\alpha(i-1)$, which is then frequency labeled during the period t_2. At the end it is back transferred to $CO(i-1)$, which continues to evolve under the $C^\alpha - CO$ coupling, for the next δ_4 period, and at time point d, the relevant density operator is

$$\rho_d = -2 N_{iz} CO_{(i-1)y} \cos(\omega_{N_i} t_1) \cos(\omega_{C_{i-1}^\alpha} t_2) \cos(\pi J_{C^\alpha C^\beta} t_2) \sin^2(\pi J_{C^\alpha CO}\delta_4) \sin(2\pi J_{NCO}\delta_1) \sin(2\pi J_{NH}\delta_3)$$

$$(6.120)$$

Magnetization is now on $CO_{(i-1)}$. During the subsequent part of the pulse sequence, the magnetization retraces the path evolving under the various couplings, and at time point e, the relevant density operator leading to observable magnetization is given by

$$\rho_e = H_{ix}^N \cos(\omega_{N_i} t_1) \cos(\omega_{C_{i-1}^\alpha} t_2) \cos(\pi J_{C^\alpha C^\beta} t_2) \sin^2(\pi J_{C^\alpha CO}\delta_4) \sin^2(2\pi J_{NCO}\delta_1) \sin^2(2\pi J_{NH}\delta_3)$$

$$(6.121)$$

The amide proton magnetization is then detected during "t_3" under ^{15}N decoupling.

After three-dimensional Fourier transformation, this leads to a spectrum schematically shown in Fig. 6.60.

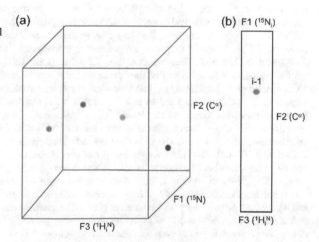

Fig. 6.60 (a) Schematic spectrum of three-dimensional CT-HN(CO)CA. (b) The F2-F3 cross-section at a particular ^{15}N chemical shift along F1. Here only the sequential connections i to $(i-1)$ are seen. Different colors are used to distinguish between the residues

6.7.4   The HN(C)N Experiment

This is a counter part of HNN in the same manner as HN(CO)CA is a counter part of the HNCA experiment, providing the directionality to the sequential assignment process of the backbone atoms along the polypeptide chain. The pulse sequence for this experiment is shown in Fig. 6.61, and the magnetization transfer pathway is shown in Fig. 6.62.

Note that the flow of the magnetization is similar to that in HN(CO)CA till the point reaches $C^\alpha(i-1)$. The C^α's are not frequency labeled, and the magnetization is transferred directly to the ^{15}N of residues i and $(i-1)$. This involves an additional

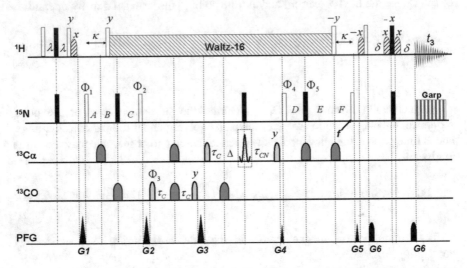

Fig. 6.61 Pulse sequence for the HN(C)N experiment. Narrow (hollow) and wide (filled black) rectangular bars represent nonselective 90° and 180° pulse, respectively. Narrow lobe (light blue) and wide lobe (gray) on carbon channel indicate selective 90° and 180° pulse, respectively. Unless indicated otherwise, the pulses are applied with phase x. Proton decoupling using the Waltz-16 decoupling sequence with field strength of 6.25 kHz is applied during most of the t_1 and t_2 evolution periods, and ^{15}N decoupling using the Garp-1 sequence with field strength 0.9 kHz is applied during acquisition. The ^{13}C carrier frequency for pulses, respectively, on $^{13}C^\alpha$ and ^{13}CO channels are set at 54.0 ppm and 172.5 ppm. The strength of the $^{13}C^\alpha$ pulses (standard Gaussian cascade Q3 (180°) and Q5 (90°) pulses) is adjusted so that they cause minimal excitation of carbonyl carbons. The 180° ^{13}CO-shaped pulse (width 200 μs) had a standard Gaussian cascade Q3 pulse profile with minimal excitation of $^{13}C^\alpha$. The delays are set to $\lambda = 2.7$ ms, $\kappa = 5.4$ ms, $\delta = 2.7$ ms. The delay τ_{CN} used for the evolution of one-bond and two-bond $^{13}C^\alpha - ^{15}N$ coupling is around 12–16 ms and must be optimized. The delay τ_C in the pulse sequence used for $^{13}C^\alpha - ^{13}C'$ (refers to carbonyl, CO, carbon) coupling evolution is 4.5 ms. The values for the individual periods containing t_1 are $A = t_1/2$, $B = T_N$, and $C = T_N - t_1/2$. The values for the individual period containing t_2 are $D = T_N - t_1/2$, $E = T_N$, and $F = t_1/2$. The delay $2T_N$ is set to 24–28 ms, and $\Delta = \tau_{CN} - \tau_C$. Phase cycling for the experiment is $\Phi_1 = 2(x), 2(-x)$; $\Phi_2 = \Phi_3 = x, -x$; and $\Phi_4 = x$; $\Phi_5 = 4(x), 4(-x)$ and receiver $= 2(x), 4(-x), 2(x)$. The frequency discrimination in t_1 and t_2 has been achieved using states-TPPI phase cycling of Φ_1 and Φ_4, respectively, along with the receiver phase. The gradient (sine bell shaped; 1 ms) levels are optimized between 30% and 80% of the maximum strength of 53 G/cm in the z-direction. These destroy the unwanted transverse magnetization components

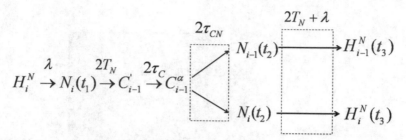

Fig. 6.62 Schematic of magnetization transfer pathway through the HN(C)N pulse sequence. Here C' refers to carbonyl (CO) carbon

transfer period, $2\tau_{CN}$. The ^{15}N are frequency labeled in the evolution time t_2, and finally the magnetization is transferred to amide protons of residues i and $(i-1)$, which are then detected during the detection period t_3.

The experiment can be analyzed using the product operator as has been done in the previous cases. The final relevant density operator at the start of the detection (t_3) is given by

$$\sigma_f = \left\{2H_{iz}N_{iy}\cos\left(\omega_{N_i}t_2\right)\Gamma_2\Gamma_4 - 2H_{(i-1)z}N_{(i-1)y}\cos\left(\omega_{N_{(i-1)}}t_2\right)\Gamma_3\Gamma_5\right\}$$
$$\sin\left(2\pi J_{C^aCO}\tau_C\right)\Gamma_6\Gamma_1\cos\left(\omega_{N_i}t_1\right) \tag{6.122}$$

where

$$\Gamma_1 = \sin\left(2\pi J_{C^aCO}\tau_C\right)\sin\left(2\pi J_{NCO}T_N\right)$$

$$\Gamma_2 = \sin\left(2\pi^2 J_{C^aN}\tau_{CN}\right)\cos\left(2\pi^1 J_{C^aN}\tau_{CN}\right)$$

$$\Gamma_3 = \sin\left(2\pi^1 J_{C^aN}\tau_{CN}\right)\sin\left(2\pi^2 J_{C^aN}\tau_{CN}\right)$$

$$\Gamma_4 = \cos\left(2\pi^1 J_{C^aN}T_N\right)\sin\left(2\pi^2 J_{C^aN}T_N\right)$$

$$\Gamma_5 = \sin\left(2\pi^1 J_{C^aN}T_N\right)\cos\left(2\pi^2 J_{C^aN}T_N\right)$$

$$\Gamma_6 = \cos\left(2\pi J_{C^aC^\beta}\tau_{CN}\right) \tag{6.123}$$

τ_C, τ_{CN}, and $2T_N = A + B + C = D + E + F$ are the delays as indicated in the pulse sequence. The resultant data after the three-dimensional Fourier transformation yields the three-dimensional HN(C)N spectrum.

$$S(t_1, t_2, t_3) \overset{3D-FT}{\rightarrow} S\left[F1\left(^{15}N\right), F2\left(^{15}N\right), F3\left(H^N\right)\right] \tag{6.124}$$

The first term in Eq. (6.122) gives rise to the diagonal peak ($F1 = F2 = {}^{15}N_i$, $F3 = H^{Ni}$) in the three-dimensional spectrum. The second term yields the cross-peak ($F1 = {}^{15}N_i$, $F2 = {}^{15}N_{(i-1)}$, and $F3 = H^{Ni-1}$). A schematic representation of the

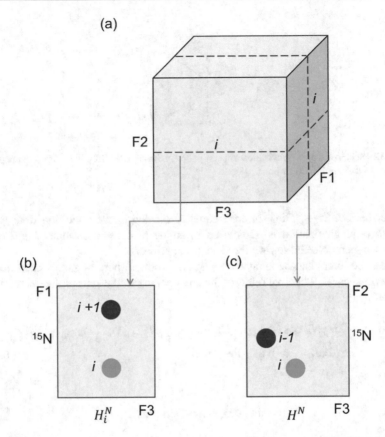

Fig. 6.63 (a) Schematic representation of the three-dimensional spectrum of HN(C)N. (b) Schematic representation of $F1 - F3$ plane at a particular $F2 = N_i$. (c) Schematic representation of $F2 - F3$ plane at $F1 = N_i$

three-dimensional spectrum is shown in Fig. 6.63a, and in Figs. 6.63b and 6.63c, the $F1$–$F3$ plane at $F2 = {}^{15}N_i$ and the $F2$–$F3$ plane at $F1 = {}^{15}N_i$ are shown, respectively.

Clearly, both the $F1$–$F3$ and $F2$–$F3$ cross-section planes provide directionality in sequential connections from the residue "i." The peaks also carry sign patterns as in the case of HNN experiment. The transfer efficiencies will be dictated by various coefficients ($\Gamma's$) in the respective terms. The different delays (τ_C, τ_{CN}, T_N) have to be optimized as before in the case of HNN. τ_c is generally set to ~4.5 ms, and τ_{CN} and T_N are typically set to ~12–15 ms.

Here again, the glycine residues make a special contribution because of the lack of C^β carbon and consequent absence of evolution under $C^\alpha - C^\beta$ coupling. This results in special patterns for glycine residues as well as for those which are adjacent to glycines. Considering the various possibilities of triplets of residues involving glycines, the expected peak patterns can be calculated as in the case of HNN. These are shown schematically in Fig. 6.64.

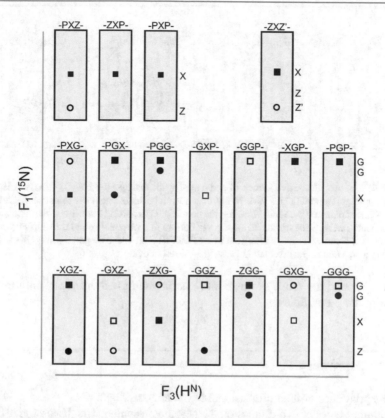

Fig. 6.64 Schematic patterns in the $F1$–$F3$ planes at the $F2$ chemical shift of the central residue in the triplets mentioned on the top of each panel, in the HN(C)N spectra for various special triplet sequences. X, Z, and Z' are any residue other than glycine (G) and proline (P). Squares are the diagonal peaks and circles are the sequential peaks. Filled and open symbols represent positive and negative signals, respectively. In all cases the peaks are aligned at the $F3$ (H^N) chemical shift of the central residue

The special features in the peak patterns in the HN(C)N and HNN spectra generate the so-called checkpoints which help greatly for sequential resonance assignments in proteins. Sections of experimental HN(C)N and HNN spectra demonstrating the sequential walk through a stretch of polypeptide chain are shown in Fig. 6.65a, b, respectively.

In the HNN and HN(C)N experiments, glycine residues served to provide checkpoints for sequential resonance assignments. Simple modifications of these experiments have been described where alanines and serines/threonines also produce distinctive peak patterns, similar to glycines. These experiments have provided the foundation for many more developments, which have enabled rapid and unambiguous assignments in different kinds of protein systems, including folded, unfolded, intrinsically disordered, and partially folded proteins. A complete description of

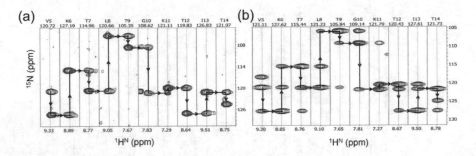

Fig. 6.65 (**a**) An illustrative stretch of sequential walk through the HN(C)N (**a**) and HNN (**b**) spectra of ubiquitin protein (1.6 mM, 76 aa). A sequential peak in one plane joins the diagonal peak in the adjacent plane on the right. Note that the panels of G10 and K11 constitute the checkpoints in this sequential walk. The numbers at the top and bottom in each panel A and B identify the F_2 (^{15}N) and F_3 (^{1}H^N) chemical shifts, which help in the identification of the diagonal peaks. Black and red contours represent the positive and negative peaks, respectively

these is beyond the scope of this book. The intention here has been only to give a flavor of the possibilities.

6.8   Summary

- The principles of multidimensional NMR are described.
- Different types of two-dimensional NMR are presented. The discussion is limited to some commonly used experiments.
- Illustrative elaborate product operator calculations are shown for some standard experiments. Some three-dimensional experiments are also described in some detail as illustrations.

6.9   Further Reading

- Principles of NMR in one and two dimensions, R. R. Ernst, G. Bodenhausen, A. Wokaun, Oxford, 1987
- High Resolution NMR Techniques in Organic Chemistry, T. D. W. Claridge, 3rd ed., Elsevier, 2016
- NMR Spectroscopy: Basic Principles, Concepts and Applications in Chemistry, H. Günther, 3rd ed., Wiley, 2013
- Understanding NMR Spectroscopy, J. Keeler, Wiley, 2005
- Protein NMR Spectroscopy, J. Cavanagh, N. Skelton, W. Fairbrother, M. Rance, A, Palmer III, 2nd ed., Elsevier, 2006

6.10 Exercises

6.1 In a two-dimensional NMR experiment, which of the following statement is correct?
  (a) Data is explicitly collected during two independent time variables t_1 and t_2.
  (b) Data is explicitly collected only during t_2.
  (c) Data is explicitly collected only during t_1.
  (d) The spectrum is generated by frequency selective excitations along the two frequency axes (F_1 and F_2).

6.2 Fourier transformation of a complex NMR signal S (t_1, t_2) leads to
  (e) absorptive line shapes along both frequency axes
  (f) dispersive line shapes along both frequency axes
  (g) absorptive line shape along F_1 and dispersive line shape along F_2
  (h) mixed line shape along both frequency axes

6.3 If SW is the spectral width along the F_1 dimension of the two-dimensional spectrum and carrier is placed at the center of the spectrum, then in the TPPI method of quadrature detection, the dwell time along t_1 dimension is equal to
  (a) 1 SW
  (b) 1/2 SW
  (c) 1/4 SW
  (d) 2 SW

6.4 Given the pulse sequence,

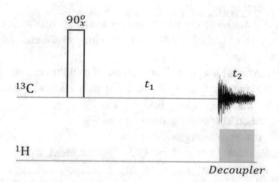

the F_2 axis of the two-dimensional spectrum for the molecule $CH_3\text{-}CH_2\text{-}CH_2\text{-}Cl$ will show
(a) 3 singlets
(b) 2 triplets and 1 quartet
(c) 1 quartet and 1 triplet
(d) 1 quartet and quintet

6.5 In a heteronuclear (C-H) spin echo experiment shown in the figure, the F_1 axis displays

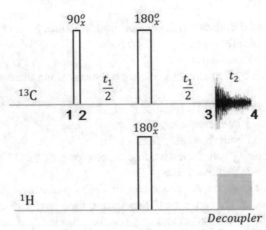

(a) carbon multiplets (J_{CH}) orthogonal to F_2 axis
(b) no multiplets
(c) multiplets tilted by 45° with respect to F_2 axis
(d) fine structure along with chemical shift of individual carbon nuclei

6.6 In a homonuclear two-dimensional J-resolved spectrum,
(a) F_1 axis has chemical shifts and F_2 axis has coupling constants
(b) F_1 axis has coupling constants and F_2 axis has chemical shifts and coupling constants
(c) F_1 axis has coupling constants and F_2 axis has chemical shifts
(d) F_1 axis has chemical shifts and coupling constants and F_2 axis has chemical shifts

6.7 In a two-dimensional homonuclear J-resolved experiment, the peaks have
(a) absorptive line shape along F_1 and dispersive line shape along F_2
(b) absorptive line shape along both F_1 and F_2
(c) dispersive line shape along both F_1 and F_2
(d) mixed line shapes along both F_1 and F_2

6.8 In a two-dimensional homonuclear COSY experiment, which of the following is correct?
(a) Cross-peak arises due to magnetization transfer mediated by dipolar interaction.
(b) Cross-peak arises due to magnetization transfer mediated by J-coupling interaction.
(c) Diagonal peak arises due to magnetization transfer mediated by dipolar interaction.
(d) Diagonal peak arises due to magnetization transfer mediated by J-coupling interaction.

6.9 Diagonal peaks in a COSY spectrum
    (a) have no fine structure
    (b) have fine structure with dispersive line shape along both frequency axes
    (c) have fine structure with absorptive line shape along F_1 axis and dispersive line shape along F_2 axis
    (d) have fine structure with dispersive line shape along F_1 axis and absorptive line shape along F_2 axis

6.10 The diagonal peak for a two-spin system ($I = 1/2$) in a two-dimensional COSY spectrum will have the fine structure

    (a) $\begin{bmatrix} + & - \\ - & + \end{bmatrix}$

    (b) $\begin{bmatrix} + & + \\ + & + \end{bmatrix}$

    (c) $\begin{bmatrix} + & - \\ + & - \end{bmatrix}$

    (d) $\begin{bmatrix} + & + \\ - & - \end{bmatrix}$

6.11 The cross-peak for a two-spin system ($I = 1/2$) in a two-dimensional COSY spectrum will have the fine structure

    (a) $\begin{bmatrix} + & - \\ - & + \end{bmatrix}$

    (b) $\begin{bmatrix} + & + \\ + & + \end{bmatrix}$

    (c) $\begin{bmatrix} - & - \\ - & - \end{bmatrix}$

    (d) $\begin{bmatrix} + & + \\ - & - \end{bmatrix}$

6.12 In the cross-peak in a two-dimensional COSY spectrum, the line shapes along the F_1 and F_2 dimension will be (*abs*: absorptive line shape; *dis*: dispersive line shape)

    (a) $\begin{bmatrix} abs & dis \\ dis & abs \end{bmatrix}$

    (b) $\begin{bmatrix} abs & abs \\ dis & dis \end{bmatrix}$

    (c) $\begin{bmatrix} dis & abs \\ dis & abs \end{bmatrix}$

    (d) $\begin{bmatrix} abs & abs \\ abs & abs \end{bmatrix}$

6.13 In a two-dimensional COSY spectrum, for a linear AMX spin system, with $J_{AM} > J_{MX}$, the M spin fine structure in the AM cross-peak is
   (a) $[+ - + -]$
   (b) $[+ + - -]$
   (c) $[+ - - +]$
   (d) $[+ + + +]$

6.14 In a two-dimensional COSY spectrum of a three-spin AMX system, the fine structure in a cross peak
   (a) is determined by the relative magnitude of the chemical shifts
   (b) is determined by the relative magnitudes of the active and passive coupling constants
   (c) does not depend upon the signs of the coupling constants
   (d) is entirely determined by the passive couplings

6.15 In a DQF-COSY spectrum of a two-spin system ($I = 1/2$),
   (a) both the diagonal and cross peak have the same fine structure and line shapes
   (b) the diagonal peak has antiphase structure and dispersive line shape, while the cross peak has antiphase structure and absorptive line shape
   (c) both the diagonal and cross peak have in-phase structure and absorptive line shape
   (d) the diagonal has in-phase structure and dispersive line shape and cross peak have antiphase structure and absorptive line shape

6.16 Phase cycling in the DQF-COSY experiment
   (a) helps to improve the signal-to-noise ratio
   (b) helps in selection of coherence transfer pathway
   (c) helps to remove artefacts of pulse imperfections
   (d) helps to improve the resolution in the spectrum

6.17 In a two-dimensional NOESY experiment, the cross peak arises
   (a) between J-coupled protons
   (b) between protons coupled by dipolar interaction
   (c) between chemically equivalent protons
   (d) between magnetically equivalent protons

6.18 The intensity of a cross-peak between two protons separated by distance "r," in a two-dimensional NOESY spectrum, is proportional to
   (a) r
   (b) $1/r$
   (c) $1/r^3$
   (d) $1/r^6$

6.19 The CT-COSY experiment achieves
   (a) homonuclear broadband decoupling along the F_1 dimension
   (b) homonuclear broadband decoupling along the F_2 dimension
   (c) selective decoupling along the F_1 dimension
   (d) selective decoupling along the F_2 dimension

6.20 In CT-COSY experiment,
  (a) J-coupling evolution does not occur during the t_1 period
  (b) J-coupling evolution occurs for the same time Δ for all the t_1 increments
  (c) chemical shift evolves through the constant time Δ
  (d) J-coupling evolution occurs for the periods $\Delta - t_1$, and the chemical shift evolution occurs for the period t_1

6.21 In the following pulse sequence,

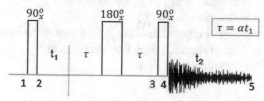

  (a) chemical shifts appear scaled up in the indirect dimension
  (b) J-values appear scaled up in the indirect dimension
  (c) J-values appear scaled down in the indirect dimension
  (d) both J-values and chemical shift are scaled up in the indirect dimension

6.22 In the given pulse sequence,

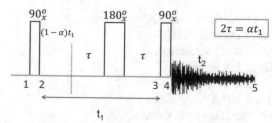

  (a) chemical shifts are scaled up in the indirect dimension
  (b) chemical shifts appear scaled down in the indirect dimension
  (c) J-values appear scaled up in the indirect dimension
  (d) J-values appear scaled down in the indirect dimension

6.23 In the TOCSY experiment, which of the following statements are true?
  A. There is in-phase transfer of coherence.
  B. There is relay of magnetization.
  C. Transfer efficiency is worse than that in INEPT transfer.
  D. Spin lock leads to isotropic Hamiltonian.
  (a) All the statements are true.
  (b) A, B, C are true.
  (c) A, B, D are true.
  (d) B, C, D are true.

6.24 In heteronuclear COSY experiment with the pulse sequence

$$90_x(^1H) - t_1 - 90_x(^1H, {}^{13}C) - t_2 - (\text{acquisition}),$$

which of the following is true?
(a) The C-H cross-peak has no fine structure.
(b) The C-H cross-peak has fine structure along F_2 alone.
(c) The C-H cross-peak has fine structure along F_1 alone.
(d) The C-H cross-peak has fine structure along both F_1 and F_2.

6.25 In a HSQC spectrum, which of the following statements are true?
    A. There is no fine structure in the cross peaks
    B. Signal-to-noise is much superior compared to direct X detection experiment
    C. Signal-to-noise ratio is inferior to direct X detection experiment
    D. Cross-peaks have fine structure along the F_1 axis
    (a) A and B
    (b) B and C
    (c) C and D
    (d) only A
6.26 In the HMQC spectrum, identify the correct statement.
    (a) The HX cross-peaks have no fine structure,
    (b) The HX cross-peaks have mixed phases resulting from H-H coupling evolutions,
    (c) The resolution is superior compared to HSQC spectrum,
    (d) The experiment takes less time than HSQC,
6.27 In a two-dimensional HSQC-TOCSY spectrum,
    (a) TOCSY causes relay along the F_2 axis.
    (b) TOCSY causes relay along the F_1 axis.
    (c) TOCSY leads to amplitude alteration HX cross peaks.
    (d) TOCSY leads to a phase alteration of HX cross peaks.
6.28 For a three-dimensional NMR experiment, recorded with 256, 512, and 1024 data points along the t_1, t_2, and t_3 axes, respectively, with acquisition time of 0.2 s and relaxation delay of 1 s, the total acquisition time with four scans for each FID will be approximately
    (a) 1.5 days
    (b) 3.5 days
    (c) 7.3 days
    (d) 11 days
6.29 In a three-dimensional experiment,
    (a) the evolution time t_1 and t_2 are incremented simultaneously
    (b) the evolution time t_1 and t_2 are incremented independently
    (c) t_1 is increment synchronously with t_3
    (d) t_2 is increment synchronously with t_3

Reference

Sanjay C. Panchal, Neel S. Bhavesh, and Ramakrishna V. Hosur (2001) Improved 3D triple resonance experiments, HNN and HN(C)N, for HN and 15N sequential correlations in (13C, 15N) labeled proteins: Application to unfolded proteins. J Biomol NMR 20:135–147

Appendix

7

Contents

7.1   Appendix A1: Dipolar Hamiltonian 278
7.2   Appendix A2: Chemical Shift Anisotropy 280
       7.2.1   Principal Axes and Principal Values 280
7.3   Appendix A3: Solid-State NMR Basics 282
       7.3.1   Magic Angle Spinning (MAS) 283
       7.3.2   Cross Polarization 284
7.4   Appendix A4: Selection of Coherence Transfer Pathways by Linear Field Gradient
       Pulses 286
7.5   Appendix 5: Pure Shift NMR 290
       7.5.1   Pseudo-Two-Dimensional Data Acquisition 290
       7.5.2   Real-Time Data Acquisition 293
       7.5.3   Homonuclear Band-Selective Decoupling 293
       7.5.4   Zangger-Sterk Real-Time Homonuclear Broadband Decoupling 295
       7.5.5   PSYCHE Homonuclear Broadband Decoupling 296
7.6   Appendix A6: Hadamard NMR Spectroscopy 298

There are six appendices here. These are intended to provide some greater awareness to students on the scope of NMR spectroscopy, in general, since in the present text the focus has been almost entirely on high-resolution NMR or solution-state NMR. In this context, some justice is given to the course, by including some basic aspects of solid-state NMR which is an important subject in itself, in the form of three Appendices, A1–A3. These deal with dipole-dipole interaction (Appendix A1), chemical shift anisotropy (Appendix A2), and magic angle spinning and cross-polarization (Appendix A3). Dipole-dipole interaction and chemical shift anisotropy are also present in the solution-state but get averaged out to zero in solution due to tumbling motions and thus do not contribute to the spectral features. This has been mentioned in Chap. 2 in the text, and accordingly reference is made there to Appendices A1 and A2. However, these interactions do contribute to relaxation processes in the solution state. It is the dipolar interaction which enables double-quantum, zero-quantum, and single-quantum transition probabilities which are

© The Author(s), under exclusive license to Springer Nature Switzerland AG 2022
R. V. Hosur, V. M. R. Kakita, *A Graduate Course in NMR Spectroscopy*,
https://doi.org/10.1007/978-3-030-88769-8_7

crucial for relaxation and also for nuclear Overhauser effects (NOE), as discussed in Chaps. 1 and 4. Moreover, the coverage in the text with regard to solution-state NMR is also limited to what can be covered in about a semester; this includes very basic topics as well as some extremely useful developments such as multidimensional NMR, which every student should know. Therefore, an advanced topic relevant for multidimensional NMR in the solution state, namely, gradient-based selection of coherence transfer pathways, is included as Appendix A4. Field gradients are also used to dephase coherences and thereby destroy unwanted magnetization components in a given pulse sequence. Such gradients are referred to as "crusher gradients." These are also used in water suppression pulse sequence which is described in Chap. 3, and hence a reference is made there for Appendix A4. Similarly, advanced topics such as pure shift NMR and Hadamard NMR, which are also relevant for solution state NMR, are included for student's benefit as Appendices A5 and A6, respectively. Pure shift NMR achieves homonuclear broad band decoupling during data acquisition and thus produces very high resolution in the spectra. Hadamard NMR invokes a new concept of encoding and decoding of NMR signals by special sequence of data acquisition and processing dictated by the so-called Hadamard matrix. So these appendices provide motivation to the students to pursue in this exciting field of NMR. In each appendix, suggestions are made for further reading.

7.1   Appendix A1: Dipolar Hamiltonian

Nuclear spins, which are magnetic dipoles, interact through space, and this interaction energy between two dipoles with magnetic moments μ_1 and μ_2 joined by vector r is given by

$$E_{\text{dipolar}} = \frac{\mu_0}{4\pi}\left[\frac{\mu_1 \cdot \mu_2}{r^3} - 3\frac{(\mu_1 \cdot r)(\mu_2 \cdot r)}{r^5}\right] \tag{7.1}$$

The Hamiltonian H_{DD} corresponding to such an interaction between two spins I and S, which are oriented with respect to the magnetic field B_0 as indicated in Fig. 7.1, is given by the following expression:

$$H_{\text{DD}} = \frac{\mu_0}{4\pi}\hbar^2\gamma_S\gamma_I\frac{1}{r^3}(A + B + C + D + E + F) \tag{7.2}$$

$$A = (1 - 3\cos^2\theta)I_zS_z$$

$$B = -\frac{1}{4}(1 - 3\cos^2\theta)(I_+S_- + I_-S_+)$$

$$C = -\frac{3}{2}(\sin\theta\cos\theta)e^{-i\varnothing}(I_zS_+ + I_+S_z)$$

Fig. 7.1 Schematic
representation of a spin-pair *IS*
orientation in the external
magnetic field (B_0)

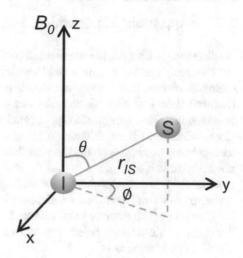

$$D = C^* = -\frac{3}{2}(\sin\theta\cos\theta)e^{i\varnothing}(I_z S_- + I_- S_z)$$

$$E = -\frac{3}{4}\sin^2\theta\, e^{-i2\varnothing}\, I_+ S_+$$

$$F = -\frac{3}{4}\sin^2\theta\, e^{-i2\varnothing}\, I_- S_-$$

This dipolar Hamiltonian has two parts: (i) a spatial part which is dependent on θ and $\varnothing$ and (ii) a spin part which depends on the angular momentum operators. Under the high field approximation, this Hamiltonian gets simplified. The terms *A–F* listed above represent the so-called dipolar alphabet. In solutions, due to rapid tumbling motions, the dipolar interaction averages to zero and hence will not contribute to the energy levels. However, the terms *B–F* which contain raising and lowering operators do cause transitions between energy levels and hence play key roles in relaxation processes. *B* causes zero-quantum transitions, *C* and *D* cause single-quantum transitions, and *E* and *F* cause double-quantum transitions. In the presence of high external fields, the Zeeman interaction is the most dominating (high field approximation), and those interactions which contribute little to the energy eigenvalues can be dropped. This is termed as *secular approximation*. Then only certain parts of the dipolar Hamiltonian need to be retained. In the heteronuclear case, only *A* is retained, and, in the homo-nuclear case, both *A* and *B* are retained. As both these terms have $(1 - 3\cos^2\theta)$ dependence, the spectral features will depend on the orientation of the internuclear vector with respect to the B_0 field.

Further Reading
- Principles of NMR in one and two dimensions, R. R. Ernst, G. Bodenhausen, A. Wokaun, Oxford, 1987
- Spin Dynamics, M. H. Levitt, 2nd ed., Wiley 2008

7.2   Appendix A2: Chemical Shift Anisotropy

As discussed in Chap. 2, the chemical shift of a nucleus arises due to the screening of the externally applied magnetic field by electronic clouds surrounding the nucleus. Currents induced in the electronic clouds by the external field produce local fields (induced fields) which influence the net field experienced by the nucleus under consideration. Every magnetic dipole (nuclear spin) is surrounded by an electronic cloud which can have different orientations depending upon the shape of the molecule, and accordingly, the induced field by the external field can have different orientations; it is not necessary that the induced field will be always parallel to the external field. Hence, the chemical shielding (σ) will be different depending on the symmetry of the molecule and the disposition of the electronic clouds with respect to the direction of the external field; this leads to the so-called angular dependence of the chemical shift; this is referred to as chemical shift anisotropy (CSA). The CSA Hamiltonian is defined as

$$\mathcal{H}_{CS} = I.\sigma.\gamma \, B_0 \tag{7.3}$$

σ in Eq. 7.3 will be a tensor. The components of the induced field, assuming the external field to be along the z-direction, are given by

$$\begin{bmatrix} B_{i,x} \\ B_{i,y} \\ B_{i,z} \end{bmatrix} = - \begin{bmatrix} \sigma_{xx} & \sigma_{xy} & \sigma_{xz} \\ \sigma_{yx} & \sigma_{yy} & \sigma_{yz} \\ \sigma_{zx} & \sigma_{zy} & \sigma_{zz} \end{bmatrix} \begin{bmatrix} 0 \\ 0 \\ B_0 \end{bmatrix} \tag{7.4}$$

The σ matrix is called the shielding tensor. The elements of the matrix have the following meaning. $\sigma_{xz} B_0$ would represent the component of the induced along the x-direction at the site of the nucleus i, if the external field were applied along the z-direction; $\sigma_{xx} B_0$ would represent the component of the induced field along the x-axis, if the external field were applied along the x-direction; and so on.

7.2.1   Principal Axes and Principal Values

In a given molecule, which may or may not have any symmetry in its shape, for every nuclear site, it is possible to define three orthogonal axes, such that, if the external field is along any of those directions, the induced field is parallel to that axis. These directions are referred to as the principal axes at that nuclear site. These may be represented by a different symbol for the axes, namely, X, Y, and Z. In other words, if the external field were applied along the X or Y or Z axis, then the induced fields would be parallel to these and would be given by $\sigma_{XX} B_0$, $\sigma_{YY} B_0$, and $\sigma_{ZZ} B_0$, respectively. The elements, σ_{XX}, σ_{YY}, and σ_{ZZ}, are called the principal values of the chemical shift tensor at that nuclear site. Clearly, the principal axes and the induced field would be different at different nuclear sites which would be dictated by the symmetry of the molecule and orientation of the electronic cloud around that site. The mean of the three principal values is termed as the isotropic chemical shift.

$$\sigma_{iso} = (\sigma_{XX} + \sigma_{YY} + \sigma_{ZZ})/3 \tag{7.5}$$

In solutions, the orientation dependence at every nuclear site will be mostly averaged out due to rapid tumbling motions, and then we observe the isotropic chemical shift. For slow motions, or at high magnetic fields, or in oriented systems, or in solids, such an averaging may not occur completely, and then one starts to see CSA effects in the NMR spectra.

In the principal axes system (PAS), the shielding tensor will be diagonal.

$$\sigma^{PAS} = \begin{bmatrix} \sigma_{XX} & 0 & 0 \\ 0 & \sigma_{YY} & 0 \\ 0 & 0 & \sigma_{ZZ} \end{bmatrix} \tag{7.6}$$

Often a different notation involving deshieding δ is used:

$$\delta^{PAS} = -\sigma^{PAS} \tag{7.7}$$

and the isotropic shift, shift anisotropy, and asymmetry are defined

$$\delta_{iso} = \frac{1}{3}\left(\delta_x + \delta_y + \delta_z\right) \tag{7.8}$$

$$\zeta = \delta_z - \delta_{iso} \tag{7.9}$$

$$\eta = \frac{\delta_x - \delta_y}{\zeta} \tag{7.10}$$

The nuclear spin interacts with the induced field as per the Zeeman interaction, and the Hamiltonian $\mathcal{H}_{CS}$ can be written as

$$\mathcal{H}_{CS} = \gamma\sigma_{xz}I_xB_0 + \gamma\sigma_{yz}I_yB_0 + \gamma\sigma_{zz}I_zB_0 \tag{7.11}$$

Under the secular approximation, only the last term needs to be retained. As stated earlier, the elements of the chemical shift tensor (σ) depend on the orientation of the molecule with respect to the external field, and for the secular part, one obtains

$$\sigma_{zz} = \frac{1}{3}\left\{\sum_j \sigma_j + \sum_j (3\cos^2\theta_j - 1)\sigma_j\right\} \tag{7.12}$$

In Eq. 7.12, j runs through XX, YY, and ZZ, and θ_j refers to the angle made by the individual principal axis with the external field. The first term in the above equation is the isotropic chemical shift, and the second term represents the orientation dependence at the particular nuclear site and contributes to the *anisotropy*.

Further Reading
- Spin Dynamics, M. H. Levitt, 2nd ed., Wiley 2008
- Introduction to solid-state NMR: anisotropic interactions, J. Titman, School of Chemistry, University of Nottingham

7.3   Appendix A3: Solid-State NMR Basics

For the characterization of materials or molecules which have solubility issues, NMR experiments will have to be carried out with powders or single crystals, and this is then referred to as "Solid-state NMR." These NMR spectra look very broad compared to the solution-state NMR spectra, as some of the interactions which get averaged out to zero in solution remain in the solid-state samples, and that leads to complexities. The general solid-state NMR Hamiltonian (H_{SS}) is given by

$$\mathcal{H}_{SS} = \mathcal{H}_Z + \mathcal{H}_{RF} + \mathcal{H}_{CSA} + \mathcal{H}_{DD} + \mathcal{H}_J + \mathcal{H}_Q \qquad (7.13)$$

These refer to the Zeeman interaction (Z), radiofrequency interaction (RF), chemical shift anisotropy interaction (CSA), dipolar interaction (DD), spin-spin interaction (J), and quadrupolar interaction (Q). Among these, $\mathcal{H}_Z$ and $\mathcal{H}_{RF}$ arise from external interactions with the spin system, while others arise from internal interactions. The quadrupolar interaction is generally seen in nuclei with spin >1, and these are typically observed in material science research. For most part of chemistry and biology research, one deals with spin values of ½. Zeeman interaction is the strongest among all the interactions (high-field approximation), and under these conditions, the other interactions can be dropped. This is the so-called secular approximation, and the Hamiltonian takes a simpler form. In Appendix A1 the dipolar Hamiltonian has been defined. The secular form of $\mathcal{H}_D$ for the homonuclear case (spins are labeled as 1 and 2) is given by

$$\mathcal{H}_{DD} = \frac{d}{2}\left(1 - 3\cos^2\theta\right)\left(3I_{1z}I_{2z} - \mathbf{I_1}.\mathbf{I_2}\right) \qquad (7.14)$$

where θ is the angle between the internuclear vector and the external field.

In the solid state, the internuclear vector for a pair of nuclei in a given molecule would have different orientations with respect to the external field in different portions of the sample, and thus the contribution of dipolar interaction to the energy levels of the spins acquires spatial dependence. This leads multiple resonance frequencies for any given spin resulting in extensive line broadening. Often the line widths span a range of several kHz.

Similarly, under the secular approximation, the CSA Hamiltonian, $\mathcal{H}_{CSA}$, can be written as

$$\mathcal{H}_{CSA} = \gamma I_z B_0 \sigma_{zz} \qquad (7.15)$$

where σ_{zz} is given by the equation

Fig. 7.2 Wide-line NMR spectrum of glycine powder sample. (Reproduced from Journal of Molecular Structure, 830, 145 (2007) with the permission of Elsevier Publishing)

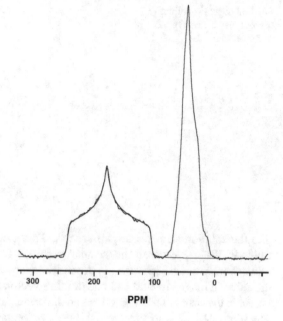

$$\sigma_{zz} = \frac{1}{3} \left\{ \sum_{j} \sigma_j + \sum_{j} (3 \cos^2 \theta_j - 1) \sigma_j \right\} \qquad (7.16)$$

Here σ_js are the principal components of the shielding tensor σ, and θ_js are the angles made by the principal axes with the external field B_0 (see Appendix A2).

The angular dependencies of $\mathcal{H}_{CSA}$ and $\mathcal{H}_{DD}$ seen in Eqs. 7.14 and 7.15 are responsible for large line widths seen in the solid-state NMR spectra. A wide-line NMR spectrum of a powder sample of a simple molecule, such as glycine, is shown in Fig. 7.2.

From such a spectrum, it becomes almost impossible to extract site-specific information. Enormous efforts are now focused on sharpening the lines by removing the DD and CSA interactions from the Hamiltonian. The most common technique employed for such purposes is the so-called magic angle spinning. This will average out the angular dependencies in the DD and CSA interactions.

7.3.1 Magic Angle Spinning (MAS)

Magic angle spinning is an extremely useful technique to remove the anisotropic contributions to the NMR spectra, namely, dipolar (DD) and chemical shift aniso-tropic (CSA) interactions, and obtain spectra with high resolution, which look almost

Fig. 7.3 Spinning of the
rotor at magic angle to the
external field

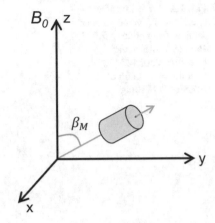

like the solution-state (isotropic) spectra. This can be understood qualitatively as
follows. We notice that the secular parts of both dipolar and chemical shift
interactions contain $(3\cos^2\theta - 1)$ dependence where θ represents the angle between
the external magnetic field and one or other vectors of the spin system (internuclear
vector in the case of DD interaction and the principal axis vector in the case of CSA).
The idea of MAS is to average out these components. The solid powder sample is
packed inside a cylindrical rotor (see Fig. 7.3) which is then aligned at an angle to the
external field. The rotor is then spun about the long axis of the cylinder at a high
speed. As a result of this spinning, there will be an averaging of various components
of the vectors, and the only component that survives is the one which is parallel to
the axis of rotation. Therefore, if this angle is adjusted to be 54.74° (β_M in the figure),
then the $(3\cos^2\theta - 1)$ term vanishes. This angle is called as the magic angle. Under
these conditions, one observes only the isotropic chemical shifts, dipolar coupling is
removed, and thus the spectrum will have sharp lines.

The sample spinning rate has to be larger than the magnitude of the anisotropic
interactions. If it is less than that rate, then sidebands will appear in the spectrum,
which will be separated by the spinning rate in Hz. A typical spectrum showing the
improvement in the resolution is shown in Fig. 7.4.

7.3.2 Cross Polarization

In solid-state NMR, cross polarization is an important technique to enhance the
sensitivity of an insensitive (low γ) and dilute nucleus (e.g., ^{13}C, ^{15}N, ^{31}P, etc.) by
transferring polarization from a more abundant and sensitive nucleus (high γ, e.g.,
^{1}H or ^{19}F). The experimental scheme for achieving this is shown in Fig. 7.5.

Transfer of polarization occurs due to the "mixing" of the I (^{1}H) and S (^{13}C)
magnetizations and is mediated by dipolar coupling between the two spins during
the spin-lock period (referred to as contact time). An efficient transfer requires that

Fig. 7.4 Resolution enhancement by MAS (RO refers to the spinning frequency). The individual chemical shifts can be read out from the spectrum at the spinning frequency of 13 kHz. In other spectra the additional peaks seen indicate the spinning sidebands. (Reproduced from Polymers for advanced technologies, 27, 1143 (2016) with the permission of Wiley Publishing)

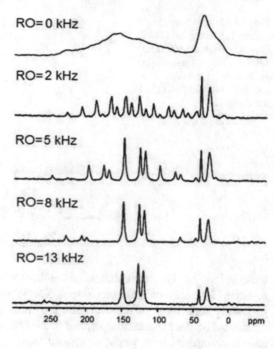

Fig. 7.5 An experimental sequence of cross polarization

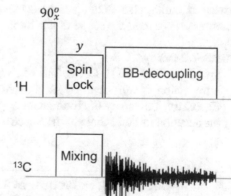

the precession frequencies of the two nuclei (abundant spin I and diluted spin S) in the rotating frames created by the spin-lock are identical (Eq. 7.17):

$$\gamma_I B_I = \gamma_S B_S \qquad (7.17)$$

B_I and B_S are the amplitudes of the radiofrequency fields applied to I and S spins, respectively. This condition is termed as Hartmann-Hahn condition. This is shown by the energy level changes in going from lab frame to rotating frame (during spin-

Fig. 7.6 Energy levels in the lab and in rotating frame when Hartmann-Hahn condition is satisfied. The ω's refer to the precession frequencies. ^{1}H is the abundant spin I and ^{13}C is the diluted spin S

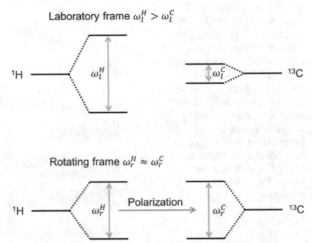

lock) in Fig. 7.6. Under conditions of the most effective transfer, the gain in signal-to-noise (S/N) ratio of the detected spin S is (γ_I/γ_S).

During acquisition of the ^{13}C data, the ^{1}H spin is decoupled, and CSA is also eliminated by the MAS technique, and this will result in sharp lines with higher S/N ratio as compared to acquisition without cross polarization. There are also many advanced multi-pulse techniques to achieve heteronuclear decoupling, but their discussion is beyond the scope of this book.

Further Reading
- Spin Dynamics, M. H. Levitt, 2nd ed., Wiley 2008
- Introduction to solid-state NMR: anisotropic interactions, J. Titman, School of Chemistry, University of Nottingham
- Introduction to Solid-State NMR Spectroscopy, M. Duer, Wiley 2010

7.4   Appendix A4: Selection of Coherence Transfer Pathways by Linear Field Gradient Pulses

In many multipulse experiments in NMR, we come across different types of coherences, such as single-quantum coherences, double-quantum coherences, zero-quantum coherences, triple-quantum coherences, etc. and also z-magnetization. These are often described in terms of change in the magnetic quantum number m. The coherences are also defined by another parameter known as the coherence order, p, which is more useful in devising strategies for selecting certain pathways of magnetization transfer through a given pulse sequence. Single-quantum coherences (SQC) are characterized by $p = +1$ or -1, and double-quantum coherences are $p = +2$ or -2. Zero-quantum coherences have $p = 0$, and finally, z-magnetization is also described by $p = 0$. A more precise definition of the coherence order of a particular coherence (represented by a part of the density operator as σ^P) is given by its response to a phase of the RF pulse that created it. Explicitly, if the RF pulse is given a phase shift of $\varnothing$, this is equivalent to a rotation around the z-axis—a coherence of order p acquires a phase shift of $-p\varnothing$.

$$\exp\left(-i\varnothing F_z\right)\sigma^{(p)}\exp\left(i\varnothing F_z\right) = \exp\left(-ip\varnothing\right)\sigma^{(p)} \tag{7.18}$$

Coherences of order p acquires a phase of $-p\varnothing$:

$$\sigma^{(p)} \xrightarrow{z-\text{rotation by } \varnothing} \sigma^{(p)}\exp(-ip\varnothing) \tag{7.19}$$

For example, for a system of single spin ½, the transition from β- state ($m = -1/2$) to the α-state ($m = 1/2$) will correspond to $p = +1$; this transition will be represented by the I^+ operator, and a phase change of ϕ for the excitation pulse will cause a phase shift $-\phi$ in the operator. The reverse transition will have $p = -1$. Similarly, for a two spin ½ system, transition from the $\alpha\alpha$ state to the $\alpha\beta$ state will have $p = -1$, and transition from $\alpha\alpha$ to $\beta\beta$ state will have $p = -2$, and so on. Note that z-magnetization and zero-quantum coherences which correspond to $p = 0$ will not be affected by these phase changes. This has been the basis of selection of coherence transfer pathways in many multipulse experiments. As one goes through the pulse sequence, the various coherences keep accumulating the phase changes caused by various pulses through the pulse sequence and, for the desired coherence transfer pathway, should add up to zero (the same phase as the receiver). The pathways for which this condition is not satisfied will not be observed. It can happen in principle that more than one pathway will get selected for a given choice of phase changes and a judicious choice of combination of phases and their cycling, while signal averaging would be required to fix on one particular pathway. This strategy is called 'phase cycling' for the selection of coherence transfer pathways.

Linear field gradients along the z-axis provide a very efficient method for coherence selection. When a linear field gradient (G_z) is applied along the z-axis, molecules at different locations experience different magnetic fields ($B_0 + zG_z$)— where B_0 is the main external field of the spectrometer which is along the z-axis— and hence their nuclei (e.g., protons in water) will precess at different frequencies:

$$\omega_{iz} = |\gamma_i(B_0 + zG_z)| \tag{7.20}$$

Here, $\gamma_i z G_z$ represents the change in the precessional frequency of nucleus i at location z. If such a gradient is applied after the coherences are created, then they all acquire different phases through the duration of the gradient, and this depends, firstly, on the location of the molecule along the z-axis and then on the coherence order. If the gradient is applied for a period τ (gradient pulse of width τ), the additional phase ($\varnothing^p$) acquired by coherence of order p is given by

Fig. 7.7 Different phases acquired by spins at different locations due to the linear field gradient G_z

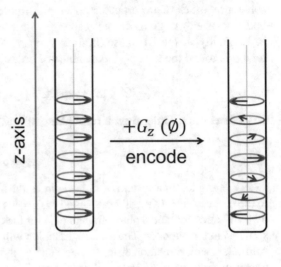

Fig. 7.8 Refocusing of the magnetization at the end of the field gradient

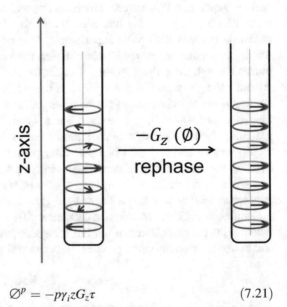

$$\varnothing^p = -p\gamma_i z G_z \tau \qquad\qquad (7.21)$$

This is pictorially depicted in Fig. 7.7.

Following this, if a reverse gradient (the sign of the gradient is changed) is applied for the same time duration, then all the spins will acquire additional phase given by

$$\phi^p = -p\gamma_i z \left(-G_z\right)\tau = p\gamma_i z G_z \tau \qquad\qquad (7.22)$$

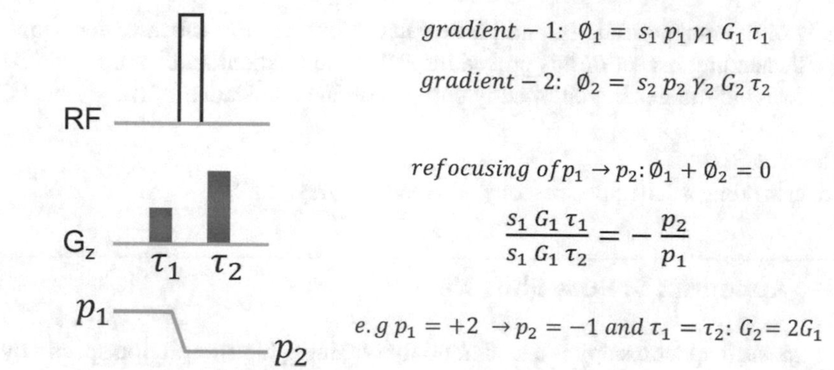

$$gradient - 1: \; \emptyset_1 = s_1 \, p_1 \, \gamma_1 \, G_1 \, \tau_1$$

$$gradient - 2: \; \emptyset_2 = s_2 \, p_2 \, \gamma_2 \, G_2 \, \tau_2$$

$$refocusing \; of \, p_1 \rightarrow p_2 : \emptyset_1 + \emptyset_2 = 0$$

$$\frac{s_1 \, G_1 \, \tau_1}{s_1 \, G_1 \, \tau_2} = -\frac{p_2}{p_1}$$

$$e.g \; p_1 = +2 \; \rightarrow p_2 = -1 \; and \; \tau_1 = \tau_2 : G_2 = 2G_1$$

Fig. 7.9 Schematic demonstration of the selection of a pathway involving two particular coherence orders p_1 and p_2 at two parts of the pulse sequence. The factor s refers to a shape factor characterizing the gradient pulses

Thus, the net phase at the end of 2τ period will be zero at every location and for every coherence. In other words, all the coherences of order p will refocus to their original position at the end of the gradient. This is depicted pictorially in Fig. 7.8.

Clearly, a gradient strength of a particular value can refocus a coherence of one particular order. In other words, a particular combination of G_z and $-G_z$ cannot refocus $p = 1$ and $p = 2$ at the same time, if both of them are present at the beginning. This is the basis of gradient-based selection of coherence orders. For example, if a coherence of order 2 has to be retained, then the second gradient can either have half the strength of the first gradient or be applied for half the duration of the first gradient. In either cases, only the $p = 2$ coherence will refocus, and all other coherences will be eliminated by complete dephasing.

In a multi-pulse experiment, coherence transfers occur at different steps of the sequence, and different coherence orders evolve in different periods. Phase changes caused by the different gradients to each of the coherence orders will have to be calculated and summed up till the detection step. The total phase has to be zero (i.e., the same as the receiver phase) for detection. By suitably adjusting the gradients at different parts of the pulse sequence, one can choose the pathway for the flow of magnetization through the pulse sequence. This is exemplified schematically in a simple manner in Fig. 7.9.

$$\sum_i \emptyset_i = 0 \tag{7.23}$$

$$\emptyset_i = s_i \, p_i \, \gamma_i \, G_i \, \tau_i \tag{7.24}$$

The gradient-based selection procedure is very simple to implement. Many of the unwanted coherences at different stages can be eliminated right away by applying the so-called crusher gradients or purge gradients. This facilitates keeping track of the flow of the magnetizations, and phase calculations become much easier. However, a demanding aspect of this procedure is that the gradient hardware should be perfect to avoid the generation of eddy currents on the application of the gradients.

Further Reading
- Understanding NMR Spectroscopy, J. Keeler, Wiley, 2005

7.5   Appendix 5: Pure Shift NMR

Pure shift NMR spectroscopy is a technique for obtaining high-resolution spectra by employing broadband/band-selective homonuclear decoupling. As a result, each chemical site in a molecule exhibits only one line in its spectrum. This situation is similar to a ^{13}C NMR spectrum recorded under heteronuclear decoupling conditions. Heteronuclear decoupling is relatively simple since the resonance frequency of the observed nucleus is very far from the decoupled nucleus, and consequently, two independent radio frequency channels can be employed to excite the two nuclei simultaneously without any interference between the two. On the other hand, achieving homodecoupled spectra, i.e., observing and decoupling the sample nuclei through applying continuous RF and acquiring the ^{1}H, is almost impossible as in the heteronuclear decoupling cases. However, homodecoupled spectra can be recorded using the following different strategies.

1. The separation of chemical shifts and J-couplings (two-dimensional NMR-based experiments, namely, J-Res and constant-time experiments). These have been discussed in Chap. 6.
2. The observation of active spins while inverting their coupled partners. Here the J-coupling evolution of the so-called active spin under coupling to passive spins is effectively refocused, while the data is being acquired. One of the cases of heteronuclear decoupling and BIRD schemes have been discussed in Chap. 3. In homonuclear schemes also, this phenomenon can be used, as in the cases of ZS (Zangger-Sterk) and HOBS (homonuclear band-selective).
3. Homonuclear broadband decoupling using low-flip angle pulse sequences (anti-z-COSY and PSYCHE (pure shift yielded by chirp excitation) schemes).

In the following, we discuss the second and the third categories of experiments. These methods can be implemented either in pseudo-two-dimensional or real-time homodecoupling acquisition modes.

7.5.1   Pseudo-Two-Dimensional Data Acquisition

Figure 7.10 represents the schematic of a pseudo-two-dimensional homodecoupled pulse sequence. Applications of pseudo-two-dimensional homodecoupling can be implemented with ZS, HOBS, and PSYCHE pulse schemes. In this experiment, the

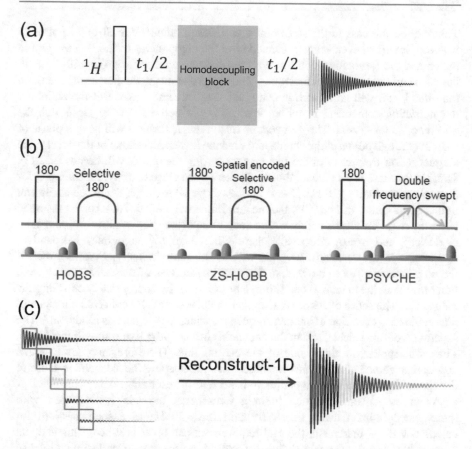

Fig. 7.10 Schematic of a pseudo-two-dimensional version of a homodecoupling pulse sequence (**a**), where the homodecoupling block can be replaced with any one of the schemes given in (**b**). Concatenating the homodecoupled FID blocks results in pure shift FID, which is given in (**c**)

total FID is collected as concatenated chunks, each of which is a part of a J-coupling refocusing evolution during a time period labelled as t_1. Finally, the one-dimensional spectrum is obtained by the Fourier transformation of the whole FID so acquired.

The J-coupling refocusing element involves the combined application of soft and hard 180° pulses at the middle of the t_1 evolution time; note that this is a spin echo sequence. The soft pulse is applied to the active spin only. Therefore, at the end of the t_1 period, that is, at the time of the echo, the J-coupling evolution is refocused. Now, the data collection in the chunk starts at a short time before the echo and ends after the same amount of time after the echo. During this, a short period of the chunk J-evolution does happen (first in the process of refocusing and then while dephasing after the echo). At the beginning of the chunk, the two vectors of the active spin doublet will be oriented with respect to the refocusing axis at an angle of $\pi J \Delta/2$ [$2\pi^*$ $(J/2)^*(\Delta/2)$] on either side of the axis, and thus the phase difference between the two vectors of the "active spin" doublet will be $\pi J \Delta$ where Δ is the length of the chunk.

This will be the case at the last point also in each chunk. For all other points in between, it will be even smaller. Considering the components of the vectors, within the chunk, the J-evolution will cause an in-phase component to grow with time till the echo and decrease again thereafter. The antiphase magnetization component, on the other hand, will go to zero at echo and then increase again. But the antiphase magnetization component is not an observable magnetization as its trace with the density operator is zero. The two vectors of the magnetization will have a phase of $\pi J\Delta/2$ at the end points of the chunk, and thus the lowest amplitude of the observable magnetization from each of the two vectors during the chunk will be $\cos(\pi J\Delta/2)$. Thus, if the phase is very small, the amplitude will be close to unity; for example, for a coupling constant of 10 Hz, if Δ is chosen to be 10 ms, then the phase at the end points of the chunk will be $9.0°$, and the amplitude will be 0.9876. It can be assumed to be the same for all the points within each chunk (the very small variation can be neglected), and hence, effectively, there will be no (or negligible) J-dependent modulation in the detected FID. This will be the same in every chunk. Further, if the amplitude is close to unity, then there will be no loss of magnetization as well, other than that due to relaxation. This can be achieved by appropriately choosing the value of Δ. The series of echoes that appear in the sequentially collected chunks will amount to the generation of another frequency, which will appear as a sideband. The separation of this sideband from the chemical shift of the active spin will be $(1/2\Delta)$. They will appear on either side of the central line. Therefore, one has to make appropriate choice of Δ for maximizing signal and keeping the sideband as close to the central line as possible so that there is no loss of resolution.

As stated, several of such indirect increments have to be recorded with incremented indirect dwell times. By utilizing ~50 Hz as a spectral-width for calculating the t_1 increment, the said requirement can be established. This is equal to only 20 ms of chunk per each indirect evolution time; hence, collecting a total of 30–40, such homodecoupled data chunks yields ~600 to ~800 ms of FID signal without any homonuclear scalar coupling information. Indeed, 0.6 s of FID length leads to a digital spectral resolution equivalent to ~1.5 Hz, which is sufficiently adequate as there is no demand to observe any scalar coupling multiplets in the homodecoupled spectra.

It is an essential condition that the chemical shift evolution must be continuous among the chunks, whereas the scalar coupling evolution has to be refocused in the middle of the chunks. For example, if "n" chunks are collected in a total acquisition time of "aq," the first chunk duration is "aq/2n," and the remaining chunks are equal to "aq/n." Therefore, the scalar coupling is refocused at the middle of each chunk (from the second onward). Then, concatenating all the chunks results in a complete homodecoupled FID.

Even though this kind of pseudo-two-dimensional homodecoupling pulse schemes produce pretty clean spectra, experiments demand relatively long data recording times compared to the conventional one-dimensional schemes, and for concatenating the resultant homodecoupled chunks, a special software is required. The development of real-time homodecoupled pulse sequences has circumvented these drawbacks, wherein spectra acquisition and processing are very similar to that of the conventional NMR methods.

7.5.2   Real-Time Data Acquisition

Figure 7.11 illustrates the real-time homodecoupled pulse sequence. In this pulse sequence, the periodic interruption of FID with homodecoupling blocks produces pure shift NMR spectra. As described for the pseudo-two-dimensional pure shift NMR, a real-time version of pure shift NMR scheme also needs ~10–20 ms of FID chunking, followed by the application of homodecoupling block again and a short duration of data acquisition. This process continues for ~30–40 FID chunks in a single step; hence, concatenating 30 homodecoupled chunks of 20 ms duration results in a ~600 ms of pure shift FID. The resultant spectra require only the conventional FT processing, and data recording is also in regular mode. Therefore, in real-time mode, a significant decrease in experimental time and an inherent ease of implementation make it a very routinely usable experiment. The relaxation losses occur during the selective refocusing pulses, and concatenating two homodecoupled chunks with different relaxation properties enhances the line widths in real-time experiments compared to the pseudo-two-dimensional modes.

These two types of data acquisition schemes, namely, pseudo-two-dimensional and real-time pulsing, have been demonstrated in various kinds of homodecoupling schemes, viz., band-selective decoupling, Zangger-Sterk (ZS), PSYCHE, and BIRD decoupling.

7.5.3   Homonuclear Band-Selective Decoupling

Figure 7.12 represents the real-time homodecoupled band-selective pulse sequence; it is often referred to as a BASH or HOBS pulse sequence. As in other homodecoupling methods, the spins selected refocus in the HOBS scheme. This method works well for the molecules that have nicely separated bands of resonances,

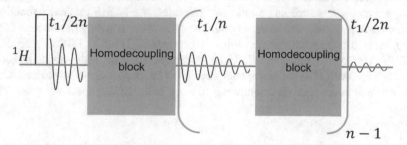

Fig. 7.11 Schematic of a real-time pure shift NMR pulse sequence, wherein the number of homodecoupling interruptions is equal to n. The rectangular pulse is a hard 90° pulse

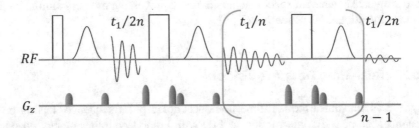

Fig. 7.12 Schematic of the real-time homonuclear band-selective (HOBS) pulse sequence. The narrow and wide rectangles are the hard 90° and 180° pulses, respectively. The Gaussian shaped pulse is the 180° refocusing pulse. In the present pulse sequence, "n" homodecoupling interruptions are used for a total acquisition time of "t_1"

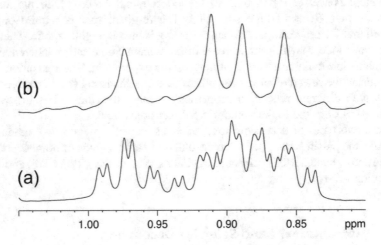

Fig. 7.13 Comparison of an experimental HOBS (**b**) and a conventional (**a**) one-dimensional spectra recorded on testosterone molecule dissolved in DMSO-D_6

and only one type of proton can be used for the experiment. Otherwise, if two scalarly coupled spins are selected together for a HOBS experiment, the scalar coupling between those two selected spins appears in the resultant HOBS spectrum. Thus, for molecules such as proteins and peptides, the HOBS method's functional performance is expected to be very good. The HOBS experiments have been utilized along with different multidimensional experiments, HSQC, TOCSY, NOESY, ROESY, HSQMBC, etc. The HOBS spectrum recorded on testosterone molecule dissolved in DMSO-D6 solvent is compared with the conventional spectrum. A vast improvement in the spectral resolution is noticed in the HOBS spectrum (Fig. 7.13).

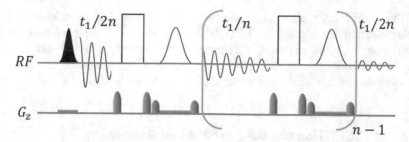

Fig. 7.14 Schematic of the real-time Zangger-Sterk (ZS) homodecoupled pulse sequence. The filled and open Gaussian shapes represent the soft 90° and 180° pulses, respectively. The rectangular shape is the hard 180° refocusing pulse. In the present pulse sequence, "n" homodecoupling interruptions are used for a total acquisition time of "t_1"

7.5.4 Zangger-Sterk Real-Time Homonuclear Broadband Decoupling

Figure 7.14 represents the real-time broadband version of the homodecoupled pulse sequence, which is also known as the Zangger-Sterk (ZS) scheme. In general, the homonuclear broadband decoupling pulse sequences are referred to as HOBB methods. The ZS homonuclear broadband decoupling method is the first-ever developed RF-based broadband decoupling method to obtain ultrahigh-resolution ^{1}H NMR spectrum, and that is one of the widely explored pure shift NMR methods. The ZS method utilizes the concepts of MRI, wherein slice-selective pulses are used in the presence of weak pulsed-filed gradients. In the ZS method, when a linear pulsed-field gradient is applied along the sample's z-axis, the different parts of the NMR sample experience different magnetic field strengths. Therefore, as a consequence, positional-dependent frequency shifts ($\Delta\omega = \gamma G_s$) along the sample volume can be established. The terms s, γ, and G correspond to the position in the tube along the sample length, the gyromagnetic ratio of the excited nucleus, and the pulsed-field gradient strengths. It is interesting to know that when a selective pulse is applied alone without pulsed-field gradients, only that particular resonance is excited/ refocused in the NMR spectrum; on the other hand, when the selective pulses are applied along with the gradients, the whole spectrum can be excited, but each resonance is excited from a thin slice of the sample. Hence, in general, in these experiments, spectral sensitivity is very inferior, which is a severe limitation of these ZS experiments. As is known, homodecoupling in broadband fashion can be achieved by combining the slice-selective refocusing pulses with a conventional hard 180° pulse. Overall, the spins on resonance to the slice-selective pulse get refocused, and the off-resonance spins get inverted. If any set of scalarly coupled spins are selected together for the refocusing, the efficacy of homodecoupling deteriorates; in such cases, small bandwidth pulses have to be used, but utilizing small bandwidth pulses cause small thin slice selection, then more relaxation loses, hence lower sensitivity. Overall, ZS methods' working performance is inadequate for the steroid kind of strongly coupled spin systems. In this concern, highly sophisticated pulse sequences (PSYCHE) have also been developed; they use low-flip angle pulses and work well even for the strongly coupled spin systems. The applications of the ZS scheme have been explored in all kinds of two-dimensional homonuclear correlation experiments.

7.5.5   PSYCHE Homonuclear Broadband Decoupling

Pure shift yielded by CHIRP pulse excitation is known as PSYCHE (pure shift yielded by chirp excitation) methodology (Fig. 7.15). This method is analogous to the anti-z COSY experiment, which uses the pulse sequence $90\text{-}t_1\text{-}\beta\text{-}t\text{-}\beta\text{-}acquire$ (t_2). The utilization of low-flip angle pulses (β) in the anti-z COSY results in only anti-diagonal signals, while suppressing the information of both the diagonal and cross-peaks, which is schematically shown in Fig. 7.16. The anti-diagonal signal sensitivity is inversely proportional to the flip angle of the pulses. In this pulse scheme, the cross-peak signals diminish at the low-flip angles (otherwise, they appear as recoupling artifacts), and signal sensitivity is reduced, which is a trade-

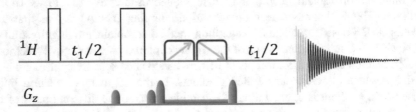

Fig. 7.15 A schematic representation of the pseudo-two-dimensional version of the PSYCHE pulse sequence. The narrow and wide rectangles represent the hard 90° and 180° pulses, respectively. The pure shift dwell increment is equal to t_1

Fig. 7.16 A schematic representation of an anti-z COSY spectrum. It shows only anti-diagonal peaks and simplified cross-peaks when compared with the conventional COSY spectra

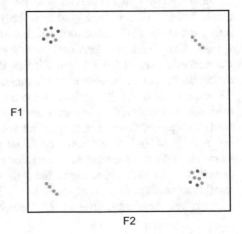

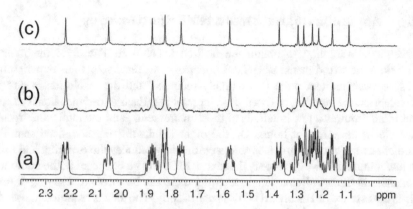

Fig. 7.17 A comparison of experimental PSYCHE (**c**), real-time ZS (**b**), and conventional (**a**) one-dimensional spectra recorded on estradiol molecule dissolved in DMSO-D$_6$

off between spectral purity and sensitivity. Since cross-peaks remain in the anti-z COSY, this method cannot readily be used as a pure shift NMR method in complex molecule cases. In this concern, a novel method, PSYCHE, has been proposed, wherein directly the hard low-flip angle pulses are replaced with the low-flip angle CHIRP-shaped pulses. This strategy results in the suppression of all the zero-quantum and cross-peak terms due to the spatiotemporal averaging phenomenon; therefore, pure shift spectra can be obtained from the PSYCHE methodology. The data is acquired in the pseudo-two-dimensional mode. Applications of the PSYCHE methodology have been explored for almost all the kinds of two-dimensional NMR methodologies that are useful for small molecules.

Figure 7.17 compares the quality of broadband pure shift NMR spectra recorded on estradiol molecule dissolved in DMSO-D$_6$ solvent, at different experimental conditions, viz., conventional, real-time ZS, and pseudo-two-dimensional version of PSYCHE. As expected, due to the significant overlapping of scalar couplings belonging to very closely separated chemical shifts, it is not easy to obtain precise chemical shift information. While such resolution issues can be resolved in the real-time ZS experiment, however, for this steroid molecule, estradiol, since chemical shifts are very closely separated, selective refocusing pulses of long durations (small bandwidths) need to be used. Hence a little increase in the line broadening is observed for the pure chemical shifts due to the significant relaxation losses during the selective pulses. On the other hand, using PSYCHE broadband homodecoupling resulted in clearly pure shift NMR spectra, but the data has to be recorded in pseudo-two-dimensional modes.

Further Reading
- Pure shift NMR, K. Zangger, Progress in Nuclear Magnetic Resonance Spectroscopy, 86–87 (2015) 1–20.
- PSYCHE pure shift NMR spectroscopy, M. Foroozandeh, G. A. Morris, M. Nilsson, 24 (2018) 13988–14000.

7.6   Appendix A6: Hadamard NMR Spectroscopy

Conventional sampling to obtain the desired spectral resolution in the indirect dimensions of two-dimensional NMR spectroscopy results in long experimental times, as each indirect point is separately recorded and the resolution is directly dependent on the number of dwell time increments. Each two-dimensional experiment often requires a few hundreds of dwell increments, and the total experimental times are in the range of hours. On the other hand, utilizing advanced sampling schemes such as nonuniform sampling significantly reduces the experimental times to a few minutes. However, if only that many FIDs can be collected as the number of resonances in the one-dimensional spectrum, there can be substantial saving of time. This is the essence of Hadamard NMR spectroscopy. The data is recorded in pseudo-two-dimensional modes. The main requirement for recording the data in Hadamard NMR spectroscopy is that the resonances should be well resolved.

In Hadamard NMR spectroscopy, in each FID, all the resonances of interest are simultaneously excited as per the elements in the rows of the Hadamard matrix. Therefore, the size of the Hadamard matrix which is a square matrix equals the number of resonances to be excited. For example, a Hadamard matrix of order four is given in Scheme 7.1, and this H_4 matrix works only for the four/less than four resonances. Wherein all the four resonances need to be acquired as per the signs in the rows of the H_4 matrix ("+" signifies a positive peak, and "–"signifies a negative peak). This type of sign encoding in Hadamard NMR spectroscopy can be achieved by using shaped pulses with appropriate excitation profiles at the defined resonance positions. Thus, as per Fig. 7.18, four FIDs are collected having the excitation profiles as per the four rows in the matrix. This is termed as Hadamard encoding. Next, each resonance can be decoded by performing proper algebraic operations on the four FIDs as per the four columns in the matrix; for example, the first resonance is the additive result of all the four rows of H_4 encoded data sets. The second resonance can also be similarly decoded by just performing row (1) + row (2) – row (3) – row (4). Similarly, the third resonance will be row (1) – row (2) + row (3) – row (4), and the fourth resonance will be row (1) – row (2) – row (3) + row (4). This whole process is termed as Hadamard decoding.

Based on the requirement (the number of resonances), an appropriate Hadamard matrix size must be selected. Different sizes of Hadamard matrices are available, and they can be denoted as H_2 and H_n ($n = 4m$, m is an integer), wherein the subscript represents the matrix size. The matrices H_2 and H_8 are also shown in Scheme 7.2. The advantage of such an exercise, in addition to the time saving, will be an improvement in the SNR by a factor $\sqrt{n}$ at each site, where n is the size of the Hadamard matrix,

Scheme 7.1 H_4 Hadamard matrix

$$H_4 = \begin{bmatrix} + & + & + & + \\ + & + & - & - \\ + & - & + & - \\ + & - & - & + \end{bmatrix}$$

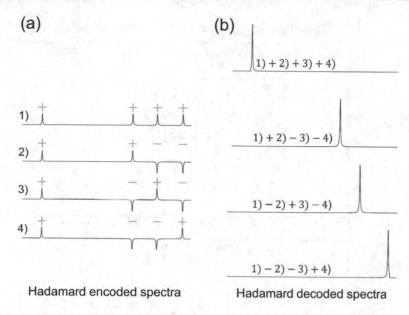

Fig. 7.18 Schematic of Hadamard encoded (**a**) and decoded (**b**) NMR spectra for a four-spin combination

Scheme 7.2 H_2 and H_8 Hadamard matrices

$$H_2 = \begin{bmatrix} + & + \\ + & - \end{bmatrix} \qquad H_8 = \begin{bmatrix} + & + & + & + & + & + & + & + \\ + & - & + & - & + & - & + & - \\ + & + & - & - & + & + & - & - \\ + & - & - & + & + & - & - & + \\ + & + & + & + & - & - & - & - \\ + & - & + & - & - & + & - & + \\ + & + & - & - & - & - & + & + \\ + & - & - & + & - & + & + & - \end{bmatrix}$$

when compared with the corresponding single resonance-selective excitation experiment. Hadamard NMR spectroscopy applications have been demonstrated for COSY, NOESY, TOCSY, HSQC, HMBC, etc., NMR spectroscopy techniques to minimize the experimental times to a great extent.

The following figures demonstrate the application of this technique in different two-dimensional spectra (Fig. 7.19). Notice here that in these spectra, Fourier transformation is done only along the direct dimension. Hadamard encoding is done only for the chosen frequencies in the indirect dimension.

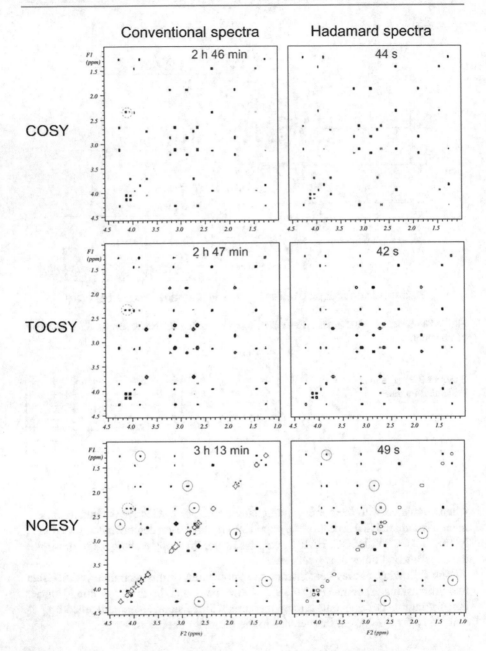

Fig. 7.19 Conventional and Hadamard NMR spectra recorded on strychnine molecule. The corresponding experimental times are given at each spectrum. (Reproduced from Journal of Magnetic Resonance, 162, 300 (2003), with the permission of Elsevier Publishing)

Further Reading

- Hadamard NMR spectroscopy, Progress in Nuclear Magnetic Resonance Spectroscopy, 42 (2003) 95–122

Correction to: Multidimensional NMR Spectroscopy

Correction to: Chapter 6 in:
R. V. Hosur, V. M. R. Kakita, *A Graduate Course in NMR Spectroscopy*,
https://doi.org/10.1007/978-3-030-88769-8_6

Chapter 6 was inadvertently published with some errors in the equations and figures.

Original page 232

$$I_{kx} \xrightarrow{\mathcal{H}_{et}} I_{kx}\left(\frac{1+\cos 2\pi Jt}{2}\right) + I_{lx}\left(\frac{1-\cos 2\pi Jt}{2}\right)$$

$$+ \left(2I_{ky}I_{lz} - 2I_{ly}I_{kz}\right)\sin\frac{2\pi Jt}{2} \tag{6.74}$$

$$I_{lx} \xrightarrow{\mathcal{H}_{et}} I_{lx}\left(\frac{1+\cos 2\pi Jt}{2}\right) + I_{kx}\left(\frac{1-\cos 2\pi Jt}{2}\right)$$

$$+ \left(2I_{ly}I_{kz} - 2I_{ky}I_{lz}\right)\sin\frac{2\pi Jt}{2} \tag{6.75}$$

Correction page 232

$$I_{kx} \xrightarrow{\mathcal{H}_{et}} I_{kx}\left(\frac{1+\cos 2\pi Jt}{2}\right) + I_{lx}\left(\frac{1-\cos 2\pi Jt}{2}\right) + \left(I_{ky}I_{lz} - I_{ly}I_{kz}\right)\sin 2\pi Jt \tag{6.74}$$

$$I_{lx} \xrightarrow{\mathcal{H}_{et}} I_{lx}\left(\frac{1+\cos 2\pi Jt}{2}\right) + I_{kx}\left(\frac{1-\cos 2\pi Jt}{2}\right) + \left(I_{ly}I_{kz} - I_{ky}I_{lz}\right)\sin 2\pi Jt \tag{6.75}$$

The updated original version for this chapter can be found at
https://doi.org/10.1007/978-3-030-88769-8_6

Original page 253

(i) Under the influence of ^{15}N chemical shifts for a period t_1.
(ii) Under the influence of ^{15}N-^{1}H coupling which refocuses antiphase ^{15}N magnetization to produce in-phase magnetization.

Page 254

$$- 2H_{iz}^N N_{iy} \rightarrow N_{ix} \qquad (6.96)$$

Correction pages 253 and 254

(i) Under the influence of ^{15}N chemical shifts for a period t_1.

$$- 2H_{iz}^N N_{iy} \rightarrow - 2H_{iz}^N (N_{iy} \cos (\omega_{Ni} t_1) - N_{ix} \sin (\omega_{Ni} t_1)) \qquad (6.96)$$

(ii) Evolution under the influence of ^{15}N-^{1}H coupling is effectively refocused and ^{15}N magnetization remains anti-phase with respect to ^{1}H.

Original (Fig. 6.55) page 258

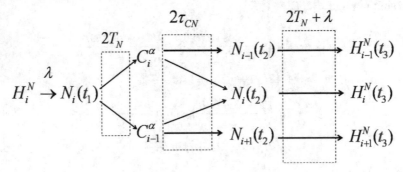

Correction (Fig. 6.55) page 258

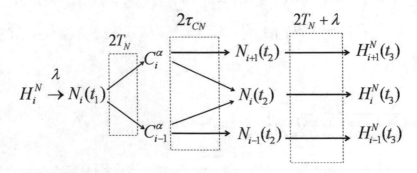

The corrections have now been incorporated.

Solutions to Exercises

Chapter 1

1.1 (c)	1.2 (d)	1.3 (c)	1.4 (c)
1.5 (a)	1.6 (a)	1.7 (b)	1.8 (c)
1.9 (b)	1.10 (c)	1.11 (a)	1.12 (d)
1.13 (c)	1.14 (a)	1.15 (b)	1.16 (d)

1.17

The transition probability from state $|I, m>$ to $|I, m'>$, all of which are orthonormal to each other, is given as per Eq. 1.71:

$$P = \gamma^2 H_1{}^2 |< m'|\hat{I}_x|m >|^2$$

For a spin ½ ($I=½$), m takes values ½ (α-state) and $-½$ (β-state).
The transition probability from α to β is proportional to

$$P_{(\alpha \to \beta)} \propto |< \alpha|\hat{I}_x|\beta >|^2 = < \alpha \,|\, \hat{I}_x \,|\, \beta >< \beta \,|\, \hat{I}_x \,|\, \alpha >$$

It follows from Box 1

$$I_x \,|\, \alpha > = \frac{1}{2}(I^+ + I^-) \,|\, \alpha > = \frac{1}{2} |\, \beta >$$

$$I_x \,|\, \beta > = \frac{1}{2}(I^+ + I^-) \,|\, \beta > = \frac{1}{2} |\, \alpha >$$

Therefore,

$$< \alpha|\hat{I}_x|\beta > = \frac{1}{2} < \alpha \,|\, \alpha > = \frac{1}{2}$$

$$< \beta|\hat{I}_x|\alpha > = \frac{1}{2} < \beta \,|\, \beta > = \frac{1}{2}$$

© The Author(s), under exclusive license to Springer Nature Switzerland AG 2022
R. V. Hosur, V. M. R. Kakita, *A Graduate Course in NMR Spectroscopy*,
https://doi.org/10.1007/978-3-030-88769-8

Therefore,

$$P_{(\alpha \to \beta)} \propto \frac{1}{4}$$

Similarly,

$$P_{(\beta \to \alpha)} \propto |< \beta|\widehat{I_x}|\alpha >|^2 =< \beta \,|\, \widehat{I}_x \,|\, \alpha >< \alpha \,|\, \widehat{I}_x \,|\, \beta >$$

$$P_{(\beta \to \alpha)} \propto \frac{1}{4}$$

Therefore, the upward and downward transition probabilities are identical.

1.18

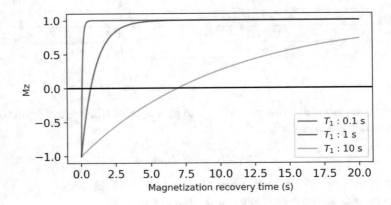

1.19

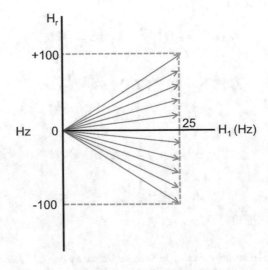

1.20

The Bloch equations in the rotating frame are given by Eqs. 1.86–1.88; these can be derived by substituting Eq. 1.85 into Eqs. 1.82–1.84.

1.21

$$v = -M_o \frac{\gamma H_1 T_2}{1 + T_2{}^2 \Delta \omega^2}$$

$h^{\max}$ will beat $\Delta \omega = 0$,

$$h^{\max} = -M_o \gamma H_1 T_2$$

And

$$\frac{h^{\max}}{2} = \frac{-M_o \gamma H_1 T_2}{2}$$

$$\frac{-M_o \gamma H_1 T_2}{2} = -M_o \frac{\gamma H_1 T_2}{1 + T_2{}^2 \Delta \omega^2}$$

$$\frac{1}{2} = \frac{1}{1 + T_2{}^2 \Delta \omega^2}$$

$$\Rightarrow 2 = 1 + T_2{}^2 \Delta \omega^2$$

$$\Rightarrow 1 = T_2{}^2 \Delta \omega^2$$

$$\Rightarrow \Delta \omega = \frac{1}{T_2}$$

Therefore, the width at half height,

$$2\Delta \omega = \frac{2}{T_2}$$

Similarly,

$$\frac{h^{\max}}{n} = \frac{-M_o \gamma H_1 T_2}{n} = -M_o \frac{\gamma H_1 T_2}{1 + T_2{}^2 \Delta \omega^2}$$

$$\Rightarrow n = 1 + T_2{}^2 \Delta \omega^2$$

$$\Rightarrow \Delta \omega^2 = \frac{n-1}{T_2{}^2}$$

$$J(\omega) = k\frac{\tau_c}{\left(1 + \omega^2\tau_c^2\right)}, \text{ where k is a constant.}$$

$$\frac{\mathrm{d}J}{\mathrm{d}\tau_c} = k\left[\frac{\left(1 + \omega^2\tau_c^2\right) - \tau_c(2\omega^2\tau_c)}{\left(1 + \omega^2\tau_c^2\right)^2}\right]$$

$$= k\left[\frac{\left(1 - \omega^2\tau_c^2\right)}{\left(1 + \omega^2\tau_c^2\right)^2}\right]$$

$$\frac{\left(1 - \omega^2\tau_c^2\right)}{\left(1 + \omega^2\tau_c^2\right)^2} = 0$$

$$\omega^2\tau_c^2 = 1$$

$$\tau_c = \frac{1}{\omega}$$

$$J(\omega) = k\frac{\tau_c}{\left(1 + \omega^2\tau_c^2\right)}, \text{ where k is a constant.}$$

$$J(\omega)_{\max} = k\tau_c$$

At half the maximum, $J(\omega) = \dfrac{k}{2}\tau_c$

Therefore, $\dfrac{k}{2}\tau_c = k\dfrac{\tau_c}{\left(1 + \omega^2\tau_c^2\right)}$

$$\Rightarrow \frac{1}{2} = \frac{1}{\left(1 + \omega^2\tau_c^2\right)}$$

$$\Rightarrow \omega^2 = \frac{1}{\tau_c^2}$$

$$\Rightarrow \omega = \frac{1}{\tau_c}$$

The interaction of the RF applied along the z-axis with the magnetic moment will be given by

$$\mu.H_1 \cos{(\omega_0 t)} = \mu_z H_1 \cos{(\omega_0 t)} \tag{S1.1}$$

The Zeeman interaction between magnetic moment and main field is given by

$$\mu.H_0 = \mu_z H_0 \tag{S1.2}$$

Thus, in the presence of the RF, the total interaction is

$$\mu_z(H_0 + H_1 \cos{(\omega_0 t)}) \tag{S1.3}$$

This will result in the oscillation of the energy levels of the spin system.

Considering Eq. S1.1, the formula for the transition probability (Eq. 1.71) will become

$$P = \gamma^2 H_1{}^2 |< m' |\hat{I}_z| m >|^2 \tag{S1.4}$$

Clearly, this cannot cause a change in the state, and hence there can be no transitions between the energy levels.

Chapter 2

2.1 (c)	2.2 (b)	2.3 (c)
2.4 (d)	2.5 (a)	2.6 (c)
2.7 (d)	2.8 (b)	2.9 (a)
2.10 (b)	2.11 (d)	2.12 (a)
2.13 (c)		

2.14

For a two-spin system, the isotropic Hamiltonian is

$$\mathcal{H} = \omega_1 I_{1z} + \omega_2 I_{2z} + 2\pi J I_1.I_2 \tag{S2.1}$$

$$= \omega_1 I_{1z} + \omega_2 I_{2z} + 2\pi J \left[I_{1x} I_{2x} + I_{1y} I_{2y} + I_{1z} I_{2z} \right] \tag{S2.2}$$

and the F_z operator is

$$F_z = I_{1z} + I_{2z} \tag{S2.3}$$

The commutator $[F_z, \mathcal{H}]$ can be written as

$$[F_z, \mathcal{H}] = \left[F_z, \left(\omega_1 I_{1z} + \omega_2 I_{2z} + 2\pi J\{I_{1x}I_{2x} + I_{1y}I_{2y} + I_{1z}I_{2z}\}\right)\right] \quad \text{(S2.4)}$$

$$= \omega_1[F_z, I_{1z}] + \omega_2[F_z, I_{2z}] + 2\pi J\{[F_z, I_{1x}I_{2x}] + [F_z, I_{1y}I_{2y}] + [F_z, I_{1z}I_{2z}]\} \quad \text{(S2.5)}$$

$$[F_z, I_{1z}] = 0; [F_z, I_{2z}] = 0 \quad \text{(S2.6)}$$

Now,

$$[F_z, I_{1x}I_{2x}] = [(I_{1z} + I_{2z}), I_{1x}I_{2x}] = [I_{1z}, I_{1x}]I_{2x} + I_{1x}[I_{2z}, I_{2x}]$$
$$= I_{1y}I_{2x} + I_{1x}I_{2y} \quad \text{(S2.7)}$$

Similarly,

$$[F_z, I_{1y}I_{2y}] = [(I_{1z} + I_{2z}), I_{1y}I_{2y}] = [I_{1z}, I_{1y}]I_{2y} + I_{1y}[I_{2z}, I_{2y}]$$
$$= -I_{1x}I_{2y} - I_{1y}I_{2x} \quad \text{(S2.8)}$$

$$[F_z, I_{1z}I_{2z}] = [(I_{1z} + I_{2z}), I_{1z}I_{2z}] = [I_{1z}, I_{1z}]I_{2z} + I_{1z}[I_{2z}, I_{2z}] = \mathbf{0} + \mathbf{0} = \mathbf{0} \quad \text{(S2.9)}$$

Substituting Eqs. S2.6–S2.9 into Eq. S2.5, we get

$$[F_z, \mathcal{H}] = 0 + 0 + I_{1y}I_{2x} + I_{1x}I_{2y} - I_{1x}I_{2y} - I_{1y}I_{2x} + \mathbf{0} = 0 \quad \text{(S2.10)}$$

2.15

The following spectrum is simulated at a field strength of 400 MHz.

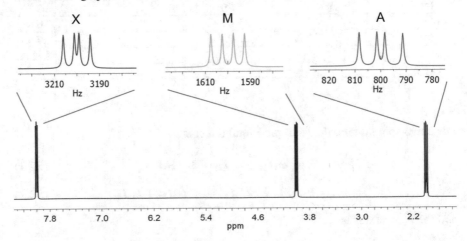

2.16

Hint: Use the given peak frequencies.

Chemical shifts: (A) 400 Hz, (M) ~1000 Hz, and (X) 1400 Hz

Scalar coupling constants: $J_{AM} = J_{MX} = 7$ Hz

2.17

Hint: Use the given peak frequencies.

Chemical shifts: (A) 21.99 ppm, (M) ~71.98 ppm, and (X) 122.39 ppm

Scalar coupling constants: $J_{AH} = 120$ Hz, $J_{MH} = 145$ Hz, $J_{XH} = 184$ Hz

2.18

Hint: Use Table 2.6.

2.19

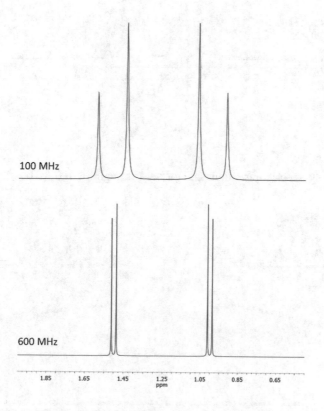

2.20

The following spectrum is simulated at a field strength of 400 MHz.

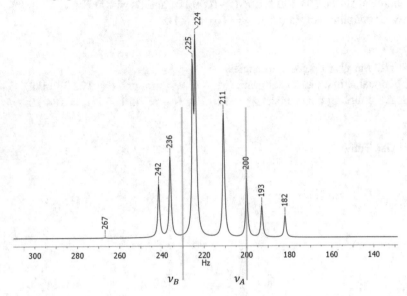

2.21

The following spectrum is simulated at a field strength of 400 MHz.

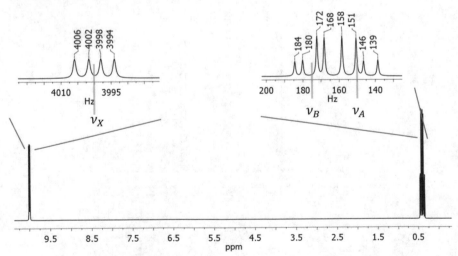

Chapter 3

3.1 (c)	3.2 (c)	3.3 (b)
3.4 (c)	3.5 (b)	3.6 (c)
3.7 (c)	3.8 (b)	3.9 (b)
3.10 (b)	3.11 (b)	3.12 (c)
3.13 (d)	3.14 (d)	3.15 (c)
3.16 (c)		

3.17

Let τ be the dwell time and $\nu_{\max}$ be the maximum frequency in Hz.

Therefore, $\tau = \frac{1}{2\nu_{\max}}$.

The evolution during the dwell time adds a phase of $2\pi\nu\tau$ (at frequency ν).

So, for the maximum frequency, the phase added will be $2\pi\nu_{\max}\tau = \pi$.

Thus, at half the maximum frequency, the phase added will be $\frac{\pi}{2}$.

So, if we have a positive absorptive line shape in the absence of the delay, then the introduction of a delay of a one dwell time causes a dispersive line shape at half the maximum frequency and a negative absorptive line at the maximum frequency.

3.18

$$\omega_{\text{eff}} = \sqrt{\gamma^2 B_1^2 + \Omega^2} \tag{1}$$

Flip angle for T_p,

$$\beta = 2\pi\omega_{\text{eff}} T_p \tag{2}$$

Given,

$$2\pi\gamma B_1 T_p = \frac{\pi}{2}$$

$$T_p = \frac{1}{4\gamma B_1}$$

From Eqs. (1) and (2),

$$\beta^2 = 4\pi^2 \omega_{\text{eff}}^2 T_p^2$$

$$= 4\pi^2 \left(\gamma^2 B_1^2 + \Omega^2\right) T_p^2$$

$$= 4\pi^2 \left(\gamma^2 B_1^2 + \Omega^2\right) \cdot \frac{1}{16\gamma^2 B_1^2}$$

If $\beta = 2\pi$, then

$$4\pi^2 = 4\pi^2 \left(\gamma^2 B_1^2 + \Omega^2 \right) \cdot \frac{1}{16\gamma^2 B_1^2}$$

$$16\gamma^2 B_1^2 = \gamma^2 B_1^2 + \Omega^2$$

Therefore, offset $\Omega = \sqrt{15}\gamma B_1$
If $\beta = \pi$, then

$$\pi^2 = 4\pi^2 \left(\gamma^2 B_1^2 + \Omega^2 \right) \cdot \frac{1}{16\gamma^2 B_1^2}$$

$$4\gamma^2 B_1^2 = \gamma^2 B_1^2 + \Omega^2$$

Therefore, offset $\Omega = \sqrt{3}\gamma B_1$

3.19

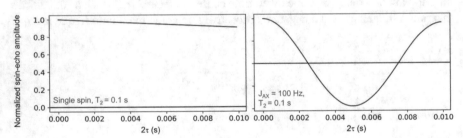

3.20

$$t = \frac{1}{4v_{\max}}$$

$$t = \frac{1}{4 \times 5000} s$$

$$t = 50\ \mu s$$

3.21

$$T_1 = \frac{\tau_{\text{null}}}{\ln 2}$$

$$T_1 = \frac{2\ s}{0.693} = 2.89\ s$$

Chapter 4

4.1 (a)	4.2 (b)	4.3 (a)
4.4 (c)	4.5 (d)	4.6 (b)
4.7 (a)	4.8 (a)	
4.9 (b)	4.10 (a)	4.11 (b)
4.12 (c)		

Chapter 5

5.1 (c)	5.2 (c)	5.3 (a)
5.4 (b)	5.5 (a)	5.6 (c)
5.7 (b)	5.8 (c)	5.9 (b)
5.10 (b)	5.11 (d)	5.12 (b)
5.13 (b)	5.14 (c)	5.15 (d)
5.16 (b)	5.17 (b)	5.18 (d)
5.19 (c)	5.20 (a)	5.21 (d)
5.22 (a)	5.23 (c)	5.24 (a)
5.25 (a)		

5.26

$2I_xS_y$ and $2I_zS_z$ in the eigenbasis of the weak coupling Hamiltonian can be calculated by using direct products of one-spin matrix representations:

$$2I_xS_y = 2 \times \frac{1}{2} \times \frac{1}{2} \begin{bmatrix} 0 & 1 \\ 1 & 0 \end{bmatrix} \otimes \begin{bmatrix} 0 & -i \\ i & 0 \end{bmatrix} = \frac{1}{2} \begin{bmatrix} 0 & 0 & 0 & -i \\ 0 & 0 & i & 0 \\ 0 & -i & 0 & 0 \\ i & 0 & 0 & 0 \end{bmatrix}$$

$$2I_zS_z = 2 \times \frac{1}{2} \times \frac{1}{2} \begin{bmatrix} 1 & 0 \\ 0 & -1 \end{bmatrix} \otimes \begin{bmatrix} 1 & 0 \\ 0 & -1 \end{bmatrix} = \frac{1}{2} \begin{bmatrix} 1 & 0 & 0 & 0 \\ 0 & -1 & 0 & 0 \\ 0 & 0 & -1 & 0 \\ 0 & 0 & 0 & 1 \end{bmatrix}$$

5.27

Consider $\alpha = x$, $\alpha' = y$, $\beta = y$, and $\beta' = z$:

$$
\begin{aligned}
\left[2I_\alpha S_{\alpha'}, 2I_\beta S_{\beta'}\right] &= 4\left[I_x S_y I_y S_z - I_y S_z I_x S_y\right] \\
&= 4\left[I_x I_y S_y S_z - I_y I_x S_z S_y\right] \\
&= 4\left[I_x I_y S_y S_z - I_y I_x S_y S_z + I_y I_x S_y S_z - I_y I_x S_z S_y\right] \\
&= 4\left[\left[I_x, I_y\right]S_y S_z + I_y I_x\left[S_y, S_z\right]\right] \\
&= 4\left[I_z S_y S_z + I_y I_x S_x\right] \\
&= \left[I_z \begin{bmatrix} 0 & i \\ i & 0 \end{bmatrix} + \begin{bmatrix} -i & 0 \\ 0 & i \end{bmatrix} S_x\right] \\
&= \frac{1}{2}\left[\begin{bmatrix} 1 & 0 \\ 0 & -1 \end{bmatrix} \otimes \begin{bmatrix} 0 & i \\ i & 0 \end{bmatrix} + \begin{bmatrix} -i & 0 \\ 0 & i \end{bmatrix} \otimes \begin{bmatrix} 0 & 1 \\ 1 & 0 \end{bmatrix}\right] \\
&= \frac{1}{2}\left[\begin{bmatrix} 0 & i & 0 & 0 \\ i & 0 & 0 & 0 \\ 0 & 0 & 0 & -i \\ 0 & 0 & -i & 0 \end{bmatrix} + \begin{bmatrix} 0 & -i & 0 & 0 \\ -i & 0 & 0 & 0 \\ 0 & 0 & 0 & i \\ 0 & 0 & i & 0 \end{bmatrix}\right] = 0
\end{aligned}
$$

Hence, the commutation is proved. This can be verified for other combinations too.

5.28

(a) For a $R_x(\pi)$, the pulse on $\widehat{I}_z$ will have the following effect:

$$
\rho = R_x(\pi)\widehat{I}_z R_x^{-1}(\pi)
$$

$$
= \frac{1}{4}\begin{bmatrix} 0 & -i \\ -i & 0 \end{bmatrix}\begin{bmatrix} 1 & 0 \\ 0 & -1 \end{bmatrix}\begin{bmatrix} 0 & i \\ i & 0 \end{bmatrix}
$$

$$
= \frac{1}{2}\begin{bmatrix} -1 & 0 \\ 0 & 1 \end{bmatrix}
$$

$$
= -\widehat{I}_z
$$

So, the z-magnetization is inverted by the π pulse along the x-axis.

(b) For a $R_y(\pi)$, the pulse on $\widehat{I}_z$ will have the following effect:

$$\rho = R_y(\pi)\widehat{I}_z R_y^{-1}(\pi)$$

$$= \frac{1}{4} \begin{bmatrix} 0 & -1 \\ 1 & 0 \end{bmatrix} \begin{bmatrix} 1 & 0 \\ 0 & -1 \end{bmatrix} \begin{bmatrix} 0 & 1 \\ -1 & 0 \end{bmatrix}$$

$$= \frac{1}{2} \begin{bmatrix} -1 & 0 \\ 0 & 1 \end{bmatrix}$$

$$= -\widehat{I}_z$$

So, the z-magnetization is inverted by the π pulse along the y-axis.

Chapter 6

6.1 (b)	6.2 (d)	6.3 (b)
6.4 (a)	6.5 (a)	6.6 (b)
6.7 (d)	6.8 (b)	6.9 (b)
6.10 (b)	6.11 (a)	6.12 (d)
6.13 (b)	6.14 (b)	6.15 (a)
6.16 (b)	6.17 (b)	6.18 (d)
6.19 (a)	6.20 (b)	6.21 (b)
6.22 (b)	6.23 (c)	6.24 (d)
6.25 (a)	6.26 (b)	6.27 (a)
6.28 (c)	6.29 (b)	

Printed in the United States
by Baker & Taylor Publisher Services